The Surface Climates of Canada

Climate plays an intimate role in landscape pattern and diversity in Canada. Focusing on climatic processes at and near the earth's surface and on how these processes interact with the natural and human-modified landscapes, *The Surface Climates of Canada* goes beyond mere descriptive climatology to reveal the two-way interrelations between Canada's surface and its climate.

In the opening chapters contributors lay out the large-scale context of the physical climate of Canada. They introduce the processes, balances, and dynamic links between the surface and atmosphere that create and maintain the diversity of surface climates found in Canada as well as describing the physical processes that operate near the ground's surface. Individual chapters are dedicated to snow and ice – the almost universal surface cover in Canada – and the other major natural surface environments of Canada: ocean and coastal zones, freshwater lakes, wetlands, arctic islands, low arctic and subarctic lands, forests, and alpine environments. The final part of the book considers those surface environments that have been strongly influenced by human activity, such as agricultural lands and urban environments, and examines the prospects for future climate change.

The Surface Climates of Canada brings together for the first time a wide range of scholarship by leading Canadian climatologists. It will be an indispensable tool for understanding Canada's surface climates and the processes responsible for their creation and control.

W.G. BAILEY is professor of geography, Simon Fraser University.
TIMOTHY R. OKE is professor of geography, University of British Columbia.
WAYNE R. ROUSE is professor of geography, McMaster University.

CANADIAN ASSOCIATION OF GEOGRAPHERS
SERIES IN CANADIAN GEOGRAPHY
Cole Harris, General Editor

Canada's Cold Environments
Hugh M. French and Olav Slaymaker, editors

The Changing Social Geography of Canadian Cities
Larry S. Bourne and David F. Ley, editors

Canada and the Global Economy
The Geography of Structural and Technological Change
John N.H. Britton, editor

The Surface Climates of Canada
W.G. Bailey, Timothy R. Oke, and Wayne R. Rouse, editors

The Surface Climates of Canada

EDITED BY
W.G. BAILEY,
TIMOTHY R. OKE,
AND WAYNE R. ROUSE

McGill-Queen's University Press
Montreal & Kingston • London • Buffalo

© McGill-Queen's University Press 1997
ISBN 0-7735-0928-3 (cloth)
ISBN 0-7735-1672-7 (paper)

Legal deposit fourth quarter 1997
Bibliothèque nationale du Québec

Printed in Canada on acid-free paper

McGill-Queen's University Press acknowledges the
financial support of the Government of Canada through
the Canadian Studies and Special Projects Directorate of the
Department of the Secretary of State of Canada.

This book has been published with the help of a grant from the
Humanities and Social Sciences Federation of Canada, using funds
provided by the Social Sciences and Humanities Research Council
of Canada.

Canadian Cataloguing in Publication Data

Main entry under title:
    The surface climates of Canada
(Canadian Association of Geographers series in Canadian geography)
Includes bibliographical references and index.
ISBN 0-7735-0928-3 (bound) –
ISBN 0-7735-1672-7 (pbk.)
    1. Microclimatology—Canada. 2. Canada—Climate.
I. Bailey, William G. II. Oke, T.R. III. Rouse, W.R.
GB447.s87 1997   551.6′6′0971   c97-900782-8

This book was typeset by Typo Litho Composition Inc.
in 10/12 Times.
Cover illustration: *Summer Clouds* by Tom Thomson.
Oil on board, 21.6 cm × 26.7 cm, private collection.

*To Professor F. Kenneth Hare,
distinguished scholar, teacher, and
researcher, a friend to many, and
an inspiration to us all*

Contents

Tables   ix

Figures   xi

Symbols, Units, and Quantities   xvii

Preface   xxi

PART ONE   SURFACE CLIMATE CONCEPTS

1 Canada's Climate: An Overall Perspective /
F. Kenneth Hare   3

2 Surface Climate Processes / Timothy R. Oke   21

3 Spatial Variability in Surface Climates /
Douw G. Steyn, Hans-Peter Schmid,
John L. Walmsley, and John D. Wilson   44

4 Winter and Snow / John W. Pomeroy and
Barry E. Goodison   68

PART TWO   NATURAL SURFACE CLIMATES

5 Oceans and the Coastal Zone / Owen Hertzman   101

6 Freshwater Lakes / William M. Schertzer   124

7 Wetlands / Nigel T. Roulet, D. Scott Munro, and
Linda Mortsch   149

8 The Arctic Islands / Ming-Ko Woo and
Atsumu Ohmura   172

9 The Low Arctic and Subarctic / Wayne R. Rouse,
Richard L. Bello, and Peter M. Lafleur   198

10 Alpine Environments / Ian R. Saunders, D. Scott Munro,
and W. G. Bailey   222

11 Forest Environments / J. Harry McCaughey,
    Brian D. Amiro, Alexander W. Robertson, and
    David L. Spittlehouse   247

PART THREE   MANAGED AND CHANGING
SURFACE CLIMATES

12 Agricultural Surfaces / Terry J. Gillespie   277

13 Urban Environments / Timothy R. Oke   303

14 Climatic Change / L.D. Danny Harvey   328

15 Epilogue / R. Ted Munn   352

    Glossary   359

    Contributors   365

    Index   367

Tables

4.1 Winter conditions for selected stations in Canada 69

4.2 Relative snow water retention on various landscape types in an open grassland environment 87

7.1 Attributes of the major wetland classes, as defined by the Canadian Wetland Classification System 153

7.2 Components of the surface radiation balance for various wetlands 157

7.3 Thermal properties of peat soils 159

7.4 Rate of evaporation from various wetlands 163

8.1 Diurnal amplitudes of air temperature for selected stations of the Arctic Islands 176

8.2 Variations in precipitation at Hot Weather Creek, Ellesmere Island 179

8.3 Energy balance of glaciers 192

8.4 Climate at the equilibrium line of Arctic Islands glaciers 193

10.1 Average daily total radiation components for ice and snow on Peyto Glacier 233

10.2 Average daily total heat balance components for three glacier sites on Peyto Glacier over ten-day periods 238

11.1 Stresses on the forest ecosystem 271

13.1 Flux ratios for daytime at the Sunset residential site in Vancouver 314

14.1 Selected general circulation–modelling groups that have simulated effects of an increase in CO_2 334

Figures

1.1   The westerlies in mid-troposphere 4
1.2   The westerlies in summer 6
1.3   Schematic of a travelling mid-latitude cyclone 10
1.4   Mean daily global solar radiation in January and July
      and annual mean 12–13
1.5   Mean annual net radiation 15
1.6   Mean annual measured precipitation 17
1.7   Mean annual calculated evapotranspiration 18
1.8   Mean annual runoff ratio 18
1.9   Mean annual dryness ratio 19
2.1   Schematic representation of the earth's surface boundary
      layers 23
2.2   Hypothetical sequence of weather for two days 26
2.3   Schematic depiction of radiation, energy, and water
      balance components 29
2.4   Typical variation of air temperature in the surface layer
      through a fine day 30
2.5   Typical shape of the variation of wind speed with height 31
2.6   Height variation of the mean and instantaneous
      velocity 33
2.7   Typical form of mean microclimatic profiles 36
2.8   Schematic depiction of vertical profiles and vertical fluxes
      in the planetary boundary layer 37
2.9   Relationship between proportion of net radiation used for
      evaporation and soil moisture content 40
2.10 Schematic depiction of mean vertical profiles and vertical
      fluxes in and above a plant canopy layer 41
2.11 Variation of the height of the planetary boundary layer 42
3.1   Surface inhomogeneities in the Canadian landscape 48
3.2   Flow adjustment to a step change in surface
      roughness 49
3.3   Growing thermal internal boundary layer 51

3.4   Suggested classification of different types of land cover 53
3.5   Horizontal profile of surface radiation temperature and hot and cold plumes 54
3.6   Schematic illustration of the roughness layer 54
3.7   Wind speed profiles over various types of terrain 56
3.8   Depiction of drainage wind activity in Canyon Creek 57
3.9   A well-developed sea-breeze circulation 59
3.10   Flow zones and reduction in wind speed over a windbreak 61
3.11   Snow fences in southern Alberta 62
3.12   Cloud patterns associated with surface controls on convection 63
3.13   Modelled and observed wind fields and trajectories 64
4.1   North American snow climatology 71
4.2   Variations in annual September–October snow-covered area over North America 72
4.3   Macroscale distribution of mean annual snowfall in Canada from 1951 to 1980 73
4.4   Mesoscale distribution of mean annual snowfall in southern Ontario 74
4.5   Disposition of winter snowfall in a forest environment 77
4.6   Interception by forest as a function of snowfall 79
4.7   Intercepted snowload of a black spruce 80
4.8   Vertical profiles of blowing snow mass concentration 81
4.9   Transport rate of blowing snow 82
4.10   Variation in snow depth and *SWE* near a spruce tree 83
4.11   Relationship between leaf area index and *SWE* 84
4.12   Winter evolution of *SWE* under various forest stands 85
4.13   Effect of land use on snow depth and density 86
4.14   Effect of topography on *SWE* 86
4.15   Mean annual mass of snow transported and sublimated as a function of fetch 88
4.16   Soil heat flux below a shallow snowpack 92
4.17   Variation in albedo of melting snowpacks 93
4.18   Fluxes comprising radiation balance of melting snowpacks 94
4.19   Energy and melt fluxes for a melting snow patch in the Arctic 96
5.1   Seasonal climatologies of sea-ice extent in the Arctic, 1953–84 102
5.2   Aerial view along the north coast of Ellesmere Island in late June 103
5.3   Ocean surface currents for the Arctic, Atlantic, and North Pacific oceans 105
5.4   Atlantic coastal average sea-surface temperatures in July and January 108–9

5.5   Radiation data for Sable Island, NS 111
5.6   Climate comparison of Vancouver, BC, Yarmouth, NS, and Regina, Sask. 115
5.7   Climate comparison within the three coastal zones of Canada 116
5.8   Climate variability within the coastal zone of the Maritimes 117
5.9   Mean monthly growing degree days above 5°C for six stations in the coastal zone 119
5.10   Depiction of fog events over the western North Atlantic in summer 120
6.1   Large lakes and major drainage basins in Canada 125
6.2   Monthly mean surface water temperatures of the Great Lakes 127
6.3   Examples of measurement systems deployed on large lakes 129
6.4   Energy and water balance components of lakes 130
6.5   Lake surface albedo 131
6.6   Diurnal variation of radiation balance components for Lake Ontario 132
6.7   Seasonal variation of radiation balance components for Lake Ontario 134
6.8   Long-term monthly mean energy balance for the Great Lakes 135
6.9   Heat storage in the Great Lakes turbulent 137
6.10   Great Lakes: long-term monthly mean turbulent heat fluxes 139
6.11   Monthly mean evaporation from the Great Lakes 141
6.12   Energy balance and evaporation of Lake Diefenbaker and Perch Lake 143
6.13   Average annual lake evaporation in Canada 144
7.1   Percentage cover of wetlands in Canada 151
7.2   Mineral-poor, open fens 152
7.3   An ombrotrophic, raised, low-shrub bog 152
7.4   A treed bog 154
7.5   Daily total radiant energy flux density for a treed swamp 155
7.6   Summer surface energy balance for three types of wetland 158
7.7   Energy balance on a blanket bog 161
7.8   Hourly Bowen ratios values in three wetlands 161
7.9   Transect of Bowen ratio in southern Hudson Bay Lowland 162
7.10   Average surface energy balance for sedge fens 164
7.11   Daylight pattern of resistances for a treed swamp 165
7.12   Diurnal pattern of surface resistance 166

7.13 Elevation of the surface and water table in a subarctic
      fen 167
7.14 Storage of carbon in the world's soils 168
8.1   Location of Canada's Arctic Islands 173
8.2   Interior of Banks Island 173
8.3   Midnight on 25 May 1979 at Eidsbotn, Devon
      Island 174
8.4   Mean hourly global solar and net radiation at Alert
      and Iqaluit 175
8.5   Mean daily air temperatures at a coastal and an inland
      station 178
8.6   Vertical distribution of air temperature for glacier and
      tundra stations 180
8.7   Aerial view of the southern coastal zone of Devon Island
      in early August 181
8.8   Melting of snow in McMaster basin 182
8.9   Hourly global solar radiation, net radiation, air
      temperature, snow temperatures, melt, and daily albedo,
      at a site near Resolute 184
8.10 Large "ploughshares" in the snow on a valley slope 185
8.11 Changes in snow temperature in McMaster basin 186
8.12 Large spatial contrasts in heat and water balances between
      snow-covered and snow-free sites 187
8.13 Energy and heat balance components at tundra and beach
      sites 189
8.14 Aerial view of the north coast of Ellesmere Island in early
      August 190
8.15 Aerial view of glaciers in late August 191
9.1   Geographical regions in the low arctic and subarctic
      regions 199
9.2   Landscapes near treeline 200
9.3   Aerial views of lowland and wetlands 201
9.4   Seasonal course of temperature and energy balance
      components at a high subarctic site 203
9.5   Summer energy and water balances of tundra and forest
      sites 204
9.6   Comparative energy and water balances and surface and
      atmosphere controls of a sedge-fen 205
9.7   Comparative energy and water balances of dryland tundra
      and wetland tundra 207
9.8   Ratio of solar radiation received at the forest floor to that
      above the canopy of subarctic forests 210
9.9   Ratio of tundra to subarctic forest Bowen ratios 212
9.10 Conceptual model of evapotranspiration sources in a
      subarctic forest 213
9.11 Energy balance of a shallow tundra lake 216

9.12   Pairwise comparison of diurnal temperature changes in eight tundra ponds 217

9.13   Priestley-Taylor evaporation coefficient over a shallow tundra lake 219

10.1   A winter view of the Cascade Mountains in southern British Columbia 223

10.2   View north from Peyto Peak in October 223

10.3   Patterned ground on a fellfield tundra surface 224

10.4   A glacier ablation zone in summer 224

10.5   Generalized physiographic divisions of the Canadian Cordillera 226

10.6   Thermally generated winds in mountain terrain 227

10.7   Hourly radiation balances for different alpine surfaces 230

10.8   Five-day sequence of albedo changes that attend melting of snowcover 232

10.9   Net radiation for selected alpine surfaces 234

10.10 Hourly energy balance for snow-covered tundra in winter 236

10.11 Surface energy balances of Peyto Glacier 237

10.12 Tundra energy balances and aerodynamic and surface resistances in summer 239

10.13 Generalized energy balance diagram for alpine surfaces in summer 241

11.1   Forest regions of Canada 248

11.2   Diurnal radiation and energy balance regimes for a mixed forest at PNFI 250

11.3   Energy balance regime for a Douglas-fir canopy in British Columbia 253

11.4   Diurnal net fluxes of carbon dioxide in a Douglas-fir stand 255

11.5   Evolution of carbon after logging an old-growth Douglas-fir forest 256

11.6   Typical profiles of mean wind and momentum flux within a dense spruce forest 257

11.7   Typical temperature profiles within forests during day and night and related heat fluxes 258

11.8   Examples of time-series data of wind speed and temperature within forests 260

11.9   Comparison of snow accumulation and melt in clearcuts and forests 262

11.10 Daily net radiation and evapotranspiration from a partially closed aspen-alder-willow stand 264

11.11 Comparison of modelled and measured mean root-zone water content for a thirty–year-old stand of Douglas fir 265

11.12 Diurnal course of evapotranspiration and canopy conductance from a young Douglas-fir stand 266

11.13 Wave forest at Spirity Cove, Newfoundland 270

12.1   Diurnal variation of the radiation balance components for a maize crop 278

12.2   Sensible and latent heat fluxes from a leaf may be quantified using the Ohm's law analogy. 279

12.3   Solar radiation, air temperature, and leaf temperature in a maize crop 280

12.4   Distribution of agricultural activities in Canada 281

12.5   Climatic moisture index and growing season classes for Canada 282

12.6   Energy fluxes for unstressed and water-stressed alfalfa 286

12.7   Winds at the top of a corn crop and from a standard 10-m tower nearby 287

12.8   Energy fluxes for a Prairie wheat crop 289

12.9   Diverse agricultural landscape found in the Annapolis Valley 290

12.10 A temperature transect down the slopes of the Beaver Valley 291

12.11 Frequency of various classes of soil temperatures 293

12.12 A surface mulch created by growing rye as a winter cover crop 294

12.13 Rows of trees planted perpendicular to the prevailing wind 295

12.14 The relationship between minimum relative wind speed and optical porosity of windbreaks 296

12.15 Corn yield may be enhanced by windbreak protection. 297

12.16 Irrigation permits intensive fruit farming in the Okanagan Valley 298

13.1   Distribution of the "misery" index across Canada 304

13.2   Idealized vertical structure of the urban atmosphere 305

13.3   Idealized flow around buildings 306

13.4   A wind-tunnel model of downtown Vancouver 307

13.5   Changes in wind profile across a city 308

13.6   Ratio of incoming radiant fluxes at industrial and rural sites 309

13.7   Radiation balance at rural and urban sites 311

13.8   Daily variation of anthropogenic heat flux density in a suburb 312

13.9   Ensemble hourly average energy balance in a suburb of Vancouver 313

13.10 Flux density ratios in a suburb 313

13.11 Differences of average energy fluxes at rural and suburban sites 315

13.12 Energy balance at rural and urban sites 315

13.13 Diurnal energy balances of an urban canyon and a lawn 316

13.14 Diurnal variation of surface conductance in a suburb 317

13.15 Urban effects on humidity in Edmonton 318

13.16 Urban canyon layer heat island in Winnipeg and Montreal 320

13.17 Temporal variation of heat island in Edmonton 320

13.18 Maximum heat island intensity in Canadian cities 321

13.19 Urban boundary layer heat island structure in Montreal 323

14.1 Variation in globally averaged surface air temperature 332

14.2 Change in surface air temperature over Canada 333

14.3 Change in winter in Canada for a doubling of CO_2 335

14.4 Change in summer in Canada for a doubling of CO_2 336

14.5 Change in summer soil moisture in Canada for a doubling of CO_2 337

14.6 Ecoclimatic provinces in Canada at present and for a doubling of CO_2 344

15.1 Profiles of air temperature and elevation up Yonge Street, Toronto 354

Symbols, Units, and Quantities

ITALIC LETTERS (UPPER CASE)

B	–	Bowen ratio (climatological)
B_i	–	Thermal quality of snow
C	J m^{-3} °C^{-1}; –	Heat capacity; runoff ratio ($= N/P$)
CMI	–	Climate moisture index
D	–; –	Bulk transfer coefficient; dryness ratio ($Q^*/L_v P$)
D_W	mm or kg m^{-2} s^{-1}	Ground water drainage
DF	mm	Crop water deficit
E	mm or kg m^{-2} s^{-1}	Evapotranspiration
E_I	mm or kg m^{-2} s^{-1}	Evaporation of intercepted water
E_{eq}	mm or kg m^{-2} s^{-1}	Equilibrium evaporation
F	y	Flushing time of lakes
F_C	g CO$_2$ m^{-2} s^{-1}	Flux density of carbon dioxide
GDD	°C d	Growing degree day
GW	mm	Ground water
I	mm	Infiltration
K	m^2 s^{-1}	Eddy diffusivity
$K\downarrow$	W m^{-2}	Global solar radiation
$K\uparrow$	W m^{-2}	Reflected solar radiation
K^*	W m^{-2}	Net solar radiation
K_D	W m^{-2}	Diffuse solar radiation
K_S	W m^{-2}	Direct-beam solar radiation
L	m	Obukhov length; length scale
$L\downarrow$	W m^{-2}	Incoming longwave radiation
$L\uparrow$	W m^{-2}	Outgoing longwave radiation
L^*	W m^{-2}	Net longwave radiation
L_f	J kg^{-1}	Latent heat of fusion
L_v	J kg^{-1}	Latent heat of vaporization
M_f	mm °C^{-1} d^{-1}	Temperature index for snowmelt
N	mm	Total landscape runoff

P	mm or m s^{-1} or kg m^{-2} s^{-1}	Precipitation
PE	mm or kg m^{-2} s^{-1}	Potential evapotranspiration
$Q\!\downarrow$	W m^{-2}	Incoming solar and longwave radiation
$Q*$	W m^{-2}	Net radiation flux density
Q_A	W m^{-2}	Net advected heat energy
Q_B	W m^{-2}	Net heat storage in a plant canopy
Q_E	W m^{-2}	Turbulent latent heat flux density
Q_F	W m^{-2}	Anthropogenic heat flux density
Q_G	W m^{-2}	Sub-surface heat flux density
Q_H	W m^{-2}	Turbulent sensible heat flux density
Q_M	W m^{-2}	Heat flux density due to melt and freezing
Q_P	W m^{-2}	Heat flux density input through precipitation
Q_S	W m^{-2}	Net heat storage in a volume per unit horizontal surface area
Q_{Sa}	W m^{-2}	Heat storage in canopy air
Q_{Sp}	W m^{-2}	Heat storage in phytomass
Q_{Sw}	W m^{-2}	Heat storage in lakes
Q_Z	W m^{-2}	Heat flux density at the base of a volume of lake or ocean water
R	mm or m s^{-1} or kg m^{-2} s^{-1}; m^3	Net surface runoff; net lake discharge
S	mm	Moisture storage
S_f	–	Fractional salinity of water
SM	mm	Soil moisture
SWE	mm	Snow water equivalent
SWE_M	mm	Snow melt water
T	°C	Temperature
U	J m^{-3}	Internal energy of snowpack
V	m^3	Water volume
W	mm	Water; water discharge
WUE	kg mm^{-1}	Water-use efficiency

ITALIC LETTERS (LOWER CASE)

c	kg m^{-3} or ppmv	Concentration of carbon dioxide
d	m	Zero-plane displacement
e	Pa	Vapour pressure
g	m s^{-1}	Conductance
h	J m^{-3}; m	Snowpack enthalpy; height of vegetation canopy or windbreak
k	–	von Karman's constant
k_s	W m^{-1} °C^{-1}	Soil thermal conductivity
p	mm or Pa	Meltwater loss from snowpack base or Total atmospheric pressure
r	s m^{-1}	Resistance

s	Pa °C^{-1}	Slope of saturation vapour pressure versus temperature curve
t	s	Time
u	m s^{-1}	Horizontal (x-plane) wind speed
u_*	m s^{-1}	Friction velocity
v	m s^{-1}	Lateral (y-plane) wind speed
vdd	kg m^{-3}	Vapour density deficit
vpd	Pa	Vapour pressure deficit
w	m s^{-1}	Vertical (z-plane) wind speed
z_i	m	Depth of convective mixed layer
z_o	m	Roughness length
$z*$	m	Depth of surface roughness layer

GREEK LETTERS

α	–	Surface albedo
α_P	–	Priestley-Taylor evaporation coefficient
α_v	m^{-1}	Extinction coefficient for radiation in water
α_L	–	Reflection coefficient for longwave radiation
β	–	Bowen ratio
γ	Pa °C^{-1}	Psychrometric constant
δ	m	Height of internal boundary layer
ε	–	Emissivity
θ	°C	Potential temperature
θ_s	–	Volumetric soil moisture
$\varkappa$	–	Fraction of grid cell receiving precipitation
λ	m^2 s^{-1}	Thermal diffusivity
μ	J m^{-2} s$^{-\frac{1}{2}}$ °C^{-1}	Thermal admittance
ρ	kg m^{-3}	Density
σ	W m^{-2} K^{-4}	Stefan-Boltzmann constant
τ	Pa; kg m^{-1} s^{-2}	Momentum flux density
Ω	–	McNaughton-Jarvis coupling factor

COMMON SUBSCRIPTS (UPPER CASE)

A	Advection
B	Biomass
C	Carbon
E	Latent heat
G	Subsurface
H	Sensible heat
I	Interception
M	Momentum
P	Photosynthesis
PE	Potential evapotranspiration

S	Storage change
V	Water vapour

COMMON SUBSCRIPTS (LOWER CASE)

a	Air; aerodynamic
b	Boundary layer
c	Canopy
f	Fusion
h	Top of Canopy
i	Ice or Snow
im	Inhomogeneity
in, out	Input or output
l	Leaf
lg	Large
o	Surface value; ocean
r	Reflected
s	Soil; stomate
sm	Small
sn	Snow
v	Water vapour; vaporization
w	Liquid water
x	Horizontal, along-wind
y	Lateral, across-wind
z	Vertical height or depth

COMMON SUPERSCRIPTS

'	Instantaneous deviation from mean
–	Time average

Preface

This book is about the climate of the near-surface zone where almost all terrestrial life exists. It is a layer, only a few metres in thickness, which straddles the earth's surface. It includes an array of climates that changes from place to place and from one moment to the next with great rapidity. This small-scale variability is directly related to the nature of the surface itself – hence "surface climates." The surface of any system is critical because its properties mediate the flows of heat, water, gases, and momentum between itself and the sun and the atmosphere above. In so doing, properties such as reflectivity, conductivity, wetness, and roughness determine how much heat, water, and gas the system retains or loses. They also influence its temperature, humidity, and gas concentration, how much drag is exerted, and hence the wind speed. Full understanding of the characteristics of a surface climate requires knowledge both of the special mix of properties possessed by a surface and the exchange processes acting on it. This book attempts to express the range of that knowledge within the context of the diversity of surfaces found in Canada.

This volume grew out of the realization that many of Canada's climatologists had been studying surface environments within the country. A body of work that covered most of the ecological and regional units of Canada had not been brought together in a comprehensive fashion. Further, whereas in the past much research and attention were directed towards presentation of the spatial patterns of weather elements such as temperature, wind, and precipitation, the unifying paradigm of more recent research was the influence of the surface on climate. This work directed efforts to measurement, modelling, and prediction of the physical processes defining the surface climates of the Canadian landscape. This volume is a synthesis of these scattered inquiries.

The nature, diversity, and pattern found in the surface climates of Canada are a direct outcome of its surface geography and macroscale climatology. The climates and ecological regions of Canada are a consequence of it position in the mid- to high latitudes, which places it in the continuum of the northern hemispheric circumpolar ecosystems. The book considers Canada's landscape to be composed of a series of natural and human-modified landscapes, each with its own set of special surface climates. Most chapters are organized within a similar template. In each case, the focus is on the role of the physical nature of the surface as a control on exchanges of heat, mass, and momentum by the processes of radiation, convection, and conduction and the resulting thermal, moisture, and wind climates in the atmospheric boundary layer and just beneath the surface. An abiding theme is the close link between energy and water. Therefore the process of evaporation, which is central to both balances, is emphasized.

The book opens with four introductory chapters, on the concepts of surface climate. The first chapter is by F. Kenneth Hare, a Canadian who pioneered research in physical climatology, and to whom this book is dedicated. It lays out the large-scale context of the physical climatology of Canada. It introduces the processes, balances, and dynamic links between the surface and the atmosphere that create and maintain the diversity of the surface climates found in Canada. Chapter 2 introduces the physical processes that operate near the ground's surface and develops the conceptual frameworks, models, and nomenclature used in the rest of the book. For simplicity, it considers mainly the vertical (one-dimensional) exchanges and climate profiles existing at a simple, flat, and extensive surface in summer. In the real world, surfaces are often complex, uneven, always limited in extent, and, in much of Canada, it is the winter that dominates. Thus chapter 3 looks at how surface climates are affected by spatial patchiness, sudden changes in surface properties, and hilliness. There follows a chapter (4) dealing with snow and ice – the almost-universal surface cover, which suddenly masks the properties of the underlying landscape for much of the year.

The next seven chapters (5–11) look at the major natural surface environments of Canada. With one-fifth of the world's fresh water and a coastline that borders on three of the world's oceans (Atlantic, Arctic, and Pacific), Canada's surface incorporates a vast area of water. This is the subject of three chapters – one (5) about the oceans and their coasts, a second (6) dealing with large lakes (especially the Great Lakes) and a third (7) on the wetlands – which provide a link between the aquatic and terrestrial environments. The next two chapters examine the surface climates of the cold, high-latitude regions. Chapter 8 is devoted to the climates found in the Canadian Arctic archipelago; chapter 9 is concerned with the low arctic and subarctic region, composed of tundra, open woodland, and lake surfaces. Alpine tundra and glacier environments are examined in chapter 10, with special emphasis on the mountains of the western Cordillera, and there is a chapter (11) devoted to the climates of the large expanse of forested terrain.

The final part of the book (chapters 12–14) considers surface environments already strongly influenced by human activity and examines the prospects for future climate change. Chapter 12 deals with the transformation already wrought by agricultural development as the original surface cover has been altered to support production of grain and field and horticultural crops. Nothing so drastically disrupts the original surface cover as urban development, which is the subject of chapter 13. It shows that most Canadians live in an urban climate that they have inadvertently created for themselves. The combination of the direct climate effects of surface-cover change and the indirect effects caused by the release of radiatively active gases and aerosols by human activity has implications for all the future surface climates of Canada.

The book closes with an epilogue by R.E. (Ted) Munn, another Canadian pioneer in the study of surface climates. He reflects on developments in Canadian climatology and looks towards tomorrow's challenges.

This book arises as a result of a cooperative effort between the Canadian Association of Geographers and McGill-Queen's University Press. It is one in a series of books that presents current summaries of Canadian research on a range of geographical subjects. The focus of this book is on understanding surface climates and defining the processes responsible for their creation and control. The geographical focus is restricted to Canada, but of course the surface types are not. Thus, though this book is a national project, its content applies to many other areas in the mid- and high-latitude parts of the world. The book assumes that the reader has received an introduction to weather and climate, such as that found in some high school and all university courses in physical geography or climatology. To help bridge any gaps, we have provided a glossary of technical terms. We expect the book to be useful, in conjunction with an introductory text in microclimate, to illustrate the range of surface climates within the country. The units and dimensions used are those of the Système International d'Unités. There is no universally agreed on set of symbols for meteorological quantities. Making our selection (see "Symbols, Units, and Quantities") was all the more difficult because we include work from hydrology, limnology and oceanography.

We gratefully acknowledge the editorial assistance of Carolyn Whitehead of Simon Fraser University and the editorial and production assistance of Joan McGilvray, Susanne McAdam, and their colleagues at McGill-Queen's University Press. All graphics have been prepared by Eric Leinberger of the Cartographic Services Unit of the Department of Geography at the University of British Columbia, and we are very grateful to him for his excellent contribution. Many of the figures are based on materials that have appeared in scientific journals and books, and their sources are denoted in each caption. Credit for the photographs should be attributed to: Alberta Agriculture (Figure 3.11), W.G. Bailey (Figure 12.16), Canada Centre for Inland Waters (Figure 6.3), Terry Gillespie (Figures 12.12 and 12.13), Richard Heron (Figure 8.12), George Hunter (Figure 3.1d), Garth Lenz (Figure 3.1c), National Air Photographic Library, Ottawa (Figures 3.1a, b, f, 3.12, 5.2, 8.7, 8.14, 8.15, 9.3, and 12.9), Alexander Robertson (Figure 11.13), Nigel Roulet (Figures 7.2, 7.3, and 7.4), Wayne Rouse (Figure 9.2), Ian Saunders (Figures 10.1, 10.2, 10.3, and 10.4), Alfred Siemens (Figure 3.1e), University of Western Ontario Wind Tunnel Facility (Figure 13.4), and Hok Woo (Figures 8.2, 8.3, and 8.10).

We wish to thank the authors of the individual chapters. It has been a privilege and pleasure to work with them. This volume presents a cumulative statement of the knowledge and understanding that they and their colleagues have acquired on Canada's surface climates. They have provided the insight, effort, and information that contributed to the evolution of this volume.

We extend our warmest appreciation to Leah Skretkowicz, Midge Oke, and Margaret Rouse for their unfailing encouragement, support, and understanding.

W.G. Bailey, White Rock, British Columbia
Tim Oke, Vancouver, British Columbia
Wayne Rouse, Hamilton, Ontario

The Surface Climates of Canada

CHAPTER ONE

Canada's Climate: An Overall Perspective

F. KENNETH HARE

INTRODUCTION

This book explores Canada's surface physical climates. It concentrates on Canada's surface because the strongest exchanges take place there and because the ecological and economic impact of climate arise directly from surface processes. However, the total climate also contains a large dynamic component that is brought to Canada by winds and ocean currents. In this first chapter we explore the concept of the total climate – dynamic and physical – and relate it to other things. The rest of the book examines such relationships in the context of specific surface types: forests, lakes, soils, agricultural fields, glaciers, and others. It also considers climatic changes that may occur because of shifts in global physical climatic processes.

CLIMATE'S DYNAMIC COMPONENT

Canada lies within the northern hemisphere's westerly belt. The vortex of westerly winds that encircles the north pole covers all of Canada's territories at all times of year. This vortex is a permanent feature of the general circulation of the atmosphere. It exists because of unequal solar heating of tropical and polar belts and because of the spin of the earth around its polar axis. This combination of forcing factors – things that drive the atmospheric and oceanic engines – ensures that easterly surface winds predominate in the intertropical belt, whereas westerlies occur over middle and high latitudes of both hemispheres, at least on the average (Figure 1.1). These westerlies are sometimes called the Ferrel westerlies, in honour of William Ferrel's celebrated account of winds and storms, written in 1857.

The westerlies dominate Canada's dynamic climate. Their mean strength is greater in winter than in summer – because differences in temperature between pole and equator are also greater in winter. The latitude and altitude of the jet streams – the high-speed cores of the westerlies – also vary. In winter the strongest westerlies usually lie across the United States and the Gulf of Mexico, and in summer, across southern Canada, close to the U.S. border.

In the westerly belt, there are strong poleward temperature gradients (Figure 1.2) in the troposphere (essentially the bottom 10 km of the atmosphere). The strength of the

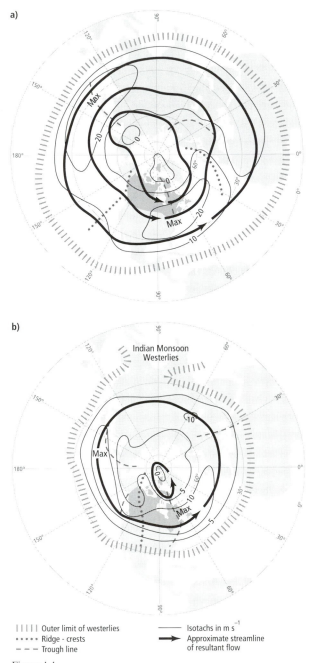

a)

b)

Indian Monsoon
Westerlies

| | | | | Outer limit of westerlies   ——— Isotachs in m s^{-1}
• • • • • Ridge - crests   ➡ Approximate streamline
– – – Trough line   of resultant flow

Figure 1.1

The westerlies in mid-troposphere (roughly 5.5 km above sea level) in
(a) January and (b) July. Heavy arrows show mean flow; lighter isolines
give mean wind speed in m s^{-1}. Average trough and ridge crests are also
shown.

Sources: Hare (1960), data from Lahey et al. (1958).

westerlies increases with height to one or more jet streams, usually at 9 to 12 km, which is also the layer in which transcontinental aircraft fly. Having to face the winds, westbound aircraft usually take an hour or two longer to cross the country than do eastbound flights. At greater heights, in the lower stratosphere (10 to 25 km), temperatures are highest in latitudes 45°N to 60°N. This means that the westerly winds decrease with height at these levels (Figure 1.2), to a maximum height of 20 to 25 km (though they are overlain by other westerlies of different origin in the colder months). On Figure 1.2 – a summer example – the summit of the westerlies is near 23 km, and the jet core lies just north of Toronto, at about 11 km.

Why do Canada's climates differ so much from region to region? Why do Vancouver, BC, and Winnipeg, Man., have such contrasted winters, given that they are in similar latitudes? Why does Canada have both arctic and temperate climates within its territories? And why are summer and winter so different?

The country's huge north–south extent guarantees a wide spread of temperatures. Pelee Island, Ont., is at 42°N latitude. Cape Columbia, NWT, is at 83°N. Southernmost Ontario had a temperate deciduous forest before it was felled, and coastal southern British Columbia used to have magnificent rainforest, as well as southern trees such as the madrona (arbutus) and the Garry oak. Yet almost a third of the country has tundra-covered and permanently frozen ground, and northern sea channels may be choked with ice year round. All this is natural to the latitude range.

Canada's physiography also influences the climate. The Western Cordillera – the complex of mountains, plateaus, and valleys in British Columbia, western Alberta, and Yukon – is an impressive barrier across the westerlies. The westerlies can surmount it, dropping a great deal of rain and snow in transit, but for much of the time do not reappear at ground level east of the Cordillera. Instead, at low altitudes (below 4 km) the way is frequently open for arctic airstreams to plunge southward, bringing the cold spells of winter and cool refreshment in summer. Also warm, humid tropical air can stream northward up the Mississippi Valley into central and eastern Canada, though in the cooler seasons it usually rises above colder air ahead, again releasing rain and snow. Occasionally it may even reach Alberta (as it did in July 1987, on the occasion of Edmonton's worst tornado). These two intrusions resemble the summer and winter monsoons of eastern Asia. Indeed, eastern Canada's arctic outpourings in late winter and spring are more persistent than the northwest monsoon of China and Japan. The never-ending interplay of airstreams and disturbances from the Pacific, the Arctic, and the tropics gives us our dramatically changeable weather.

The oceans also exert a strong influence. The Pacific is not specially warm for its latitude range, but because it stays unfrozen it gives the west coast a mild, wet, and cloudy winter. The Atlantic coast is paralleled by the southward-flowing Labrador Current, which carries pack ice and icebergs (from Greenland) through winter, spring, and early summer. East winds are cold and fog-laden in spring and summer; coastal Labrador, the island of Newfoundland, Prince Edward Island, and Nova Scotia experience a very late spring in consequence. Hudson Bay, too, retards the eastern spring, with its annual pack ice not disappearing until July or early August. In fall, the Bay does not extensively freeze until December. Together with thousands of lakes on the Laurentian Shield, the Bay gives unbroken stratocumulus cloud over eastern Canada

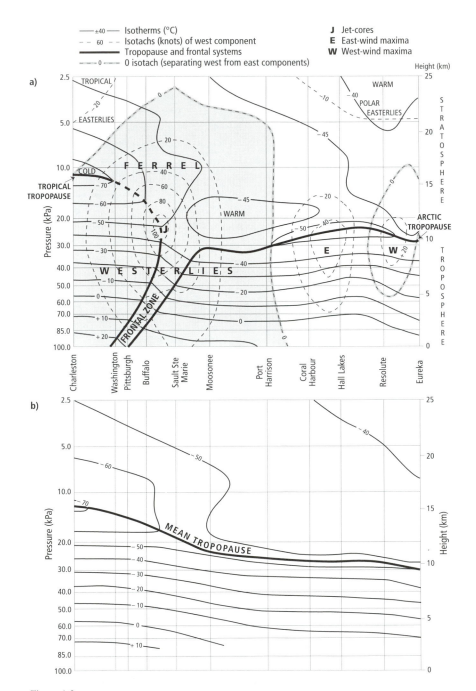

Figure 1.2
The westerlies in summer; (a) vertical cross-section for 31 July, 1958 along 80°W. Isolines show west-wind component (isotachs) and temperatures to 25 km. (b) Cross-section for July 1958 as a whole shows how latitudinal variation in position of the jet core smooths out distribution of mean temperature. *Source*: Hare (1960).

during autumnal arctic outbreaks, with heavy snowfalls in the lee of each open water body. Not until freeze-up is complete do arctic airstreams carry the clear but bitter weather of mid- and later winter (Hare and Thomas, 1979).

The westerly regime over Canada is thus much modified by the physiography of the continent. Warm and cold airstreams criss-cross the country, with cold weather alternating with mild or warm conditions. Surface winds may blow from other directions for long periods. But the westerly control is always there, even in the episodes known as blocking, when deep-seated anomalies (like gigantic tumours) deform the circumpolar westerly pattern.

THE ROLE OF DISTURBANCES

The fluctuating airstream movements just described are orchestrated by large-scale disturbances of dynamic origin, familiar to weathercast viewers as cyclones and anticyclones – regions of low and high pressure, respectively. Above 4–5 km these merge into troughs and ridges (waves) in the westerlies. Some of these disturbances are stationary or slow-moving, others fast-moving towards the east, steered by the westerly current in which they are embedded.

Among the slow-moving disturbances, the most familiar is a ridge of high pressure whose north–south axis tends to lie over the Western Cordillera or the eastern Pacific (Figure 1.1). A complementary trough of low pressure often sits over eastern Canada, between Hudson Bay, the Great Lakes, and the Atlantic shoreline. These features imply that upper winds (above 3–4 km) over central Canada should flow on the average from a little north of west. These winds in turn tend to steer disturbances from the Pacific across northern British Columbia and Alberta (where they often start to intensify) towards the Great Lakes and the Maritimes. When the eastern trough is further west than usual, disturbances from the u.s. south and midwest are often steered towards the St Lawrence Valley. The slow-moving westerly disturbances, often called long waves, thus steer the faster-travelling highs and lows and give contrasting spells of weather. When they are displaced by more assertive ridges, called blocking highs (as they often are in spring and fall), the entire scheme of things is altered for days or even weeks. Blocking may bring unseasonable thaws in winter and cool or dry weather in summer and fall.

The real weather-makers are the smaller travelling disturbances. Of these, the dominant form is the travelling cyclone, often called a depression, or just a low, to distinguish it from its smaller but violent tropical cousin, the hurricane. At least five families of cyclones affect Canada.

(1) Pacific systems are large, intense storms moving on to the BC or Alaskan coasts, as often as once every two days or less in winter. These are responsible for heavy rains on west-facing hillsides, with deep snows at higher levels. In summer most of these systems affect Alaska, but the north coast is still covered by the onshore winds and rain of their outer circulations.

(2) Alberta Lows are less intense systems that become organized over Alberta or Montana in the lee of the Rockies, usually when a trough in the overlying westerlies enters from the Pacific. They tend to move rapidly east or southeast and form a major element in the climate of central and eastern Canada. In spring and summer,

systems of a related sort may move slowly across the Prairies, giving copious rains or thundershowers. When they fail to appear, drought may be severe.

(3) Central continental cyclones are systems that form over the u.s. interior, chiefly in Colorado, Texas, Oklahoma, Nebraska, Kansas, and Missouri, and head northeastward towards the Great Lakes, the Ohio and St Lawrence valleys, or New England. These cyclones are confined largely to the cooler seasons and cause much of the stormiest, rainiest, and snowiest weather of Ontario, Quebec, and the Maritimes. They reach peak intensity when they pass off the Atlantic shore.

(4) Arctic lows are cold-cored cyclones affecting mainly northeastern Canada. They are often the persistent remains of old cyclones of the previous groups. They may drift across arctic Canada for days, bringing light precipitation, even in winter. "Cold lows," as the family is collectively called, occasionally form over southern Canada or the United States and may remain almost stationary for days, giving prolonged periods of cold, damp weather on their northern flanks (or anomalously mild weather in winter).

(5) Atlantic coast systems form over or just off the southern and eastern coasts of the United States and travel northeastward across Quebec, the Atlantic provinces, or the Grand Banks, bringing strong winds and heavy precipitation to those regions. In summer and fall some of these storms originate as tropical hurricanes, which are in the process of becoming mid-latitude cyclones. A small number of these hurricanes and their mid-latitude successors (for example, Carol and Hazel in 1954) may move inland into the Great Lakes or St Lawrence Valley areas. Most approach from east of the Appalachian Mountains, but some may travel to Canada via the Mississippi Valley.

These storm systems are plainly visible on weather maps. Typically they measure 15 to 60 degrees of latitude or longitude across. Many are roughly oval in plan, usually with the longer axis directed sw–ne, so that they appear tilted on the weather map. This feature arises from their need to carry westerly momentum poleward, as Harold Jeffreys pointed out in 1926. They also pull tropical, moist air northward and cooler, drier air southward. The various cyclone families thus carry out most of the required meridional (north–south or vice versa) transports of heat and moisture. They also generate vast amounts of kinetic energy, enabling the westerly current to maintain itself against frictional drag from the earth's rough surface.

What causes these disturbances, and how do they relate to long-established (since World War I) ideas about homogeneous airmasses separated by fronts? Any textbook of weather and climate can provide the reader with details of these Norwegian, Finnish, and Swedish ideas – the Norwegian, or Bergen school of analysis – which still heavily influence the analysis of weather maps by forecasters. The Bergen scheme viewed the atmosphere as lying over homogeneous source regions (tropical or polar, maritime or continental) long enough for huge bodies of air (airmasses) to become horizontally uniform in temperature and humidity. These airmasses would emerge from their source regions to form parts of the westerly current, within which they would meet one another along sharp boundaries called fronts. The fronts were actually gently sloping surfaces, with the warmer airmass tending to climb upward over wedges of cold air below, thereby cooling to yield clouds and precipitation. Frontal cyclones formed, in the classical model of J. Bjerknes and his colleagues, as unstable, eastward-moving waves on the frontal surface.

Though the Bergen system has many elements of truth within it, facts now available require it to be modified in several ways. First, daily upper-air soundings show that the mid-latitude westerlies are invariably baroclinic – they always contain horizontal gradients of temperature and humidity. Large, uniform airmasses are confined to the tropics and polar regions. Winds from these sources that enter the westerlies always become baroclinic, not only because they are differentially heated or cooled but because of internal dynamic processes. Hence they cease to meet the definition of an airmass. It is more realistic to speak instead of airstreams from specific source regions. The boundaries between contrasting airstreams of different origins are marked by fronts, which are zones of sharp transition, rather than actual discontinuities of temperature, humidity, and wind velocity.

Second, cyclone formation follows the traditional model less frequently than often supposed and does so mainly in preferred localities (notably in the eastern interior of the United States, and the u.s. Atlantic coast). The western interior plains of North America are one of the world's most prolific sites of cyclone formation (cyclogenesis), but the mechanism often differs from the Bjerknes model. Pressure falls over Alberta, Montana, Colorado, Oklahoma, or Texas, in a broadly baroclinic part of the upper westerly current. Such currents are inherently unstable in response to a wide class of initial perturbations. As the young cyclone moves eastward and deepens, the motion itself tends to create a cold front in the storm's southwest quadrant, and sometimes a warm front ahead of the centre. In other words, the moving and developing cyclone appears to create, or at least intensify, the frontal structures, which become more clearly defined as the storm evolves. In effect, the fronts are sharpenings of the generally baroclinic structure.

There has also been a change of view as to the mechanism of precipitation within the moving cyclone. Figure 1.3 shows a sketch of the pattern now seen as typical. It shows how air moves through the advancing cyclone wave. This view, based on extensive observation and analysis in the United Kingdom and the central United States, suggests that most precipitation falls from two broad rising airstreams, or "conveyor belts," one warm and the other cold (Browning et al., 1973; Harrold, 1973; Carlson, 1980; and, in particular, Browning and Roberts, 1994). The warm conveyor belt consists of a broad, warm, and moist airstream that moves northeastward along, above, and ahead of the cyclone's cold front. It then rises above the warm front, at the same time turning towards the east, and then southeast, with the warm air moving faster than the cyclone wave itself. The cold conveyor belt is made up of cold easterly winds that move westward north of the cyclone centre. They are drawn upward as they move, because their motion is strongly convergent, and ultimately turn back eastward as they reach the level of westerly control. Much of the steady rain or snow that falls over eastern Canada (and over the Prairies in spring and summer) can be seen on radar screens to move from the east, often from the Atlantic, not from overrunning westerlies. Between these two precipitation belts, drier air may be injected into the cyclone from the southwest, giving the entire cloud mass a comma shape on satellite images. The warm conveyor belt was identified and carefully mapped over fifty years ago by Jerome Namias and C.G. Rossby, but only recently have satellite observations and detailed temperature soundings revived interest in one of world climate's most striking features.

In addition, we now realize that the precipitation affecting Canada is often triggered by smaller mesoscale systems, only tens to a few hundreds of kilometres across (massed thunderstorms, for example, or the remarkable fall and winter snow belts in

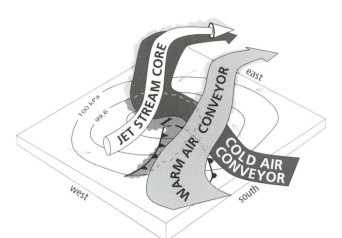

Figure 1.3
Schematic of a travelling mid-latitude cyclone, showing warm and cold
fronts of traditional analysis. Superimposed (shaded area) is the comma
cloud, caused by indrawing of three elements – a moist, warm southwesterly conveyor belt that moves faster than the low centre and ultimately
climbs over the cooler southeasterlies ahead of the warm front; a moist,
cold conveyor belt consisting of air drawn in from ahead of the cyclone's
centre, which rises to create the rain and cloud north of the centre and in
the comma cloud's hangback; and dry air pulled in to create the tuck in the
comma cloud. All are shown as trajectories with respect to the moving
centre of low pressure.
Sources: Simplified from Carlson (1980) and Browning and Roberts
(1994).

the lee of the Great Lakes, Hudson Bay, and the Gulf of St Lawrence). These are often
missed by the synoptic analysis and in many cases are hard to relate to the larger synoptic systems that dominate the weather map. Only since satellite imagery became
available have we been able to compile an adequate climatology of these mesoscale
systems.

Taken together, these factors (physiography, mean westerly flow, long-wave systems, and various disturbances) account for the astounding day-to-day variability and
seasonal contrasts of Canada's climates. They account for regional disparities and for
the element of hazard that is always present – gales, floods, tornadoes, hurricanes,
fog, ice-storms, and many more. Much of Canada's daily weather is benign, at least in
the warmer months. But Canadians still face one of the earth's most challenging environments. Climatic stress and hazard play a key role in national life.

THE PHYSICAL COMPONENTS
OF CLIMATE

How does the physical climate of Canada relate to the foregoing account? How can
we characterize the outputs and receipts of mass and energy at the land and water
surfaces?

The list of quantities exchanged or converted at or very near the surface is extensive. It includes:

- water substance, in the form of rain, snow, dew, frost, and evaporation;
- sensible and latent heat (the energy detected by thermometers, plus that involved in changes of phase of water), in the form of radiation, precipitation, and convection;
- momentum, because of the drag of the earth on the winds (or the reverse);
- kinetic energy, as a consequence of momentum;
- carbon dioxide and methane, chiefly because of processes involved in life;
- ozone – the triatomic form of oxygen;
- nitrous oxide, chiefly from soil denitrification;
- oxygen and nitrogen;
- trace gases such as helium and radon (from radioactive decay) and various organic substances;
- a wide variety of particles, including spores and pollen from plants;
- many pollutants, such as sulphur dioxide, emitted from smokestacks and deposited chiefly as sulphates in acid rain, or directly as dry deposition; oxides of odd nitrogen; and the various synthetics released by industry, such as the notorious but useful chlorofluorocarbons (CFCs).

For a variety of reasons climatologists have generally confined their attention to a few of the above fluxes – radiation, precipitation, convective fluxes of sensible heat and water, the soil heat flux, and various measures of atmospheric state, such as pressure, temperature, and humidity. They have treated the remainder as pertaining to air quality or to biological processes, and hence as lying within other professional domains.

However, in recent years the scope of climatology has been broadened to include some of these neglected processes. There is no logical reason why the list should be split up, because the physical and dynamic mechanisms, though not the chemical, are the same for all.

The Radiative Energy Exchanges

The energy balance at the earth's surface is forced by solar energy delivered as radiation; the forcing energy-drive of the entire climatic system begins when incoming solar radiation is absorbed – converted into sensible or latent heat. This happens mainly at the earth's surface. In this section I examine the time-averaged radiative fluxes over Canada, including not only solar radiation but the longwave radiation originating within the planetary system itself.

Figure 1.4 shows the mean daily solar radiation reaching the surface of Canada in midwinter (January) and midsummer (July) and the average for the whole year (McKay and Morris, 1985). The figures, standardized for the decade 1967–76, estimate the energy content of direct solar irradiance, plus diffuse solar irradiance from scattering and diffuse reflection (skylight) within the atmosphere. Losses resulting from interception by clouds are hence allowed for. The figures are based on a small number (fifty-seven) of long-term monitoring records, reinforced by modelling estimates. The observational basis is weak, because meteorologists, for historical reasons,

a)

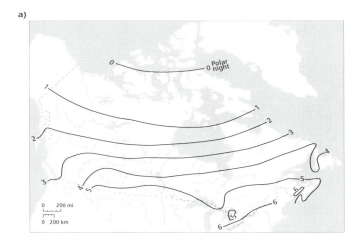

b)

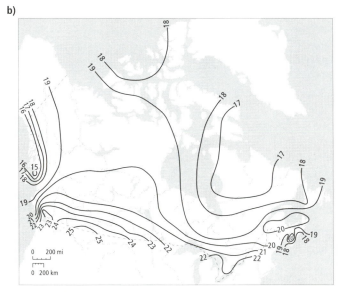

Figure 1.4
Mean daily global solar radiation in (a) January and (b) July, and (c) annual mean. Units are MJ m^{-2} d^{-1}. Equivalents in W m^{-2} are given, for twenty-four–hour averages for six representative stations in (c).
Source: McKay and Morris (1985).

have tended to neglect direct measurement of radiation (Canada, Atmospheric Environment Service, 1982).

As might be expected, Canada's surface receives only a moderate amount of solar radiation. In midsummer, southern districts receive a twenty-four–hour average of 230 to 300 W m^{-2} on a horizontal surface. This compares with about 350 W m^{-2} in sunny central California. Most parts of the tundra and boreal-forest biomes receive 200 to 230 W m^{-2}, which is quite high because the many hours of daylight offset the relative weakness of the low-angle solar beam. The sunniest area is southern Saskatchewan. In May and June, daily receipts over the high Arctic actually exceed

c)

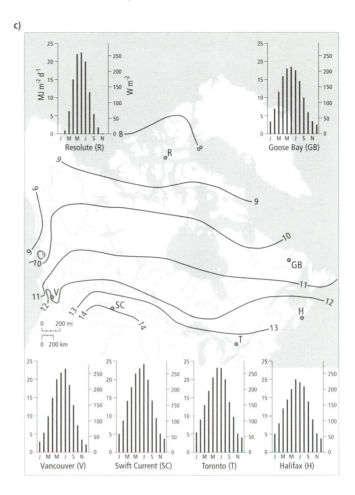

those over southern Canada, but increasing cloudiness reduces northern incomes by July and August. On the spatial average, 75 to 85 per cent of these estimated incomes is absorbed; the rest is reflected back.

In winter, the short period of daylight and low solar elevations yield low daily solar energy incomes, ranging from zero over the Queen Elizabeth Islands (in total darkness for one to three months) to about 30–50 W m^{-2} in southern districts. At this time almost the whole country is covered with snow or ice and hence highly reflective. Consequently, no more than half of the solar income, and probably much less, is absorbed.

Also of interest, but rarely observed on a continuing basis, is net radiation or radiation balance (see chapter 2). This is the absorbed solar radiation minus the net exchange of energy between the surface and atmosphere in the form of longwave radiation. Meaningful estimates of this significant quantity are hard to obtain. It is usually not clear, for example, whether net radiation increases or decreases with physiographic height. Longwave emission from the surface can be estimated reasonably well, but return radiation from the atmosphere is difficult to measure or calculate. Figure 1.5 shows an estimate for the mean annual value over Canada. The solar radiation input to this map is based on a slightly different period and observational set from those of Figure 1.4 (Hare and Hay, 1974; values derived from Hay, 1970). On the average, for southern Canada, net radiation is about one-third to one-quarter of the solar radiation received.

Figure 1.4 illustrates a general rule: there is no simple relationship between solar radiation and temperature. Thus in January, the southern Prairies receive about a third more solar radiation than does Victoria, BC, but the Prairies are nevertheless 20 to 25 Celsius degrees colder. Another illustration concerns arctic Canada. It is natural to feel that during the frigid winter, radiative cooling must be intense. In fact it is only moderate (Figure 1.5). At times the flux density (energy received or emitted per unit time and area) may even be slightly positive (the surface is warmed by net radiation). This is because the surface is near radiative and convective equilibrium. In total darkness, emissions of longwave radiation from the surface (proportional to the fourth power of the Kelvin temperature, and hence quite low in cold weather) are largely balanced by the incoming radiative and convective fluxes from the overlying air, which is usually warmer than the surface itself in the arctic winter.

There is nothing remarkable about Canada's radiation climate. The various fluxes are similar to those over other countries in similar latitudes. It is the dynamic processes that give Canada's climates such diversity and changeability. The very mild winter of the Pacific coast, for example, is maintained by warm airstreams from the Pacific Ocean. When arctic air flows westward across the mountains, as it occasionally does for a few spells each winter, bitter cold can reach even Vancouver and Victoria, BC.

Other Exchange Processes

The net radiative heating and cooling of the surface, plus the import or export of heat by travelling airstreams (advection), control diurnal and annual temperature cycles, plus day-to-day weather changes so familiar in the Canadian climate. There are, however, many other surface exchange processes that can be considered part of the physical climate. Some are convective, such as the fluxes of heat and moisture to and from the surface brought about by turbulent eddies. Others depend on conduction, such as soil heat flux; on gravitational movement, such as precipitation or water percolating through the soil; and on molecular diffusion, such as movements of vapour and trace gases in the soil. Only a few of these are known in sufficient detail for them to be shown as national distributions.

We may write the vertical energy and moisture balance equations for unit area of the earth's surface as:

Energy: $Q^* = Q_E + Q_H + Q_G,$ (1.1)

Moisture: $P = E + N + \Delta S,$ (1.2)

where Q^* is net radiation, Q_E convective latent heat flux density, Q_H convective sensible heat flux, Q_G change, if any, in stored heat in the soil, subsoil, glaciers, snowcover, lakes, and streams, P precipitation, E evaporation (or, over land, evapotranspiration, which cannot easily be separated into its components over plant-covered terrain), N percolation plus runoff (local water surplus), and ΔS change, if any, in stored water in the soil.

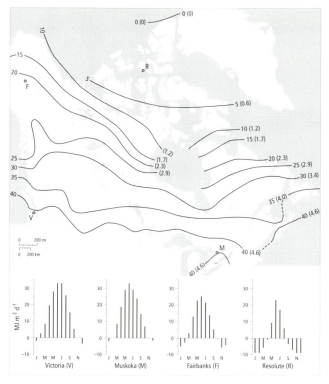

Figure 1.5
Mean annual net radiation. Units of isolines are kly y^{-1} with equivalent values in MJ m^{-2} d^{-1} in brackets.
Sources: Hare and Hay (1974), based on data from Hay (1970).

Of the terms in equations 1.1 and 1.2, precipitation, evapotranspiration, runoff, percolation, and net radiation can be measured or calculated with some confidence. In addition, one can define non-dimensional quantities that link equations (1.1) and (1.2) into a single statement of the conservation principle for water and energy. Over a significant period of time – say, several years – the storage terms in equations (1.1) and (1.2) almost vanish. If we then define the quantities

$B = Q_H/Q_E$ (the climatological Bowen ratio),

$C = N/P$ (the runoff ratio), and

$D = Q*/L_v\, P$ (the dryness ratio), (1.3)

we can write, for climatological annual means,

$D = (1 + B)(1 - C)$. (1.4)

B, the ratio of the mean annual sensible heat flux to the equivalent latent heat flux, is the climatological analogue of β, the Bowen ratio for instantaneous values. C is the runoff ratio, relating mean annual runoff to mean annual precipitation. D, first defined by Budyko (1958), and derived in its present form by Lettau (1969), is the dryness ratio – the ratio of mean annual net radiation to the energy required to evaporate all the mean annual precipitation. Equation (1.4) is the balance equation for annual mean values of the main climatic surface fluxes and may be called Lettau's law. It is also valid for other periods over which water and heat stored in the soil do not change (Hare, 1980; 1983).

There are serious difficulties in preparing valid national distributions of these quantities, especially with regard to precipitation (Hare, 1980; Lins, Hare, and Singh, 1990). First, precipitation (P) is made up of rain, hail, and snow, and inaccuracies of measurement are serious, especially for snow. Moreover, observations are made as point values in a sparse network of stations. Interpolation is hence necessary when isolines are drawn. Second, evapotranspiration (E) is not routinely measured in a satisfactory way (pan evaporation not being representative). Hence it must be estimated from empirical formulae, or from the difference ($P - N$), which is valid only when long-term mean values are used. Estimates are point values, at least in principle. Third, runoff (N) is measured as streamflow for many central and southern Canadian streams and rivers by the Water Survey of Canada and by other agencies. Results are for streamflow at gauging points. This gives the runoff from the entire basin upstream from the gauge. By subtracting one gauge's values from those of the next gauge downstream, one gets the net runoff from the area of the basin between the gauges – an areal mean value. Percolation to groundwater is generally measured only at research stations. Fourth, net radiation (Q^*) is nearly always a calculated value, and interpolation over extensive areas may be very difficult, particularly in complex terrain. This was attempted in Figure 1.5.

Preparing consistent national maps hence involves comparing quantities measured or calculated at local stations, with quantities estimated over areas. To compare point with areal values, it is necessary to average all quantities over arbitrary areas, such as 5° latitude-longitude squares. Hare (1980) gives estimated annual mean values for each such square in Canada and the United States south of 60°N.

Figure 1.6 shows the mean annual distribution of precipitation (P) over Canada, estimated for the thirty-year period 1931–60 (Hare and Hay, 1974, after data from Canada, Department of Transport, 1967, and u.s. Weather Bureau, 1968). Also shown as insets are the distributions over the months of the year at representative stations. The main features of the distribution are the very wet Cordilleran area of British Columbia and Yukon (with heavy amounts on the west face of each mountain range and much drier conditions in the valleys); the dry Prairies, and even drier Arctic; and a well-watered eastern area, roughly Ontario, Quebec, and the Atlantic provinces. Much detail has been filtered out of this figure, especially in hilly areas.

There are sharp seasonal variations in precipitation regimes. On the Pacific coast there is a pronounced late fall-winter maximum, and summer is quite dry in southern British Columbia. Over the Prairies, early summer is the time of maximum rainfall; unfortunately, its occurrence is unreliable in any given year. Over most of eastern Canada precipitation is evenly distributed by season, though its character changes with the season. Except in coastal British Columbia, cool-season precipitation is usually snowfall.

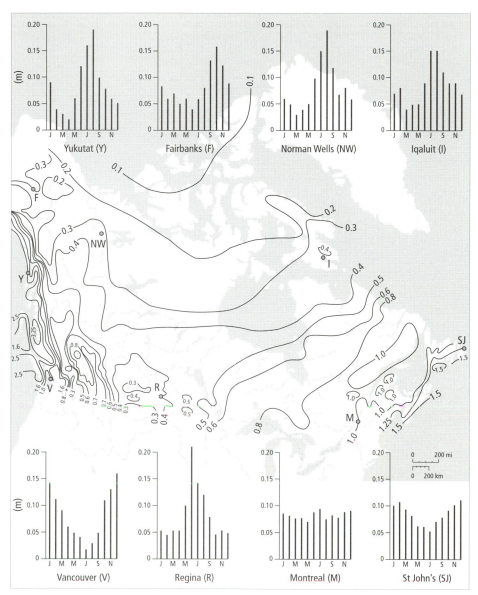

Figure 1.6
Mean annual measured precipitation (m).
Sources: Hare and Hay (1974), derived from Canada, Department of Transport (1967), and u.s. Weather Bureau (1968).

Figure 1.7 shows an estimate of the spatially averaged annual actual evapotranspiration, E, based on a comparison between the precipitation distribution (P) and the drainage basin–derived value of N, the regional runoff. Figure 1.8 shows an estimate (Hare, 1980) of the annual runoff ratio, N/P – the fraction of mean annual precipitation that actually runs off in streams and ultimately discharges to the sea. Figures 1.6

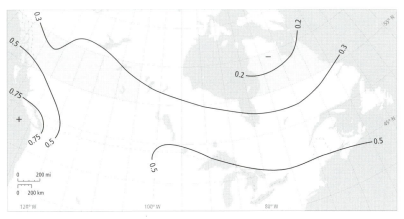

Figure 1.7
Mean annual calculated evapotranspiration (m), averaged by five-degree squares (not valid over Great Lakes or oceans). Stippled area is tentative.
Source: Hare (1980).

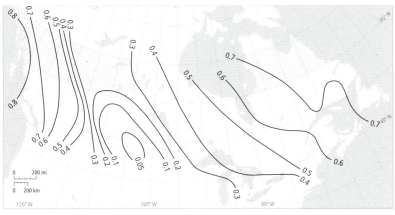

Figure 1.8
Mean annual runoff ratio (*N/P*), averaged by five-degree squares (valid for land areas only). Stippled area is tentative.
Source: Hare (1980).

through 1.8 are estimates of varying degrees of crudity. Nothing better can be yet achieved on the national scale. But much greater precision can be attempted for local sites or even entire drainage basins.

The seasonal variation in these water-related processes depends on precipitation regime (see insets for Figure 1.6) and on incidence of evapotranspiration. The latter is driven mainly by net radiation. The inset diagrams on Figure 1.4 show that in winter little energy is available for evapotranspiration, which hence has a warm-season maximum. Runoff, by contrast, is dominated in most areas by the spring snowmelt, which produces the freshet in all but the largest streams, usually in May or June.

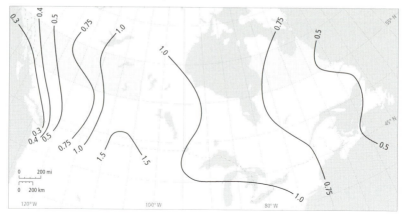

Figure 1.9
Mean annual dryness ratio ($Q*/L_vP$) for land areas, averaged by five-degree squares.
Stippled area is tentative.
Source: Hare (1980).

It clearly emerges that Canada has a well-watered Pacific sector (runoff ratio above 0.4), which is cut off sharply on the east by the Rocky Mountains. East of this barrier, the Prairies are very much drier (runoff ratios below 0.2 over most areas). Many Prairie districts have no net surface runoff, with drainage being collected into the saline ponds and sloughs so common over Saskatchewan and Alberta. From the Great Lakes eastward, however, well-watered conditions are again universal. Figure 1.9 shows the dryness ratio, D, derived from Lettau's law – see equations (1.3) and (1.4). In broad terms, a dryness ratio much above unity indicates conditions of summer moisture stress, with grassland or semi-arid ecosystems. Below a D of approximately 1.1, forest ecosystems predominate.

CONCLUSIONS

The student of surface climates must distinguish at all times between those things that depend on local mechanisms – chiefly exchanges of heat and moisture via the processes discussed in the following chapters – and those that are actually imported to the site by the winds and storms to which Canada is so prone. The total climate has physical (local) and dynamic (advective) components.

But each site on Canada's surface contributes to the overall energy and moisture balances that drive the general circulation of the atmosphere. In summer, for example, all of Canada contributes to the warming of the northern hemisphere, because heat is on balance transferred from the national land surface to the overlying atmosphere. So also, in some areas, is moisture (evaporation may exceed rainfall). In winter, Canada's surface is the recipient of heat and moisture from many overriding warm airstreams.

Thus the physical surface climate of Canada is driven by the global properties imported by the winds and ocean currents (which affect coastal areas), but in return it contributes to the driving forces behind the global climate. How each surface type fares in each of these two roles is discussed in the following chapters.

REFERENCES

Browning, K.A., Hardman, M.E., Harrold, T.W., and Pardoe, C.W. 1973. "The Structure of Rainbands within a Mid-latitude Depression." *Quarterly Journal of the Royal Meteorological Society* 99: 215–31.

Browning, K.A., and Roberts, M.N. 1994. "Structure of a Frontal Cyclone." *Quarterly Journal of the Royal Meteorological Society* 120: 1535–58.

Budyko, M.I. 1958. *Heat Balance of the Earth's Surface*. (In Russian: Leningrad: Gidrome-teoizdat.) Trans. N.A. Stepanova, MGA 13E-286, United States Weather Bureau, Washington, DC, PB131692, 141–5.

Canada, Atmospheric Environment Service. 1982. *Canadian Climatic Normals, 1951–1980, Volume 1, Solar Radiation*. Downsview, Ont.: Environment Canada.

Canada, Department of Transport. 1967. *Climatological Normals*. Cited in Hare and Hay (1974).

Carlson, T.N. 1980. "Airflow through Midlatitude Cyclones and the Comma Cloud Pattern." *Monthly Weather Review* 180: 1498–1509.

Hare, F.K., 1960. "The Westerlies." *Geographical Review* 28: 345–67.

– 1980. "Long-term Annual Surface Heat and Water Balances over Canada and the United States South of 60°N: Reconciliation of Precipitation, Runoff and Temperature Fields." *Atmosphere-Ocean* 18: 127–53.

– 1983. *Climate and Desertification: A Revised Analysis*. World Climate Applications Programme, Report WCP-44, World Meteorological Organization and United Nations Environment Programme, Geneva, 51–63.

Hare, F.K., and Hay, J.E. 1974. "The Climate of Canada and Alaska." In R.A. Bryson and F.K. Hare, eds., *Climates of North America, World Survey of Climatology*. Elsevier, Amsterdam, 11: 49–192.

Hare, F.K., and Thomas, M.K. 1979. *Climate Canada*. 2nd ed. Toronto: John Wiley & Sons.

Harrold, T.W., 1973. "Mechanisms Influencing the Distribution of Precipitation within Baroclinic Disturbances." *Quarterly Journal of the Royal Meteorological Society* 99: 232–51.

Hay, J.E. 1970. "Aspects of the Heat and Moisture Balance of Canada." PhD thesis, University of London.

Lahey, J.F., Bryson, R.A., Wahl, E.W., Horn, L.H., and Henderson, V.D. 1958. *Atlas of 500 mb Wind Characteristics for the Northern Hemisphere*. Madison: Department of Meteorology, University of Wisconsin.

Lettau, H.H., 1969. "Evapotranspiration Climatonomy: 1. A New Approach to Numerical Prediction of Monthly Evapotranspiration, Run-off, and Soil Moisture Storage." *Monthly Weather Review* 97: 691–9.

Lins, H.F., Hare, F.K., and Singh, K.P. 1990. "Influence of the Atmosphere." Chapter 2 of M.G. Wolman and H.C. Riggs, eds., *The Geology of North America*, Vol. 0-1, *Surface Water Hydrology*, 11–53. Geological Society of America, Boulder, Col.

McKay, D.C., and Morris, R.J. 1985. *Solar Radiation Analyses for Canada, 1967–1976*. Vol. 5. Ottawa: Environment Canada.

U.S. Weather Bureau. 1968. *Climate Atlas of the United States*. Washington, DC: Environmental Science Services Administration.

CHAPTER TWO

Surface Climate Processes

TIMOTHY R. OKE

INTRODUCTION

This chapter sets out the way an individual surface is affected by the large-scale external controls outlined in chapter 1 and the nature of its response. The surface is bathed by heat from the sun, buffeted by winds driven by the large-scale circulation and synoptic systems that migrate over it, and occasionally inundated with rain and snow. Some of these inputs are cyclical (for example, the daily and annual variation of solar heating), and some are quasi-random (for example, storm winds and precipitation). The response of the surface is determined by those of its properties that govern absorption, reflection, and emission of radiation; the frictional drag, channelling, and turbulent nature of airflow; the heating and cooling of the soil; and the availability of water at the surface of the soil and within plants for evaporation and transpiration. That is, we need to understand the physical microclimatology of natural systems. At the heart of this set of processes is the way in which properties such as surface albedo, emissivity, roughness, conductivity, heat capacity, soil moisture content, and leaf stomatal conductance regulate the exchanges (fluxes) of heat, water, carbon, and momentum. These properties establish characteristic budgets of radiation and momentum and the balances of heat, water, and carbon.

This chapter is a succinct introduction to the essential concepts and terms used throughout the book – a rapid introduction to physical microclimatology. Ideally we would hope that the reader is familiar with the subject through a standard text such as Arya (1988), Campbell (1977), Lowry (1988), Monteith and Unsworth (1990), or Oke (1987). Explanations of some of the more important new technical terms are gathered together in the Glossary. Parallel discussions to those in this book are to be found in the pioneering text of Munn (1966), which laid out microclimates according to surface type and the two-volume review of principles and case studies of vegetation climates edited by Monteith (1975, 1976). In many ways, this book is an attempt to extend these works in a Canadian context.

ATMOSPHERIC BOUNDARY LAYERS

The earth's surface is where most of the climatic activity that directly affects humans is generated. It is the primary site for absorption of solar energy; it is the source of most of the heat, all the water, and most of the gases and aerosols in the atmosphere; and its drag causes the wind to come to a standstill. It is thus a boundary for the atmosphere, the influences of which are transmitted up into the air above, creating the planetary boundary layer (PBL). Above this is the free atmosphere, which comprises the rest of the troposphere (Figure 2.1), but its properties derive from source regions hundreds or thousands of kilometres upwind.

The PBL is defined as the layer directly affected by surface friction, heating, and cooling. The primary means of carrying these effects upward is atmospheric turbulence. There are two sources for turbulence: thermal convection resulting from vertical stratification of temperature, and therefore of air density, and mechanical convection caused by wind shear in the air and the roughness of the ground. The height of the PBL fluctuates from day to day and day to night in response to passage of weather systems and the diurnal cycle of radiant heating (Figure 2.2, below). As a result, the influence of the surface expands from as little as 100 m on a calm night to as much as 2 or 3 km on a fine summer day.

The PBL consists of many sublayers, each nested inside another at increasingly small scales. Figure 2.1 illustrates this idea by spanning scales from a leaf to the whole PBL. Each scale interacts with that above and below, but the effects are not simply cumulative – this non-linearity in scaling is addressed by Jarvis and McNaughton (1986) for the case of evaporation. The surfaces providing these boundaries are, in descending order, the regional landscape, an individual surface type (with or without a plant canopy), a single element such as a plant, and an individual leaf.

The first division of the PBL is between an outer and an inner layer (Figure 2.1b). In the outer region, the specific nature of the underlying surface is least important, except in its ability to generate thermal convection that can reach the top of the PBL. In this layer, Coriolis turning effects caused by the earth's rotation add a deflecting force to wind directions. By day, the PBL's outer region is often dominated by convective activity and so is referred to as the mixing layer (ML). There most properties except wind direction show little variation with altitude because of the stirring (Figure 2.8a, below). The top of the PBL is often "capped" by an inversion – a layer where temperature increases with height (Figures 2.1a, b). With cooler, denser air underlying warmer air, upward motion is restricted, and so the inversion makes it more difficult for near-surface properties to reach the free atmosphere. The transition between the top of the PBL and the base of the capping inversion is called the entrainment zone. It is bombarded from below by buoyant thermals, which overshoot into the free atmosphere, are repulsed, and in so doing carry with them (entrain) some of the warmer, drier, and usually cleaner air back down into the PBL.

The characteristics of the inner region, or surface layer (SL), are dominated by the nature of the surface itself (Figures 2.1b, c). By day, the height of this layer is about 10 per cent of that of the PBL, and turbulent activity is so vigorous that the fluxes can be effectively considered to be constant with height. Over a horizontally extensive site it should therefore be possible to place instruments at any location in the SL and obtain similar values (to within approximately 5 per cent of each other). Very close to the surface, however, the effects of individual elements cause spatial variability in fluxes and climatic properties – the result of, for example, thermal and moisture plumes and turbulent wakes from spots that are sunlit or shaded, vegetated or bare (Figure 2.1d). This is the roughness layer, whose top can be thought of as the

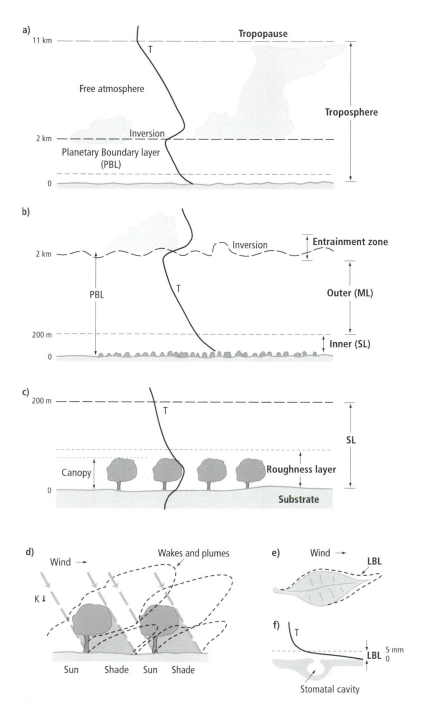

Figure 2.1
Schematic representation of the earth's surface boundary layers, including characteristic
daytime environmental temperature (T) profiles: (a) troposphere (cf. Standard Atmosphere); (b) planetary boundary layer (PBL); (c) surface layer (SL) and canopy layer;
(d) turbulent wakes and thermal plumes of individual roughness elements; (e) laminar
boundary layer (LBL) of a single leaf; and (f) at the scale of a leaf stomate.

level at which the mixing action of turbulence has blended the individual plumes and wakes into a homogeneous state, so that conditions are characteristic of the spatially averaged surface below (Figure 2.1c). The height at which full blending is achieved depends on the dimensions and spacing of the surface elements but is typically about $100\ z_o$ (Garratt, 1992), where z_o is the surface roughness length (see p. 30).

Relatively tall surface elements may be part of a more or less coherent vegetation canopy such as those of grassland, crop, or forest ecosystems. In effect the canopy forms an elevated active surface, where most of the radiation and precipitation are intercepted, water is transpired and evaporated, dewfall accumulates, and drag on airflow is exerted. This canopy layer (Figure 2.1c) has its own distinct microclimate. As a first approximation, the primary active surface can be considered to operate at a level about two-thirds up the height of the plants. For example, consider the shape of the variation of wind speed with height over such a canopy (see inset Figure 2.5a, below). If this profile is extrapolated downward into the plant stand it appears that it will come to a halt about one-third of the way down. This height is interpreted to be the location of the effective sink for momentum. It is analogous to raising the ground surface to this level, so it is called the height of zero-plane displacement (Figure 2.5, below).

Very close to individual surface elements turbulent activity ceases, due to the skin-friction layer surrounding all objects. In this laminar boundary layer (LBL) (Figure 2.1e, f), usually only millimetres thick, transport of properties is governed by motion at the molecular scale. The thickness of the LBL depends on the dimensions of the object, the wind speed, and the magnitude of the temperature difference between the object and the ambient air. All other things being equal, the LBL is thicker over large objects and in calm conditions. Since all quantities have to pass through this layer, it provides an effective buffer, or resistance, to release or uptake of heat, water vapour, carbon dioxide, and pollutants.

Most of this book is concerned with microclimates that exist at the spatial scale of the SL (say, tens to hundreds of metres in height and hundreds to thousands of metres in horizontal distance), but the controls provided by the LBL and the dynamics of the complete PBL, which might seem to lie outside these bounds, are also important. Time scales are typically less than a day, but seasonal change is often relevant.

In order to illustrate the way the microclimates of different surfaces come about, it is helpful initially to simplify things by considering conditions on a single day for a large, flat, and bare surface, such as a field of soil on the Canadian Prairies. In that way we can stress the dynamics of the daily cycle and consider all fluxes and profiles one-dimensional – in other words, only vertical differences are involved. Later we can add complications caused by seasonal change, presence of a plant or snowcover, and three-dimensional interactions resulting from changes of surface type and the uneven character of the natural landscape.

SYNOPTIC CONTROLS ON SURFACE CLIMATES

Weather systems tend to bring several linked influences to bear on surface microclimate. Figure 2.2 is a hypothetical sequence imposed on our simple soil surface over a period of two days. It starts with an anticyclone (high-pressure system) present. Such systems are typically free of cloud because of large-scale subsidence. Therefore the solar radiation input on day 1 follows a symmetric curve from sunrise to sunset, centred on local solar noon (Figure 2.2b). Fine weather conditions such as these create marked diurnal microclimatic rhythms. For example, the strong radiant heating results in a large variation of surface (and near-surface air) temperature, with its maximum in the mid-afternoon (Figure 2.2d). This heating also enhances thermal convection and the growth of the mixing layer (z_i) through the day (Figure 2.2c), which increases the coupling between the surface and the rest of the PBL. Interaction serves to mix the faster-moving air above with the slower airflow near the rigid boundary (i.e., it produces a greater downward flux of horizontal momentum). This injection of momentum into the SL causes wind speeds to increase (Figure 2.2e). For a few hours after sunrise, the growth of the PBL is relatively slow, so that the moistening role of evaporation serves to raise the humidity near the ground. Later the mixing volume greatly expands, which combines with the downward mixing of drier air to decrease humidity (Figure 2.2f). This combination of a high temperature, which gives a high saturation vapour pressure and relatively low humidity, produces a maximum vapour pressure deficit (*vpd*, or the effective "drying power" of the air) in late afternoon.

Cloudless skies at night hasten radiative cooling. A relatively shallow radiative inversion layer forms near the ground (Figure 2.2c). This causes the collapse of the mixing layer, leaving only a residual, weak, mixed layer aloft and decoupling the surface from the air above. Therefore after sunset temperature and wind speed drop, and humidity initially increases because evaporation continues for a while (Figure 2.2d, e, f). Later, dewfall progressively depletes near-surface moisture. Air temperature drops throughout the night, reaching its minimum just after sunrise.

As day 2 of Figure 2.2 shows, this classic diurnal pattern can be largely obliterated by arrival of a weather disturbance with cloud, rain, and strong winds. As air pressure drops fairly quickly, it indicates a sharp horizontal gradient of air pressure on the weather map, and as a result winds increase. Arrival of dense cloud reduces much of the solar radiation. Cloud eliminates direct beam radiation, leaving only diffuse radiation input to the surface. However, during clearing patches, a sharp return to cloudless values is possible (Figure 2.2b). Dense cloud suppresses the ability of solar heating to set the amplitude of the daytime temperature wave. It permits the characteristics of the advecting airstreams, which created the disturbance (see chapter 1), and the mechanical mixing produced by strong airflow, to influence the diurnal temperature pattern. Because of lack of heating and thermal convection, the depth of the mixing layer is relatively small (Figure 2.2c). Nocturnal cloud and strong winds both reduce radiation cooling so that little or no thermal inversion layer is likely to form, but a mechanical mixing layer may be present. The absolute value of temperatures may perhaps be higher than that on the preceding cloudless night. While the moisture content of the air may be higher than the previous day, its diurnal variation, vertical gradient, and drying power are all smaller. If the surface is wetted by rain, temperature and humidity variation is further diminished.

These simple illustrations serve to show that surface microclimates are sensitive to modulation by what Hare (chapter 1) refers to as "climate's dynamic component."

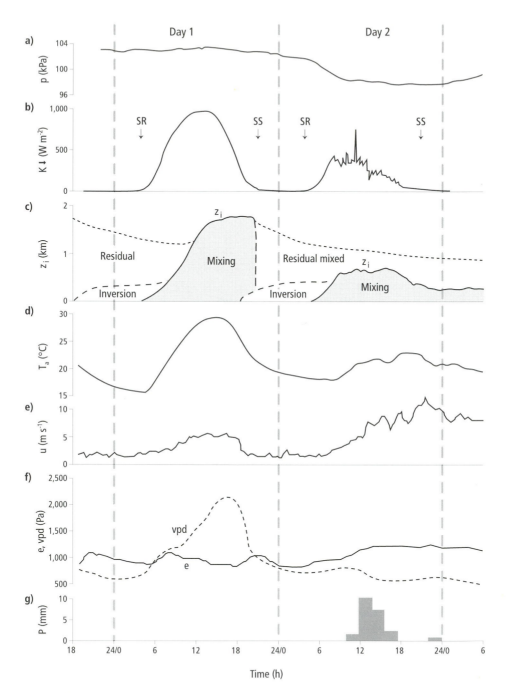

Figure 2.2
Hypothetical sequence of weather for two days, with an anticyclone followed by a day with a weather disturbance; p is total atmospheric pressure; $K\downarrow$ is solar radiation; z_i is height of the PBL; T_a, u, e, and vpd are near-surface air temperature, wind speed, vapour pressure, and vapour pressure deficit, respectively; and P is precipitation. SR and SS are sunrise and sunset, respectively.

SITE CONTROLS ON SURFACE CLIMATE

The particular mix of surface properties possessed by a site, in combination with its synoptic setting, establishes its microclimate. For land areas, the exact nature of the soil and vegetation cover are the critical attributes. At a simple, horizontal and flat site the properties of relevance to physical climatology are (after Oke, 1987):

(i) radiative – the albedo ($\alpha = K{\uparrow}/K{\downarrow}$), emissivity ($\varepsilon$), and temperature ($T_o$) of the surface and any surface geometry that influences the receipt or loss of radiation;
(ii) aerodynamic – surface roughness length (z_o), zero-plane displacement (d), and the presence of upstream obstructions to flow;
(iii) thermal – thermal properties of the substrate materials as they affect conduction of heat (k_s – thermal conductivity), temperature change (C_s – heat capacity), diffusion of temperature waves ($\lambda = k_s/C_s$ – thermal diffusivity), and the ability of the system to accept or release heat ($\mu = (k_s/C_s)^{1/2}$) – surface thermal admittance;
(iv) moisture – soil and plant water status and the abundance and species of plants as they affect availability of water for evaporation and transpiration.

These properties determine the response of the site to the flows of heat, water, and momentum brought to it by the external influences of the sun, wind, and rain. The size of these external "forcings" on climate sets bounds on energy, mass, and momentum balances at the site. The following sections sketch these relationships and provide the background and expressions used in the chapters that follow.

RADIATION BUDGET

The daily variation in the input of solar radiant energy (with wavelengths in the range of 0.15 to 4 micrometres) at the top of the atmosphere depends on earth–sun geometric relationships alone. The amount received at the surface depends on the state of the intervening atmosphere. Atmospheric constituents such as particles and gases (especially ozone and water vapour) scatter, reflect, and absorb parts of the solar beam. The result is that the amount of solar energy at the surface ($K{\downarrow}$) is less than that at the top, and instead of its all arriving as a direct parallel beam (K_S) some is diffuse (K_D), because it comes from all parts of the sky hemisphere. Cloud can be particularly effective at reducing and diffusing solar input. Without cloud about 70 to 90 per cent of the energy penetrates; with a low overcast only 10 to 20 per cent may make it, and all will be diffuse radiation (i.e., $K{\downarrow} = K_D$, Figure 2.2b).

A portion of the input ($K{\downarrow} = K_S + K_D$) is reflected from the surface ($K{\uparrow}$) and is dependent on the albedo of the surface. Since albedo values can vary from about 0.05–0.1 for water, dark soils, and coniferous forest to 0.9 for fresh snow, α is a powerful determinant of a site's climate. It is a direct control on its energetic status. The rest of the input is absorbed (K^*). This is the net solar radiation.

All objects emit radiation. At the temperatures typically found in the earth-atmosphere system, the strength of the output is much smaller than that of the sun, and it is at longer wavelengths (4 to 100 micrometres). Emission depends on the emissivity of the body and is proportional to the fourth power of the surface temperature (T_o). Emissivities of most natural surfaces cover a relatively narrow range (0.90–0.98), so this property is less important than albedo in differentiating microclimates. The atmosphere emits longwave radiation upward to space and downward to the ground ($L\downarrow$). Surfaces on the ground emit radiation upward ($L\uparrow$). Since both the emissivity and the temperature of the ground are usually greater than those of the atmosphere, output, in both day and night, normally exceeds input, so net longwave radiation (L^*) is negative. Hence L^* is a continual energy loss for most surfaces.

In summary, the net radiation for a surface (Q^*) is the algebraic sum of these streams of radiation:

$$Q^* = K^* + L^* \tag{2.1a}$$

where

$$K^* = K\downarrow - K\uparrow = K\downarrow (1 - \alpha) \tag{2.1b}$$

and

$$L^* = L\downarrow - L\uparrow \tag{2.1c}$$

where $L\downarrow$ is the incoming longwave radiation from the atmosphere, $L\uparrow = \varepsilon_o \sigma T_o^4 + L\downarrow$ $(1 - \varepsilon_o)$, and σ is the Stefan-Boltzmann constant (5.67×10^{-8} W m^{-2} K^{-4}). The sign convention is that radiative fluxes directed to the surface are positive. A schematic summary of these fluxes and their diurnal variation is shown in Figure 2.3. Clearly, by day solar inputs dominate and Q^* is positive (the surface has a surplus of radiant energy and heating prevails). At night only longwave exchanges operate; therefore $Q^* = L^*$, which is negative, and the surface cools. Clouds have a higher emissivity than the cloudless atmosphere, and cloud-base temperature is usually higher than that of the bulk of the atmosphere. Therefore arrival of cloud increases $L\downarrow$ and decreases net longwave radiant loss. As a result, all other things being the same, surface cooling is less on a cloudy night than on a cloudless night.

The radiation balance alone does not determine the diurnal variation of surface and air temperature, but there tends to be a fairly good relationship, unless advective influences dominate (Figures 2.2b, d). Daytime radiative heating leads to warming of the lowest air layers, producing a lapse profile (air temperature decreases with altitude; see the lowest portion of profile 2, and all of profile 3 in Figure 2.4a). This creates instability and generates thermal convection. Together with entrainment at the base of the overlying inversion, these cause the mixing layer to grow and the PBL to warm progressively through the day. Just before sunset, and afterward, the surface heat source is replaced by a heat sink, due to longwave cooling. The near-surface air cools first and establishes a surface-based temperature inversion (T increases with altitude; see the lowest portion of profiles 4 and 1 and the mid-portion of 2 in Figure 2.4a). This stable layer resists vertical motions and effectively decouples the surface from the rest of the PBL. The full PBL temperature structure by day is shown using T in Figure 2.1a. The same temperature profiles given in Figure 2.4a are replotted as potential temperature, θ, in Figure 2.4b (see caption for definition).

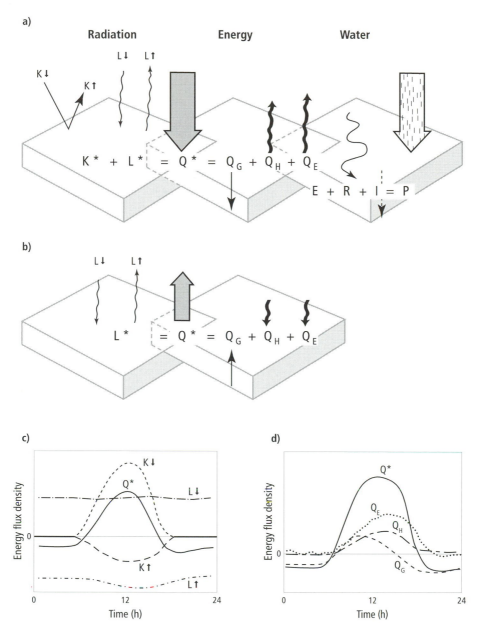

Figure 2.3
Schematic depiction of radiation, energy, and water balance components (a) by day and (b) at night, and idealized variation of (c) the radiation budget and (d) energy balance components through a day with fine weather. Inclusion of a daytime input of rain in the water balance diagram illustrates the existence of a mass system that is forced externally. Clearly rain would not have occurred on a day with such a radiation curve. A nocturnal counterpart might show a small input of dew.

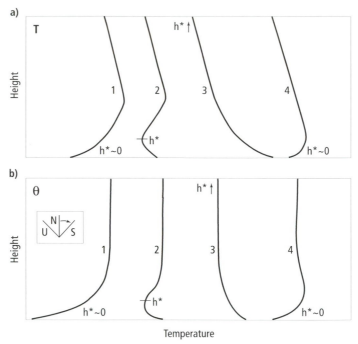

Figure 2.4
Typical variation of air temperature (T) in the surface layer through a fine day.
(a) Profiles: (1) before sunrise, (2) soon after sunrise, (3) midday and afternoon,
and (4) near sunset. (b) Same profiles plotted as potential temperature (θ).
This is a correction to T to account for the decrease of pressure and therefore
expansionary cooling (or heating) of parcels that rise (or sink). Conversion
to θ makes it possible to compare the temperatures that a parcel of air would
have if it were moved upward or downward adiabatically (without mixing with
the surrounding air). In essence the profiles in (a) are rotated clockwise by a
slope of about one Celsius degree per 100 m in height – the value of the dry
adiabatic lapse rate. This allows a simple assessment of atmospheric stability,
which is a measure of the tendency for air parcels to move vertically (see inset
– the more the θ profile leans to the right (or left), the more stable (S) or unstable
(U) is the atmosphere. If the θ profile is vertical, the layer is neutral, N.

MOMENTUM

The wind speed in the PBL is reduced by friction with the ground (Figure 2.5a). The
effect is increasingly felt as the surface is approached, so that the height variation of
the mean wind speed ($\bar{u}$) is logarithmic. In neutral stability the log wind profile law
accurately describes the shape:

$$\bar{u} = \frac{u_*}{k} \ln \left(\frac{z}{z_o} \right), \tag{2.2}$$

where u_* is friction velocity, k the von Karman constant (0.4), z the height, and z_o
the surface roughness length. Both u_* and z_o can be determined from a semi-log plot
of z versus $\bar{u}$ (Figure 2.5b) in neutral conditions. As a first approximation, z_o is typi-

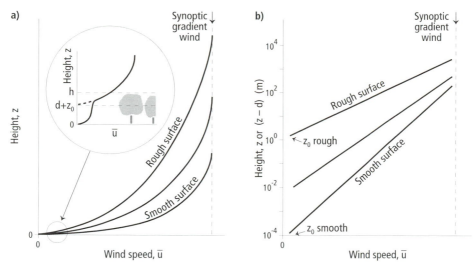

Figure 2.5
(a) Typical shape of the variation of wind speed with height in neutral stability over surfaces that are relatively smooth (for example, snow, water), moderately rough (tundra, grassland), and rough (forest, city). Height is distance above the effective momentum sink. For surfaces with tall roughness elements, a zero-plane displacement, d, must be subtracted from the height scale (inset); (b) same profiles as in (a), but plotted with a logarithmic height scale. The slope of each line is equal to u_*/k, and z_o is the height at which the line, extrapolated to $\bar{u} = 0$, intersects the height axis.

cally about one-tenth the height of the average surface elements (grass stalks, trees, and so on), but again the element density is also relevant. If a substantial vegetation canopy covers the ground, the height z in equation (2.2) is adjusted to read $(z - d)$ to account for the new zero plane (Figure 2.5a inset).

Smooth surfaces, such as open water and snow, and only moderately rough ones, such as tundra and grasslands, are therefore relatively windy, and their influence on the flow extends only a short distance above the ground. Rough surfaces such as forests and cities generate more turbulence and affect a deeper layer of air, so while the decrease of wind speed is greater, the rate of decrease with height is less steep.

The fundamental property of the flow affected by the surface is its momentum (the product of the density of air, ρ, and the mean velocity, $\bar{u}$). Friction causes the greater momentum of the flow aloft to be transported to the surface, where flow (and therefore momentum) is zero. Thus the ground is the sink for the momentum extracted from the wind by terrain roughness. The flux of horizontal momentum (τ) is always downward and by convention is considered negative (Figure 2.7, below). It can be shown that τ is the same as the frictional force of the wind per unit area of ground with units of N m^{-2}. All other things being equal, this flux is greater over rough surfaces and when the wind is strong.

There are several approaches taken to represent (and assess via observation) such turbulent fluxes. A few of those that appear elsewhere in the book are outlined here. One is to visualize the downward flux of momentum in the following manner. Turbulence consists of the erratic, almost haphazard, three-dimensional motions of air parcels of many sizes (eddies). Each eddy carries with it certain concentrations of heat, mass substances, and momentum. Since the motions are three-dimensional, it is useful to decompose the flow into three components – along-wind u, across-wind v, and vertical w velocities. Further, on an instantaneous basis, each component consists of a time mean and a fluctuating part. So the instantaneous along-wind velocity $u = \bar{u} + u'$, where the overbar represents the time average, and the prime the instantaneous fluctuation from the mean (and similarly for the v and w components). The height variation of these parts for the vertical and along-wind components is shown in Figure 2.6. Over time $\bar{w}$ goes to zero over a horizontal site, because as much air rises as falls, and therefore $w = w'$. This is not true for the u-component, because the mean wind is non-zero, so $u' = u - \bar{u}$. A positive deviation from the mean, a gust, is most likely to be associated with a downdraft, because the flow in its source region above is generally faster. Conversely a lull is delivered by an updraft. Thus the instantaneous vertical transport of u-momentum is given by the product of the momentum, $\rho u'$, and the vertical speed, w'. Over time, the mean flux of momentum is:

$$\tau = -\rho\ \overline{u'w'}, \tag{2.3}$$

where ρ is density of air and the continuous overbar indicates the time-averaged mean of the instantaneous product $u'w'$. This is the eddy correlation formulation of the momentum flux, which can be measured with matched, fast-response anemometers and a logger that can keep track of both signals accurately and perform the sums. As long as the properties of all the eddies can be monitored, this approach has great merit because it is almost free of assumptions. It relies on direct accounting of the speed of the transporting elements and their contents.

Flux-gradient formulations, in contrast, make the assumption that the strength of the momentum flux is proportional to the wind gradient. For example, using the analogy that eddies can be considered similar to molecules:

$$\tau = \rho K_M \frac{\partial \bar{u}}{\partial z}, \tag{2.4}$$

where K_M is the eddy diffusion coefficient, which has the form of Fick's molecular diffusion law; cf. equation (2.10). Since K_M is extremely variable in time and space, it is difficult to measure. Hence appeal is often made to the log law – equation (2.2) – and the fact that $\tau = \rho u_*^2$, to arrive at:

$$\tau = \rho k^2 \left[\frac{\Delta \bar{u}}{\ln(z_2/z_1)}\right]^2 \tag{2.5}$$

Alternatively, atmospheric fluxes of entities such as momentum can be likened to the flow of electrons in an electrical circuit. Ohm's law relates the flow (current) to the electrical potential applied (voltage) and the cross-section of the wire (its electrical resistance). Thus the micrometeorological Ohm's law analogy sees the flux to be equivalent to a potential difference in concentration (for momentum, $\rho\Delta\bar{u}$) divided by the resistance of the intervening atmosphere to the transport of the property:

$$\tau = \frac{\rho\Delta\bar{u}}{r_{aM}}, \tag{2.6}$$

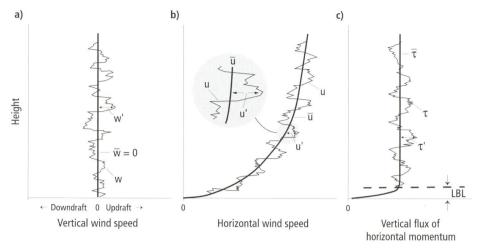

Figure 2.6
Height variation of the mean (heavy line) and instantaneous (thin) velocity for: (a) the vertical (w) and (b) horizontal along-wind (u) components of the wind, and (c) vertical flux of horizontal momentum (τ), near the ground's surface. Primed values are the differences between mean and instantaneous quantities at all heights (see inset).

where r_{aM} is the aerodynamic resistance to momentum transfer, which again can be solved using the wind profile equation (2.2):

$$r_{aM} = \frac{[\ln\,(z-d)/z_o]^2}{k^2\bar{u}} .$$
(2.7)

Recently there is a move towards use of conductance (g) instead of resistance; g is simply r^{-1}. A problem with implementing flux-gradient approaches is that as atmospheric stability departs from neutral, the shape of the wind profile becomes increasingly non-logarithmic. This necessitates evaluation of a stability parameter, such as Richardson's number (Ri) or the Obukhov length (L). These measures of atmospheric stability relate the relative roles of buoyancy (free convection) and mechanical (forced convection) forces in generating turbulence. In strong lapse conditions by day, free forces dominate, and it is unstable. In an inversion, buoyancy is suppressed, and it is stable. In windy, cloudy conditions, mechanical mixing dominates, and the temperature structure is weak, so the air is neutral. The stability measure can be used with empirical functions to correct equations such as (2.5) and (2.7) (for example, Oke, 1987: Fig. A2.8, 383).

This discussion of momentum exchange sets the framework for the turbulent transfer of heat and mass that follows.

ENERGY BALANCE

The net radiant energy surplus and deficit at the earth's surface (Q^*) obeys the law of conservation of energy through the surface energy balance:

$$Q^* = Q_E + Q_H + Q_G + Q_P + Q_B + Q_A,$$
(2.8)

where the terms are the flux densities: Q_E for convective latent heat, Q_H for convective sensible heat, Q_G for conductive heat to the substrate, Q_P for photosynthetic energy uptake or release, Q_B for heat stored in the air (and phytomass if there is a plant cover), and Q_A for horizontal heat gain or loss due to upstream influences. This is a fuller version of equation (1.1).

Yet other terms can enter in special systems – for instance, the heat associated with the melting and freezing of snow and ice (chapter 4) and the anthropogenic heat released through combustion of fuels in cities (chapter 13). In practice, Q_P and Q_B are so small in comparison with the other terms in equation (2.8) that they are usually neglected. Chapter 3 shows that Q_A is large near borders between surfaces and that surface heterogeneity opens much of the real world to micro-advection. Most studies of surface climate have been conducted at sites where such effects are minimal, so they have been conducted using a simplified balance:

$$Q^* - Q_G - Q_H - Q_E = 0. \qquad (2.9)$$

This expresses the fact that the daytime radiative surplus is used to (see Figure 2.3):

(i) warm the surface of the soil and, by conduction (Q_G), drive a temperature wave downward;
(ii) heat the PBL air by convection (Q_H), thereby propagating a temperature wave upward and causing the convective boundary layer to grow and warm;
(iii) evaporate water from the soil and other surfaces (wetted leaves, open water, and so on) and transpire water from plants (Q_E).

Conversely, the nocturnal radiative deficit at the surface is replenished by (a) upward flux of heat from the deeper layers of the substrate that are now warmer, (b) downward flux of sensible heat from the lowest layers of the air, which helps to cool the air and intensify the temperature inversion, and (c) to deposit dew intermittently on the cool surface, thereby depleting part of the SL moisture content.

The daily march of an idealized surface energy balance is given in Figure 2.3d. It shows conditions for a moist surface on a fine summer day. During the daytime, Q_G is the least significant flux density in terms of its absolute magnitude. It is relatively large for a few hours after sunrise, when conduction is favoured over convection – because thermal convection, and therefore turbulent fluxes, are constrained by the small thickness of the PBL. It peaks before Q^* does and turns negative before sunset. Q_E, in contrast, typically consumes the major portion of the available energy, especially in the afternoon, when the PBL volume is large, winds are brisk, and the air's vapour deficit is large. For such a site, with moisture available, the Bowen ratio ($\beta = Q_H/Q_E$) may be about 0.5 during the day. It may be as low as 0.1 for a wet site or open water and about 1.0 for a suburb. It can reach large whole numbers for semi-arid sites, and it is undefined for a true desert. At night, when the wind speed drops, turbulent fluxes become very small, and the nocturnal radiative drain (Q^*) is almost wholly replenished by conduction from the soil heat reservoir (Q_G).

The heat flux to and from the substrate is governed by the soil's thermal conductivity (k_s) and the vertical gradient of soil temperature ($\partial \overline{T}_s / \partial z$):

$$Q_G = -k_s \frac{\partial \bar{T}_s}{\partial z} \quad . \tag{2.10}$$

Q_G is at its maximum when the rate of change of surface temperature ($\partial T_o/\partial z$) is greatest. In the ideal case of uniform soil and sinusoidal variation of surface temperature Q_G peaks about three hours before the peak of the daily temperature wave (about 1.5 months for the annual wave). The heat storage change by a soil layer of thickness Δz is given by the soil's heat capacity (C_s) and the rate of temperature change:

$$Q_{G\,o\text{-}z} = C_s\,(\Delta \bar{T}_{o\text{-}z}/\Delta t)\,\Delta z \quad . \tag{2.11}$$

We saw earlier the three expressions for the turbulent flux of momentum in the SL:

$$\tau = -\rho\,\overline{w'u'} = \rho K_M \frac{\partial \bar{u}}{\partial z} = \frac{\rho \Delta \bar{u}}{r_{aM}} \quad .$$

Convective transport carries all entities, so that completely analagous relationships apply for the turbulent transport of sensible heat, latent heat, and carbon dioxide (F_C):

$$Q_H = C_a\,\overline{w'T'} = -C_a\,K_H \frac{\partial \bar{T}}{\partial z} = -C_a \frac{\Delta \bar{T}}{r_{aH}} \quad , \tag{2.12}$$

$$Q_E = L_v\,\overline{w'\rho_v'} = -L_v\,K_V \frac{\partial \bar{\rho}_v}{\partial z} = -L_v \frac{\Delta \bar{\rho}_v}{r_{aV}} \quad , \tag{2.13}$$

$$F_C = \overline{w'c'} \quad = -K_C \frac{\partial \bar{c}}{\partial z} \quad = -\frac{\Delta \bar{c}}{r_{aC}} \quad , \tag{2.14}$$

where each has its own diffusivity and aerodynamic resistance subscripted according to the entity and C_a is the heat capacity of air, L_v the latent heat of vaporization, ρ_v the water vapour density, and c the concentration of carbon dioxide. The accepted sign convention for these fluxes is given in Figure 2.7.

To calculate turbulent fluxes, it is helpful to eliminate the diffusion coefficients (the Ks), because they are extremely variable and difficult to measure. One typically does this by assuming that they are equal – i.e., $K_M = K_H = K_V = K_C$. Except for the case of momentum, this equality is often a reasonable assumption. That being so, if we take the ratio of equations (2.12) and (2.13) and use finite differences between the same two levels to approximate the gradients, then:

$$\beta = \frac{Q_H}{Q_E} = \frac{C_a\,\Delta \bar{T}}{L_v\,\Delta \bar{\rho}_v} = \gamma \frac{\Delta \bar{T}}{\Delta \bar{e}}, \tag{2.15}$$

where e is vapour pressure. Implementation of equation (2.15) requires measurement of temperature and humidity only at two levels. Then, using equation (2.9), we can find the turbulent fluxes from the energy balance–Bowen ratio method:

$$Q_H = \frac{Q^* - Q_G}{1 + \beta^{-1}},$$

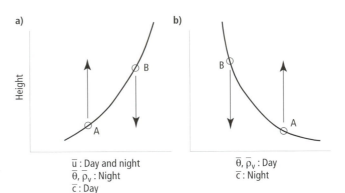

a)   b)

Height

$\bar{u}$: Day and night
$\bar{\theta}, \bar{\rho}_v$: Night
$\bar{c}$: Day

$\bar{\theta}, \bar{\rho}_v$: Day
$\bar{c}$: Night

Figure 2.7
Typical form of mean microclimatic profiles near the surface (also see
Figure 2.8). (a) Horizontal wind speed (all occasions), air temperature
and humidity at night (stable conditions with condensation), and carbon
dioxide by day (assimilation); (b) air temperature and humidity by day
(unstable with evaporation) and carbon dioxide at night (respiration). The
eddies marked A (or B) have positive (or negative) vertical velocity fluctuations w' and carry negative (or positive) fluctuations u', θ', v' and c'
in (a), and the converse is true in (b). The covariances (fluxes) $u'w'$, $w'\theta'$,
$w'\rho_v'$, and $w'c'$ in (a) are downward and considered negative; those in
(b) are upward and positive.
Source: Garratt (1992).

and

$$Q_E = \frac{Q^* - Q_G}{1 + \beta}.$$

The typical, fair-weather arrangement of fluxes and their associated vertical mean
distributions of climatic properties are summarized in Figure 2.8. By day (Figure
2.8a), the ground is a sink for momentum as a result of friction and for carbon dioxide
because of the uptake by plants in the process of photosynthesis. Hence, the vertical
profiles of wind speed and of carbon dioxide concentration decrease through the SL,
towards their respective sinks, and their fluxes are directed downward. At the same
time the ground is a source of heat and water vapour for the PBL because of radiative
warming and evapotranspiration, respectively. The profiles of temperature and humidity therefore decrease (lapse) upward, especially through the SL. Above the PBL the
air is often warmer (inversion) and drier. As a result, entrainment into the PBL acts as
a second source of heat but as a sink for water vapour. Within the ML, most properties
are approximately constant with height because of distance from the source or sink
and excellent mixing. In the case of water vapour, concentrations tend to decrease
slightly up through the ML. This is because, unlike the cases for heat and momentum,
where the top and bottom of the PBL provide matched sources or sinks, both the PBL
and the free atmosphere are sinks for water vapour by day. During the growing season, carbon dioxide increases upward in the ML.

At night (Figure 2.8b) the ground remains a sink for momentum, but also for heat
caused by radiative cooling, and sporadically for water vapour as condensation (dew-

a)

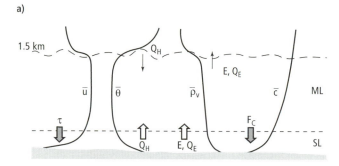

b)

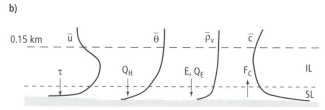

Figure 2.8
Schematic depiction of mean vertical profiles of wind speed ($\bar{u}$),
potential temperature ($\bar{\theta}$), vapour density ($\bar{\rho}_v$), and carbon dioxide ($\bar{c}$) in
the planetary boundary layer by (a) day and (b) night. Also included are
the directions of the related vertical fluxes of momentum (τ), sensible heat
(Q_H), water vapour or latent heat (E or Q_E), and carbon dioxide (F_C).

fall) occurs. Often periods of dewfall are interspersed with weak evaporation. The surface is, however, a source of carbon dioxide from the soil and plant respiration. The
nocturnal inversion layer is characterized by weak winds and poor turbulent diffusion
because of its decoupling from the upper PBL. This relative lack of mixing encourages
development of relatively strong vertical gradients in profiles. The temperature and
moisture profiles increase upward (inversion), and the wind profile commonly exhibits
an elevated, jet-like zone. Concentrations of carbon dioxide initially lapse upward from
the source, where they may be 30 to 40 ppm by volume greater during the night.

MASS (WATER AND CARBON
DIOXIDE) BALANCES

The corresponding balance of water mass at the ground surface is:

$$P - E - R - I = 0, \tag{2.16}$$

where P is input of precipitation, E loss of water to the air by evapotranspiration
(or gain by condensation as dewfall), R net surface runoff by horizontal flow (over-
land and streamflow), and I surface infiltration. This relation is analogous to that for
energy (equation 2.9) and applies to the continuous flux of water across a plane. In
practice, precipitation is not continuous, and it is difficult to specify or measure some

of the terms on an instantaneous basis. Therefore it is often more useful to specify the balance of a volume over a period such as a day or longer. Then:

$$P - E - R - D_W - \Delta S = 0,$$
(2.17)

where D_W is water drainage to layers beneath the volume and ΔS net change in soil moisture content in the volume. Over the long term, this is equivalent to equation (1.2) with $N = R + D_W$. Equation (2.17) also applies to water bodies and to irrigated and city areas if the necessary additional inflows (such as rivers and pipe-water supply) and outflows (rivers and sewers) are included.

Notice that evaporation is common to both the balances of energy and water mass (Figure 2.3a), since $E = Q_E/L_v$. Evaporation requires energy to vaporize the available water, a vapour deficit in the overlying air, and air motion to carry the vapour aloft and drier air downward. These controls are evident in the Combination Model, which combines the energy required and a formulation of the diffusion mechanism integrated from the surface to the height of measurement. The Penman-Monteith form is:

$$Q_E = \frac{s}{s + \gamma} \left[\frac{(Q^* - Q_G) + C_a \cdot v dd_a / r_{aH}}{1 + r_c / r_{aH}} \right],$$
(2.18)

where s is the slope of the curve for saturation vapour density versus temperature, γ the psychrometric constant (C_a/L_v), and r_c the surface or bulk stomatal resistance. The "energy" term incorporates the heat available to support evaporation, and the "vapour deficit" or "advective" term expresses the roles of the dryness of the air and the resistance of the surface-atmosphere system to upward diffusion. The aerodynamic resistances of the atmosphere to transport of heat and to transport of water vapour can be considered equal to each other, but both are greater than that for momentum (which is aided by pressure effects).

One way of accounting for this difference is to use different roughness lengths. Typically those for heat and vapour are about ten times smaller than that for momentum in equation (2.7):

$$r_a = \frac{\ln[(z_u - d)/z_{oM}] \ln[(z_v - d)/z_{oV}]}{k^2 \bar{u}_z},$$
(2.19)

where z_u and z_v are the respective heights of the wind speed and vapour deficit measurements and z_{oM} and z_{oV} are the roughness lengths for momentum and water vapour, respectively. It is especially useful that observations from only one height are required. The surface resistance r_c in equation (2.18) expresses the resistance of the surface to diffusion of water vapour. For vegetated surfaces this is a bulk stomatal resistance which represents the *physiological* control on transpiration (evaporation of water from leaf stomata, which open and shut in response to changes in radiation, temperature, vapour deficit, carbon dioxide, and leaf water status). For a surface wetted by rain, dew, or irrigation or for an open water surface, $r_c \approx 0$. This is referred to as potential evaporation, which is a measure of the *meteorological* control on evaporation.

The first term in equation (2.18) is usually by far the largest contributor to evaporation. Indeed, observations over extensive moist surfaces in humid climates support the simple empirical relation first forwarded by Priestley and Taylor (1972) for regional potential evaporation:

$$Q_{PE} = \alpha_P \frac{s}{s + \gamma} (Q^* - Q_G). \tag{2.20}$$

Values of the Priestley-Taylor parameter (α_P) are about 1.26. In effect this need to boost the "energy term" by about one quarter is a statement of the fact that on average PBL air maintains a surface-to-air vapour deficit due to downward mixing of dry air from the free atmosphere above. In arid climates, an α_P value of 1.74 is recommended (Shuttleworth, 1993). McNaughton and Jarvis (1983) offer another insight by expressing the relative importance of the energy and advective (vapour deficit) terms:

$$Q_E = \Omega \frac{s}{s + \gamma} (Q^* - Q_G) + (1 - \Omega) \frac{(C_a \cdot vdd_a)}{\gamma \, r_c}, \tag{2.21}$$

where Ω is a dimensionless coupling factor defined:

$$\Omega = \left[1 + \frac{\gamma}{s + \gamma} \frac{r_c}{r_a} \right]^{-1}, \tag{2.22}$$

which ranges from 0 to 1.0, and where r_a applies to the complete depth of the SL. Ω expresses the degree of coupling between the SL and the PBL. Ω becomes larger as r_a decreases and r_c increases. For well-watered, relatively smooth surfaces, r_a is large and the energy term dominates. Effectively, the SL is decoupled from the rest of the PBL by the large r_a. This partly explains why simple equations such as equation (2.20) work well over short grass and water surfaces. However, over very rough surfaces such as forests and cities, r_a is small compared to r_c and the second term becomes more important. Evaporation is then sensitive to the vapour deficit of the whole PBL, and so the SL is "coupled" to the advective influences of the PBL and larger scales. This is especially true for regions where surface moisture is not freely available. Observations suggest Ω values can be as high as 0.8 for grassland and as low as 0.2 for forests and cities, but values are highly variable and depend on mesoscale and synoptic conditions.

The nature of the exchange of carbon between the surface and the atmosphere is, in most respects, similar to that of water. Both are transfers of mass involving change of phase, which involves energy uptake or release. In the case of carbon, photosynthesis depends on the uptake of atmospheric carbon dioxide carried to the plant by convection – equation (2.14). Carbon dioxide diffuses first through the laminar boundary layer, then through the stomatal aperture to the stomatal cavity (Figure 2.1f). It then passes through the cavity wall into the chloroplasts. Biochemical reactions driven by solar radiation convert the CO_2 and water into carbohydrates and oxygen. At the same time, there is the reverse process of respiration, in which the carbohydrates are oxidized, leading to release of CO_2, water vapour, and a small amount of energy. Respiration involves plant leaves, their root systems, and soil microbial action. By day,

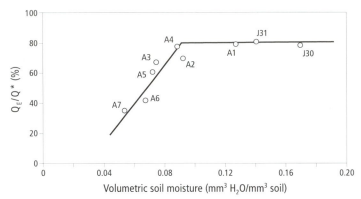

Figure 2.9
Relationship between proportion of net radiation used for evaporation (Q_E/Q^*)
by a soybean crop at Simcoe, Ont., and soil moisture content. Daily averages
starting 30 July 1974, following 20-mm rainfall on the 29th. A further 2 mm fell
on the 30th, and 2.5 mm early on 4 August. J: July; A: August.
Source: Bailey, pers. comm., 1997.

gross photosynthesis exceeds respiration; therefore plants are a net sink for CO_2. Over
time, this fact results in increases in production of dry matter. At night, only respiration occurs, and both plants and the soil are sources of CO_2. The energy requirements
involved are much smaller than for water (typically less than 2 per cent of net radiation). Hence the climatic impact is small, but the biological significance is fundamental to the existence of life on earth.

THE "REAL WORLD"

The preceding account briefly outlines the exchanges and microclimatic characteristics over an extensive, flat, and bare (or low-plant cover) surface. While the relations
covered explain the basic function of surface climate systems, the real world contains
many exceptions, some of which I will memtion and which are examined in later
chapters.

Seasonal shifts in the governing surface properties can alter the relative roles of processes. For example, over a period of days to months, the relative significance of evaporation in the energy and water balances fluctuates because of wetting (by rain or
irrigation) and drying out (as the upper soil becomes partially depleted of freely available water). Figure 2.9 shows how the role of Q_E declines as an initially moist soybean
crop starts to dry. Following rain, when water was easily available, latent heat consumed about 80 per cent of the net radiation. Within a few days, the figure dropped to
only 35 per cent. Since sensible rather than latent heat would then dominate in the
energy balance, an implication of such a shift is that the surface climate would become
warmer and drier. Notice that when rain fell on 4 August (A4) the proportion temporally reverted to one with latent heat dominant. To track conditions over an extended
period it would be necessary to keep a running water balance – equation (2.17).

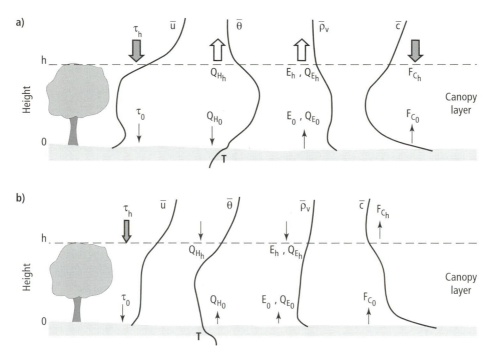

Figure 2.10
Schematic depiction of mean vertical profiles of wind speed ($\bar{u}$), potential temperature ($\bar{\theta}$), vapour density ($\bar{\rho}_v$), and carbon dioxide ($\bar{c}$) in and above a plant canopy layer, (a) by day and (b) at night. Also included are the directions of the related vertical fluxes at the height of the plant tops (h) and the ground (o).

As noted in discussion of Figure 2.1c, a complete plant canopy inserts a major climatic buffer between the ground and the sL. It effectively moves the prime plane of climatic activity (the "active surface") upward to somewhere near the top of the canopy, thereby modifying the vertical profiles of climatic elements (Figure 2.10). The active surface behaves as a sink and source for PBL fluxes, as outlined in the discussion of Figure 2.8. Within the canopy, fluxes are not of the same magnitude, or even in the same direction, as those of the active surface and sL (Figure 2.10). For example, the daytime warmth of the canopy and the shading of the ground create the possibility of a downward flux of sensible heat. The reverse occurs at night (also see Figure 11.7). Similarly, the daytime canopy sink for CO_2 creates an upward sub-canopy flux from the CO_2-rich layer near the soil floor, where respiration is continual. At night, since both the floor and the canopy are sources, both the sub- and above-canopy fluxes are directed upward. Airflow within a plant or urban canopy is light and variable. On average, the wind profile behaves as in Figures 2.5 and 2.10. In a forest, if there is a trunk zone, a minor "jet" often occurs. These general characteristics of vegetation canopies are elaborated on in chapters 11 and 12.

Snow on the surface introduces concepts of seasonal dynamics and places a buffer layer between the ground and the atmosphere. The first snowfall immediately transforms the microclimatic properties of the landscape: the albedo may increase

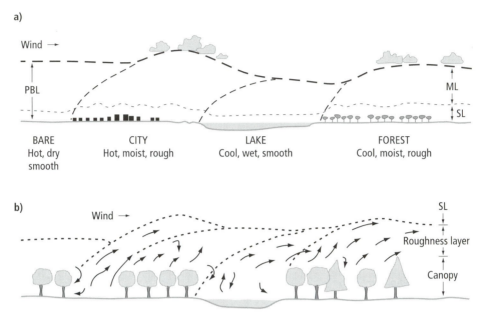

Figure 2.11
Variation of the height of the planetary boundary layer over mesoscale terrain units with contrasting surface climates. (a) Terrain units; (b) portion of the forest in (a), illustrating microscale heterogeneity where differences in the canopy and roughness layers are blended to form a more homogeneous, mesoscale surface layer.

many-fold; emissivity approaches that of a black body; if the surface is melting, its temperature and vapour density are locked at 0°C and 4.85 g m^{-3}, respectively; the surface probably becomes much smoother; an excellent insulating layer is placed over the soil; and the surface has a surface resistance for vapour loss that is either zero (when melting) or close to infinite (if totally frozen). Nothing in nature so utterly alters the surface boundary conditions for the atmosphere in such a short span. The nature of this quintessentially Canadian surface, which covers much of the country for an extended period, is the subject of chapter 4.

The blanket of snow is also a great climatic equalizer. It smothers the natural heterogeneity of the landscape. When it disappears again, the spatial variability of the mix of microclimatic properties reasserts itself. The climate of any one surface can be considered in the one-dimensional framework outlined up to this point (for example, Figure 2.1). In reality, this one-dimensional case occurs only over terrain that is sufficiently extensive in the horizontal to allow the surface effects enough time to extend throughout the full height of the PBL – so a representative new boundary layer has formed. The surface exchanges over such areas dictate the strength of the heat flux, which in turn determines the PBL's height. So, for example, the PBL extends higher over a rough, hot city than over a cool, smooth lake and is intermediate over moist vegetated areas (Figure 2.11a). At smaller scales, the variability produces a series of smaller internal boundary layers, each containing the different characteristics of more localized areas within the larger terrain unit. These may be the result of the internal spatial variability of surfaces or of the effects of a partly cloudy sky, which dapples the landscape with sunny and shaded patches. The effects of such small-scale hetero-geneity often do not survive surface-layer turbulent mixing and merge with other internal plumes (Figures 2.11b and 3.6).

In reality, even these more complex, two-dimensional depictions are greatly simpli-fied here. They are really three-dimensional phenomena, and the effects of topo-graphic relief also have to be added. That is the task of the next chapter, which shows that the lower atmosphere is continually readjusting its characteristics in response to the varying boundary conditions of different surfaces. Heating of the earth's atmo-sphere is dominantly from below. Therefore, ultimately, synoptic and global climates derive their characteristics from the integrated effects of surface climates.

REFERENCES

Arya, S.P. 1988. *An Introduction to Micrometeorology*. New York: Academic Press.

Campbell, G.S. 1977. *An Introduction to Environmental Biophysics*. New York: Springer-Verlag.

Garratt, J.R. 1992. *The Atmospheric Boundary Layer*. Cambridge: Cambridge University Press.

Jarvis, P.G., and McNaughton, K.G. 1986. "Stomatal Control of Transpiration: Scaling Up from Leaf to Region." In E.D. Ford and A. Macfadyen, eds., *Advances in Ecological Research* 15: 1–45.

Lowry, W.P. 1988. *Atmospheric Ecology for Designers and Planners*. McMinnville, Ore.: Pea-vine Publications.

McNaughton, K.G., and Jarvis, P.J. 1983. "Predicting Effects of Vegetation Changes on Tran-spiration and Evaporation." In T.T. Koslowski, ed., *Water Deficits and Plant Growth*, Vol. 7, New York: Academic Press, 1–47.

Monteith, J.L., ed. 1975. *Vegetation and the Atmosphere, Vol. 1, Principles*. London: Academic Press.

– 1976. *Vegetation and the Atmosphere, Vol. 2, Case Studies*. London: Academic Press.

Monteith, J.L., and Unsworth, M.H. 1990. *Principles of Environmental Physics*. London: Edward Arnold.

Munn, R.E. 1966. *Descriptive Micrometeorology*. New York: Academic Press.

Oke, T.R. 1987. *Boundary Layer Climates*. 2nd edn. London: Routledge.

Priestley, C.H.B., and Taylor, R.J. 1972. "On the Assessment of Surface Heat Flux and Evapo-ration Using Large-scale Parameters." *Monthly Weather Review* 100: 81–92.

Shuttleworth, W.J. 1993. "Evaporation." In D.R. Maidment, ed., *Handbook of Hydrology*, New York: McGraw Hill, 4.1–4.53.

CHAPTER THREE

Spatial Variability in Surface Climates

DOUW G. STEYN, HANS-PETER SCHMID,
JOHN L. WALMSLEY, AND
JOHN D. WILSON

INTRODUCTION

Most of chapter 2 addresses an important, but special case: the atmosphere in equilibrium with its underlying surface. In such situations the near-surface turbulent fluxes of heat, moisture, and momentum (and any other properties of interest) do not vary in the horizontal. For this condition to be met, surface properties such as temperature, moisture, and roughness, which act as sources or sinks for the fluxes, must be horizontally uniform, at least in the upwind direction. Much of the material presented in the chapters that follow also explicitly or implicitly assume that the surface is horizontally homogeneous. This is clearly not generally true. The real world is an infinitely complex mosaic of different surfaces which abruptly change in space or merge one into another (Figure 3.1, below). The central concern of this chapter is to examine the effects of horizontal inhomogeneity in surface properties on the atmosphere.

HOMOGENEITY, SCALE, AND SPATIAL VARIABILITY

At first glance, ocean surfaces far from coastlines seem relatively uniform and horizontally homogeneous, and the Canadian Prairies seem nearly so. Closer examination reveals several complicating constructs. Terms such as "variability" and "heterogeneity," on the one hand, and "uniformity" and "homogeneity," on the other, are in practice only subjective qualifications. They depend both on the case at hand and on its context. The principal idea may be best explained by borrowing from photography (also see Figure 2.1). In the context of a wide-angle photograph of the landscape, a forest, a meadow, or even a city may be perceived as one of the internally homogeneous image elements that make up the dominant structure of the (inhomogeneous) picture. However, if a telephoto lens zooms in on these elements, the overall context of the previous picture, and its inhomogeneous structure, are lost, and the elements themselves (forest, meadow, city) become the new context. If the field of view is narrowed still further, individual trees, bushes, tufts of grass, flowers, or buildings become the dominant elements. Thus when one reduces the scale, small-scale variability emerges in an area that otherwise appears homogeneous at a larger scale.

In the context of climate and surface variability, an extreme example is an ocean surface. To most land-dwellers, the oceans are simply uniform blue areas on the world map. However, when the field of view is reduced, oceans become varied surfaces. For example, in some places, spatial changes of sea surface temperatures are sharp enough to justify the term "oceanic front." Similarly, ocean roughness due to surface waves and the fluxes of heat, moisture, and momentum vary from place to place across areas that initially appear homogeneous. This inhomogeneity may be so slight, and so slowly varying in space, as to allow us to assume ocean surfaces to be homogeneous over large patches, but not in their entirety. So, if we are studying a relatively small-scale phenomenon, we may be able to justify the assumption that the oceans are homogeneous. However, for a study of a larger-scale phenomenon, we may have to explicitly account for inhomogeneity in surface properties and air–sea exchanges. Homogeneity is thus closely linked to scale. A surface that can be treated as homogeneous in relation to small-scale meteorological phenomena may be inhomogeneous when viewed from a larger perspective.

Further, consider the results from measurements made at different heights above an ocean surface. Measurements close to the sea surface monitor atmospheric and oceanic properties relatively near the point of measurement. Those made at greater height refer to those further away. Near-surface measurements thus reflect smaller-scale phenomena and more easily allow one to assume the existence of horizontal homogeneity. However, if the height of measurement is reduced to the point where it becomes comparable with the wave height, an entirely different kind of inhomogeneity comes into play. Then the measurements begin to reflect the size and speed of individual waves on the ocean surface.

If we apply the same logic to an agricultural area of the Prairies, similar features emerge. With care, one can choose a location at which measurements show an atmosphere that is fully in equilibrium with the underlying surface. For example, with measurements made a few metres above the canopy and in the centre of an extensive field of wheat, the phenomena reflected in the data have maximum horizontal scales no larger than the dimensions of the field. The effects of individual plants and intervening bare patches are blurred by the mixing effects of turbulence. But if the measurement height is increased, eventually the observations are affected by the properties of surfaces upstream of the field. The effects of different crops, bare fields, hedges, ditches, and roads can then enter the picture, and the surface represented by the data is clearly no longer homogeneous.

If the measurement height is reduced, a point is reached at which measurements are dominated by individual surface elements nearby, because the turbulent mixing has not had sufficient time to blend their separate influences into a homogeneous field. At this height, measurements can reflect only the properties of very small-scale atmospheric phenomena, and the surface cannot be considered homogeneous. In this case the data are subject to strong spatial variability: a slight horizontal movement of the position of measurement results in significant change in the observed values. Therefore, with respect to homogeneity, the scale of inquiry depends on the height of measurement.

SURFACE FORCING AND
ATMOSPHERIC EFFECTS

The surface and the atmosphere are in a relationship of mutual influence involving an intricate network of feedbacks. It is a convenient first step to consider the behaviour of the atmosphere as influenced by the surface without considering reverse influences of the atmosphere on the surface. In this simplified, one-way view of the climate system, the roughness, thermal, and moisture properties of the surface exert their influence on climate by forcing the atmosphere to respond in a particular way. Such surface forcing is transmitted vertically to the lower atmosphere by exchange processes of energy (radiation and sensible and latent heat), mass (for example, water vapour, trace gases, dust, and snow) or momentum (friction resulting from wind shear near the surface, and drag resulting from streamline curvature around topography), and it is transmitted horizontally from surface to surface by advection – transport by the mean wind.

This chapter focuses on the influence that mean flow across horizontally variable surface properties and uneven terrain (considered as *forcings*) exerts on turbulent transport and the mean flow itself (considered as *responses or effects*). In general, surface obstacles and relief affect primarily wind speed and direction (mean flow) and only secondarily influence the turbulent fluxes of heat, moisture, and momentum. In contrast, variations in the physical properties of the surface (such as roughness, albedo, and thermal and moisture properties) affect mainly turbulent fluxes in the energy and water balances and have little influence on the mean wind field.

Though this distinction between *surface forcings* and *atmospheric effects* is not universally held and is highly dependent on scale, it provides a convenient framework to organize this chapter. The initial sections illustrate the interplay of forcings and affected processes by simple examples, where only one group of atmospheric processes is affected by a dominant type of surface forcing. Later sections focus on cases where forcings and effects are combined. In practice, properties of the mean flow and of turbulence are always linked. For example, a change in topographic relief is commonly associated with changes in surface properties, so that separation of forcings from their effects on specific atmospheric processes is somewhat arbitrary. We are trying, however, to give order to an extremely complex web of causes and effects.

The structure of the earth's surface and atmospheric processes vary across a wide range of scales. As a result so do the atmospheric features that they produce. Meteorological and climatological phenomena can be divided into three broad size classes: micro, meso, and macro (Steyn et al., 1981). The microscale, which includes all surface layer phenomena, covers spatial extents from millimetres to a kilometre and time periods from a second to an hour. The mesoscale covers horizontal distances ranging from hundreds of metres to no more than a thousand kilometres and periods from tens of minutes to a few days. It includes phenomena ranging from land-sea breezes to synoptic fronts. The macroscale covers items ranging from hundreds of kilometres up to the circumference of the earth in size and time from days to years. These three classes provide a simplifying framework within which to study atmospheric phenomena. This simplification is helpful because a restriction in scale results in the need to consider only a subset of all possible processes.

As this book focuses primarily on the behaviour of the atmosphere near the surface, and therefore on smaller-scale phenomena, we restrict consideration to the micro- and mesoscales. Furthermore, we exclude climate variability due to thermodynamic forcing of flow over mountains (such as increased precipitation on windward slopes and Chinook effects in their lee). Such phenomena are well described in standard texts.

INTERNAL BOUNDARY LAYERS

Mean Flow Responses to Changing Surface Properties in Flat Terrain

The simplest form of surface inhomogeneity is an abrupt change of at least one surface physical property (other than elevation) at a border placed approximately at right angles to a strong mean flow. This essentially two-dimensional step may seem an idealized case, but many examples occur in nature – for instance, lake and ocean shorelines, field and forest borders, and urban-rural edges (Figure 3.1). In such cases, mean properties of wind speed, temperature, humidity, and any other relevant concentration have gradients in the horizontal (along-wind) direction as well as the vertical. The air upwind of the transition is presumed to be in equilibrium with the surface, but after it advects across the leading edge, the influence of the new surface propagates upward by vertical turbulent diffusion, in response to the new forcing.

The layer of modified air, called an internal boundary layer (IBL), increases in depth with distance downstream from the leading edge (Figure 3.2a). Upwind of the discontinuity, turbulent fluxes of heat, moisture, and momentum are approximately height-independent, and so the micrometeorological profiles are related to fluxes by standard one-dimensional theory (chapter 2). Any pressure effects of the leading edge that may propagate in the upwind direction are minor (as long as the leading edge does not constitute an obstacle).

An IBL grows because surface influences are advected downstream more rapidly than they are diffused vertically. There is a continuous transition from conditions close to the ground, where profiles and fluxes are almost completely adjusted to the new surface, to a height where the flow is still nearly in equilibrium with the upstream surface. However, for many purposes it is convenient to define a discrete interface – the internal boundary-layer height, $\delta(x)$ – above which the profiles or fluxes are affected by the new surface by less than, say, 1 per cent. The lowest 10 per cent of such an IBL, $\delta'(x)$, called the equilibrium layer, is assumed to be fully adjusted to the new surface (Figure 3.2a). If the discontinuity in surface forcing is purely a change in surface roughness, then a mechanical IBL (MIBL) is generated. If the forcing is thermal, caused by a discontinuity in surface temperature and/or humidity, then a thermal IBL (TIBL) is generated. In many situations it is necessary to consider both an MIBL and a TIBL – as an example, flow from a cold and relatively smooth ocean to warmer, rougher land. Garratt (1990) gives a comprehensive review of the experimental and theoretical knowledge of both micro- and mesoscale IBLs.

An MIBL is conventionally treated as a step-change in the surface roughness length z_0. Munro and Oke (1975) give an example of flow from a surface of newly planted tobacco seedlings ($z_0 = 0.085$ m) to an extensive field of mature wheat ($z_0 = 0.165$ m)

Figure 3.1
Surface inhomogeneities (patches and borders) in the Canadian landscape. Patches: (a) lakes and tributary
channels of the Mackenzie River; (b) cloud shadows cast on Fraser valley farmland (note also the edge
of sea breeze cumulus over the warmer land but cloudless skies in subsidence over the cooler water of
Georgia Strait in the background); (c) clearcut chequerboard in Wood Buffalo Park, Alta (© Garth Lenz);
(d) cultivation strips near Cowley, Alta. At larger scales there are edges between cultivated land and
grassland, between relatively smooth prairie and the rougher foothills of the main range of the Rocky Mountains (© George Hunter). Borders: (e) suburban-rural edge of Ladner, BC; (f) coastal edge between
agricultural field patches and sea near Tsawwassen, BC.

a)

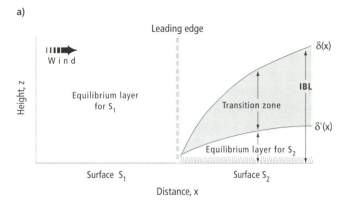

b)

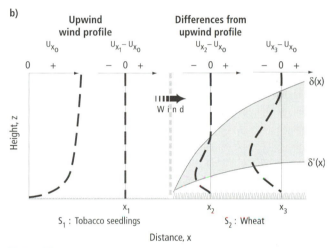

Figure 3.2
Flow adjustment to a step change in surface roughness. (a) Schematic
representation of flow adjustment from surface S_1 to a rougher surface, S_2.
An internal boundary layer (IBL) with height $\delta_{(x)}$ and in which some of
the effects of the new roughness are felt grows downwind from the leading
edge. The lowest 10 percent of the IBL ($\delta'_{(x)}$) is the equilibrium layer in
which conditions have fully adjusted to the roughness of S_2. (b) Simplified
representation of observed wind profile changes as air flows from a field
of tobacco seedlings ($z_o = 0.085$ m) to a field of wheat ($z_o = 0.165$ m) near
Simcoe, Ont. Profiles of the deviations of wind speed from an upwind
equilibrium profile (u_{x_o}) are shown, directly upwind of the leading edge
(at x_1) and at two locations downwind (x_2 and x_3).
Source: Munro and Oke (1975).

at a site near Simcoe, Ont. (Figure 3.2b). The flow upwind of the leading edge in this
schematic is in equilibrium with a relatively smooth surface. Therefore the wind profile
is logarithmic, and, because the extraction of momentum from the mean flow by friction is relatively inefficient, fairly high wind speeds persist down to close to the ground.
When flow encounters the rougher surface at the leading edge, greater friction and

hence increased turbulence produce two effects. First, kinetic energy from the mean flow is diverted into turbulent kinetic energy, slowing the mean flow. Second, greater turbulence ensures an increased transfer of momentum from the mean flow to the surface, which also decelerates the mean flow. Initially the reduction in wind speed is effective only very close to the ground, but as the air advects downwind turbulent diffusion involves greater heights, though the amount of slowing declines.

In principle, the opposite scenario, with flow from rough to smooth surfaces, results in analogous but reverse effects. The rate of adjustment to a smoother surface is slower because of the reduction in the efficiency of turbulent exchange.

Because the mean wind is both the advected quantity and the medium of advection in an MIBL, the detailed analysis of flow adjustment is a non-linear, and consequently non-trivial, problem. The results of numerical models and simplified theoretical considerations have been summarized in parametric formulae by Walmsley, Taylor, and Salmon (1989). These provide easily accessible approximations of the change in wind speed (and its vertical structure) within an MIBL. They use a similarity approach to divide the adjustment to a roughness change into two steps. First, the height of the MIBL interface for a given fetch is estimated by an iteration formula, dependent on the downwind roughness length. Second, the wind profile in the transition zone is interpolated by matching the shape of the upwind equilibrium profile above the MIBL interface to that in the new equilibrium boundary layer, where full adjustment to downwind conditions is assumed; for details, refer to Walmsley, Taylor, and Salmon (1989).

Turbulent Flow Responses to Changing Surface Properties in Flat Terrain

An example of thermal response is the IBL that develops with onshore flow across a coast, especially when the temperature of the water surface is significantly lower than that of the land (for example, the Great Lakes in springtime). The formation of a TIBL is similar to its mechanical counterpart. However, if for simplicity we neglect changes in roughness at the coastline, mean flow can be considered unaffected by the change in thermal properties of the surface. In this case, it is the turbulent response to surface forcing that is the dominant mechanism.

Development of a TIBL downwind from the leading edge at a coast is illustrated in Figure 3.3. When the neutral or stably stratified air over the cool water is advected over the warmer land, its lowest layer is heated and gains buoyancy. The resulting vertical motion associated with thermal convection enhances the turbulence and increases the upward transfer of the new surface influences. The cool-to-warm TIBL thus grows more quickly than a pure MIBL (the opposite is the case in a warm-to-cold TIBL, where vertical movements are suppressed). In a manner similar to formation of a convective mixing layer in the morning hours, the cold-to-warm TIBL grows into the overlying stable air by entrainment (Steyn and Oke, 1982). In this process, the excess buoyancy of heated air parcels carries them into the overlying inversion, where their potential energy surplus is quickly exhausted as they mix with the warmer air above. As a result, warm air from above the inversion is mixed down into the internal boundary layer.

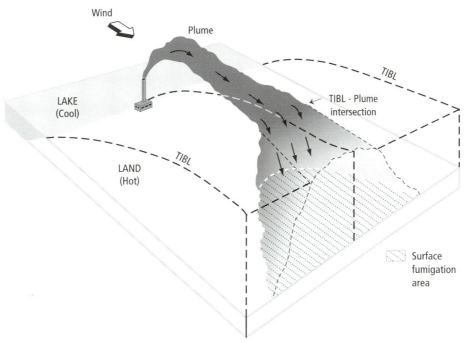

Figure 3.3
Growing thermal internal boundary layer (TIBL), as air of lake origin flows over the warmer coastal
zone during daytime. The elevated pollutant plume from the chimney stack at the coast, which is initially in
stable air, is fumigated downward when it intersects the TIBL.
Source: Modified from Portelli (1982).

This process of entrainment is the prime agent for the growth of the TIBL. It is also responsible for importing properties from the air aloft down into the TIBL and, because of the efficient mixing therein, to the surface. For example, when a pollutant plume, meandering downstream with little dilution in the stably stratified flow over a cold lake, encounters the growing TIBL over land, fumigation can result (Figure 3.3). This causes pollutants aloft to be transported to the ground in high concentrations. This is a common problem in the mid-morning along coastal plains with power plants, refineries, pulp mills, or other industrial stacks.

Fumigation of the SO_2 plume from two 200-m-high smokestacks of the Nanticoke Thermal Generating Station in a TIBL near the shores of Lake Erie in southern Ontario has been studied (Kerman et al., 1982; Portelli, 1982; Ogawa et al., 1986). With onshore flow at Nanticoke, fumigation was observed at the distance inland where the TIBL first attained the plume height. Because of plume rise, the critical TIBL height was usually about 400 m, and that height was attained between approximately 4 and 10 km inland. The location of the peak ground-level concentration moved towards or away from the power plant synchronous with changes in the height of the TIBL. Thus the ability to predict the rate of growth of the TIBL with distance inland is useful in setting stack heights and the location of new residential developments so as to minimize exposure to pollutants.

At Nanticoke, in spring, the water surface is cooler than that of the adjacent land. In the Canadian Arctic archipelago, especially in winter, the water surface can be much warmer than the surrounding ice and snow and the overlying air. Small localized areas (up to 10^2 km^2) of water, called polynyas, remain ice-free throughout the winter, despite the large temperature differences between the atmosphere and the ocean surface. A well-developed TIBL was found to originate from the upwind edge of a polynya close to Dundas Island, NWT (den Hartog et al., 1983; Smith et al., 1983). Large differences in sea-air temperatures (20 to 34 Celsius degrees), maintained by advection of cold air from the ice surface, resulted in large losses of sensible and latent heat from the water surface. The average combined energy loss was 329 W m^{-2} at a time when the net radiation gain by the water was only about 100 W m^{-2} during the day. The extra energy was provided from the heat stored in the relatively warm (approximately $-1.8°$C) sea water.

Patchy Surfaces

Though a simple step change in surface properties is a useful approximation, in reality most surfaces consist of a patchwork of surface properties (Figure 3.1). The scale of patchiness is often so fine that individual patches are smaller than the smallest scales of turbulence (for example, grains of sand and soil, and blades of grass). Such "micro-patchiness" is of little consequence to the evolution of the bulk of the atmospheric boundary layer. In the surface layer and above, the atmosphere responds to an ensemble of surface boundary conditions rather than to the individual elements that make up ecosystems such as grassland, crops, or forest. The lower, horizontal length scale necessary for patches to contribute to atmospheric inhomogeneity is about 10^{-2} to 10^{-3} m.

At the other limit, patches may be large enough to cause the entire atmospheric boundary layer to become adjusted to one (internally homogeneous) patch – for example, a large lake, prairie, or forest stand. The minimum size of patch for this to happen can be estimated using IBL concepts. Consider inhomogeneities in surface energy balance or roughness to occur at length scales L_i. When the flow encounters a change in surface condition at the edge of a patch, turbulence causes the effects to diffuse upward so that by the time the flow encounters the next patch, the internal boundary layer of the first has propagated to a height δ, appropriate to the length scale of the patch. The height δ depends on the vertical diffusion velocity (the state of turbulence in the new boundary layer) and the time for the flow to move across the patch. Briggs (1988) argues that in a fully developed convective mixed layer, the requirement that δ reach to the top of the mixed layer translates to a minimum patch size of $L_i \sim 10^3$ to 10^4 m. In less strongly convective conditions or stable stratification, both the diffusion velocity and the height of the boundary layer are decreased, so this estimate of the minimum patch size necessary to affect the whole boundary layer is conservative. Surfaces containing patches with diameter greater than 10^3 m to 10^4 m may thus be considered homogeneous, as far as boundary layer processes are concerned, except for the leading-edge zone, where the new IBL is developing. However, patchiness with characteristic lengths of a few kilometres introduces spatially variable forcing, which acts at the meso- rather than the microscale. Such

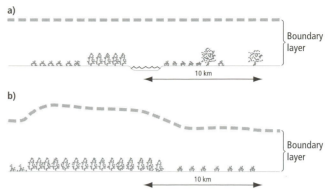

Figure 3.4
Suggested classification of different types of land cover, by ability to organize the boundary layer structure: (a) disorganized; (b) organized. The characteristic length (10 km) is chosen as ten times the typical depth of the planetary boundary layer. Note: vegetation not drawn to scale. *Source*: Shuttleworth (1988).

patches may spawn convective clouds or local circulation systems that are themselves inhomogeneous.

Shuttleworth (1988) coined the term "organized variability" for this type of large-scale patchiness, capable of affecting the planetary boundary layer in a coherent manner (Figure 3.4b). Between the scale of "micro-patchiness" and that of organized variability, there is a wide range of patch sizes, which, if taken independently, do not affect the entire boundary layer. Nevertheless, their collective effects cannot be ignored nearer the surface. This scale of patchiness Shuttleworth termed "disorganized variability" (Figure 3.4a). Patches of this size are able to organize the lower layers of the boundary layer but not its entire depth. They can create variability of surface layer climates. Hence it is the scale of patchiness most relevant to this book. This example again illustrates the principle that a definition of surface homogeneity depends on point of view.

In practice, surface variability at all scales occurs in relation to a nested structure of surface morphology or texture. A degree of organization in the texture is perceived only if there is a clear division between large-scale patches (with length scales L_{lg}) and small-scale patches (with length scales L_{sm}), so that $L_{lg} > L_{sm}$. If this is the case, the surface is considered homogeneous if the scales of inhomogeneity, L_{im}, satisfy $L_{lg} > L_{im} > L_{sm}$. Homogeneity in this sense is the spatial equivalent of temporal stationarity. In this case it means that there is no overall trend over the area of interest. This condition is characteristic of many surface types in Canada. Mixed forests, prairie, or tundra environments may stretch over hundreds of kilometres (L_{lg}) but contain moisture or temperature variations on a scale of hundreds of metres or less (L_{sm}) (Figures 3.1 and 3.5). Similarly, suburban sprawl may continue with little change in character for many kilometres but consist of houses with length scales of 10 to 20 m.

a)

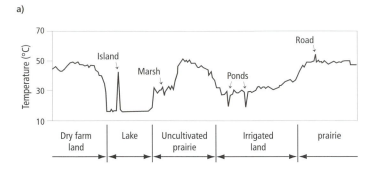

b)

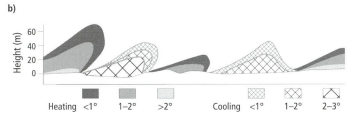

Figure 3.5
Horizontal profile of (a) surface temperature and (b) hot and cold plumes over a diverse Prairie landscape. Based on aircraft observations on the afternoon of 6 August 1968 near Brooks, Alta.
Source: Holmes (1969).

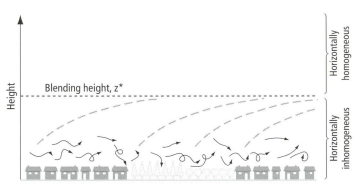

Figure 3.6
Schematic illustration of the roughness layer and its top (the blending height), where the effects of individual surface features are lost and the turbulent properties of the boundary layer become homogeneous.

If we adopt the IBL concept for two-dimensional patchy surfaces, we see that a new IBL develops every time the flow encounters a patch boundary, and characteristics imparted by the new surface boundary diffuse upward into it. With increasing height, individual IBLs are blended by turbulent mixing. Eventually, at some height z^*, the top of the roughness layer (see chapter 2), mixing may be sufficient to produce horizontal homogeneity – so that a representative spatial average of the various surface types in the area is achieved (Figure 3.6). Below z^*, the atmosphere shows the influence of sources and sinks associated with individual surface elements (Raupach and Thom, 1981). Point measurements of scalars or fluxes, such as those obtained by single sensors or eddy correlation systems, may not be representative of ensemble–mean surface layer values characteristic of a larger region (Smith, Simpson, and Fritschen, 1985). This can have serious consequences for estimates of surface energy balance components (Schmid et al., 1991). However, the vertical fluxes at z^*, though horizontally uniform, need not equal the spatially averaged surface fluxes, if there are changes in storage within the layer between the surface and z^*.

The degree to which a given surface may appear spatially homogeneous or inhomogeneous with respect to its influence on boundary-layer properties thus depends both on the surface texture and on the height. In order to estimate the height of the roughness layer, and thus find the scale at which the atmosphere assumes a more homogeneous structure, it is necessary to determine the field of view of a (hypothetical) observer, or rather of the process or property considered. This problem has long been recognized in measurements of surface radiation, where the field of view of an instrument is defined by Lambert's cosine-law and is dependent only on the height of the instrument (Schmid et al., 1991). The aim is to ensure that the instrument "sees" an area large enough to contain a representative average of surface patches. The same principle can be applied to measurements of turbulent fluxes.

During direct measurement of an atmospheric surface-layer property, which is subject to turbulent diffusion (such as temperature, humidity, or wind speed), the sensing instrument is being influenced by a specific portion of the surface upstream, called the "source area" (Schmid and Oke, 1990). Canadian groups have been prominent in developing models to estimate the extent of this two-dimensional source area for the measurement of mean variables or their vertical flux (Leclerc and Thurtell, 1990; Schmid and Oke, 1990; Schuepp et al., 1990; Wilson and Swaters, 1991; Schmid, 1993). The contributions of individual surface elements within the zone of influence are combined to produce a composite influence of the source area and are reflected in the measured signal.

Schmid et al. (1991) report the spatial variability of energy fluxes over a suburban area of Vancouver, BC. Sensible heat fluxes were measured in a relatively homogeneous residential housing area by two eddy correlation systems mounted at 28 m above the ground and separated horizontally by 0.4 to 1 km. The spatial variability was found to decrease as the size of the source area of the instruments increased, and the measurements became representative of a larger area. They show that variability may be considered negligible when the source area that accounts for 50 per cent of the influence covers at least four city blocks (each with dimensions of about 200 × 100 m), or an average of about one hundred residential house lots. This finding indicates that in nearly neutral or stable conditions the height of the roughness layer above this suburban surface may be less than 30 m, or about twice the average spacing of houses. It is considerably higher in unstable conditions.

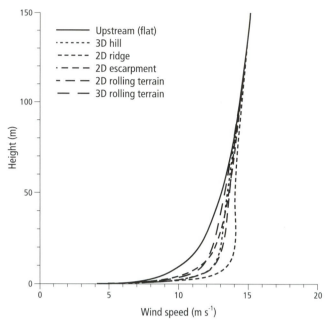

Figure 3.7
Wind speed profiles over various types of terrain, as determined by the
Guidelines model.
Source: Walmsley, Taylor, and Salmon (1989).

FLOW OVER UNEVEN TERRAIN

Response of Mean Flow to Topographic Relief

When a moving fluid, such as air, encounters an obstacle, it is either dammed up
or has to flow over or around it. Immediately upstream of the obstruction, an area of
increased dynamic pressure is created, causing a decrease in flow speed. Above the
obstacle, pressure decreases and velocity increases. A similar effect is found in flow
above and below the wing of an airplane. The wing is designed to give a greater pressure decrease on the upper side than it does on the lower, thus generating lift. Downstream of the obstacle, pressure increases again and velocity decreases. If the leading
or trailing edges of the obstacle are abrupt, or if the flow is very strong, the flow
separates from the surface of the obstacle. Regions of weak flow, stagnation, or even
reverse flow can then occur upstream or downstream from the obstacle. In a similar
fashion, wind flowing over a hill first decelerates, then accelerates, and finally decelerates again. Separation regions may occur on the windward and leeward sides of the
hill, with associated strong gustiness and turbulence.

The degree of thermal stability of the atmosphere also affects airflow over terrain.
When the surface layer is unstable or neutral, lateral deflection of flow tends to be
minimal. Therefore, if the feature has no abrupt edges, almost all of the flow goes
over the obstacle, rather than around it. As the stability of the atmosphere increases,

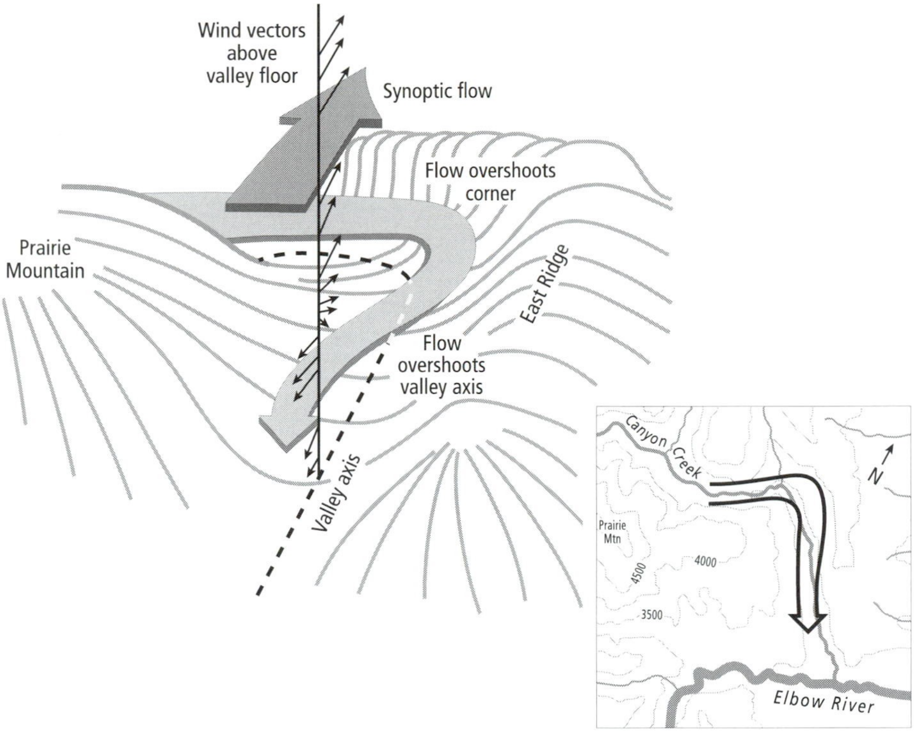

Figure 3.8
Depiction of drainage wind activity in Canyon Creek, Alta, as the flow tries to adjust to the 90° bend in the valley. Viewed looking up-valley from the confluence of the Canyon Creek and Elbow River valleys. *Source*: Sakiyama (1990).

more of the flow is deflected around the terrain feature, if that is possible. In some situations the nature of the terrain feature itself dominates over effects caused by atmospheric conditions – for instance, flow tends to be channelled by steep valley walls. In other cases, the effects of terrain and atmospheric stability interact strongly, producing diurnal upslope winds, nocturnal downslope winds, and strong leeside winds.

There is a seemingly endless variety of interactions between topographic relief and winds. The complexity of the flow depends partly on the geography of the terrain and partly on the physics and dynamics of the atmosphere. Nevertheless, the Guidelines method of Taylor and Lee (1984) and Walmsley, Taylor, and Salmon (1989) is able to provide estimates of wind speed and turbulence intensity on the top of certain idealized terrain features (such as hills, ridges, escarpments, and rolling terrain; Figure 3.7). The method can be used only if the geometry of the terrain features is simple and the slopes are not too steep.

However, the variety of real terrain is often more complex than the broad classes represented by the Guidelines method. In many cases, the relief of a given region is not amenable to treatment in a generic way, and analysis of atmospheric flows must rely on observational case studies, numerical flow modelling, wind-tunnel simulation, or a combination of all three. Sakiyama (1990) conducted an observational study of winds in the Canyon Creek and Elbow River valleys, located in the foothills of the Rocky Mountains, approximately 50 km west of Calgary, Alta. (Figure 3.8). Canyon Creek, which is deep, winding, and variable in cross-section, feeds into the broader Elbow River valley. Large mountains (2,500–3,000 m above MSL) lie to the west. The study concentrated on nocturnal drainage flows – down-valley winds, driven by radiative cooling and the associated increase in density of air. The flow in Canyon Creek is channelled and amplified by the valley geometry as air is forced through a constricted channel. Sakiyama shows that the flow is contained within the valley and reaches its maximum speed at a height of 94 m, and there is a minimum speed at a height of 218 m, in a roughly parabolic profile. Above this level, the synoptic flow continues unaffected by the relief. Cross-valley components of wind indicate "sloshing" behaviour, as the flow adjusts to a sharp bend in the valley with overshoots (Figure 3.8). The valley climate is significantly affected by these frequently recurring drainage flows. The enhanced ventilation and advection of cool and dry air, and local accelerations or decelerations near bends and constrictions, have created microclimatic niches based on the relative exposure to, or shelter from, these local winds.

LAND-SEA BREEZES

Combined Mean Flow and Turbulent Response to Changing Surface Properties in Flat Terrain

In the foregoing, we have presented separately the responses of the mean and turbulent flows to different surface forcings. This does not mean that either response always dominates over the other. As we have seen, differences in land-water surface temperature and roughness result in development of a TIBL and/or an MIBL, respectively, downstream from the coast. To this point, a well-developed mean flow across the leading edge has been assumed – for example, a regional or synoptic-scale wind. However, at the mesoscale (where differential heating occurs over length scales of 10^4 to 10^5 m), differences in surface heating between land and water surfaces themselves may lead to a characteristic mean-flow regime – the land-sea (or land-lake) breeze circulation (Figure 3.9). This is a thermally driven, diurnally reversing, mean circulation in the vertical plane, oriented across the shore.

The dynamical basis of the circulation is the differential heating of land and water surfaces. By day, the more rapid warming, convection, and thus expansion of air lying over the land quickly involve the entire mixed layer. Thus, at the same height in the upper part of the boundary layer above both land and water, pressure gradients are directed towards the water. This condition produces a weak upper flow offshore. At low levels, this redistribution induces a somewhat stronger pressure gradient from water to land (in the reverse direction) and a resultant sea or lake breeze near the surface. The two flows are linked, by subsidence over the water and uplift over the land, to form a closed circulation cell (Figure 3.9). At night the circulation reverses, because of stronger cooling over land than water, resulting in a weaker, low-level land breeze. The zone separating the moist, cool sea air from the dry, warm land air may be quite sharp and is termed the sea breeze front. It starts at the surface some distance inland and slopes upward towards the sea (Hunt and Simpson, 1982).

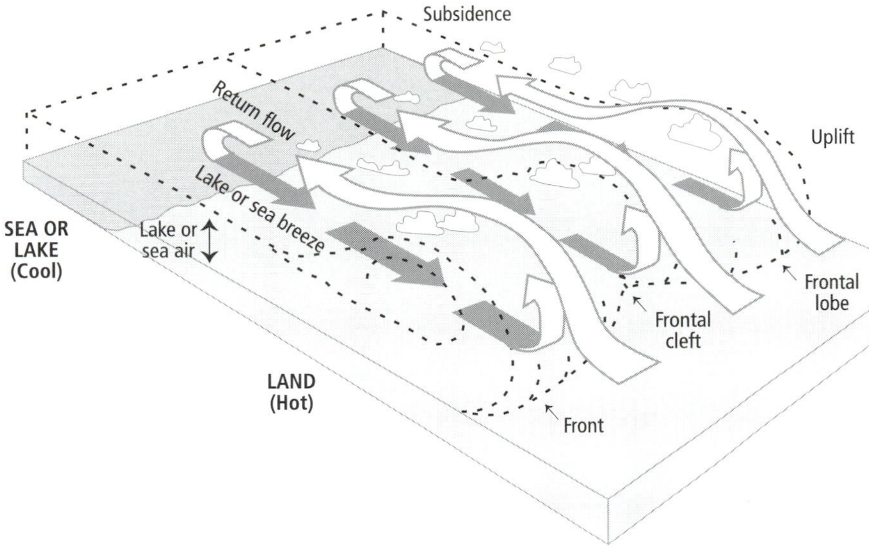

Figure 3.9
A well-developed sea-breeze circulation in the presence of a light, offshore synoptic wind. The relatively cool, moist, dense marine or lake air penetrates inland and is lifted upward at the sea- (or lake-) breeze front, forming fair-weather cumulus cloud aloft. The clouds drift back towards the water body in the return flow and dissipate in subsidence offshore. In detail the lowest layers of the marine or lake air will be modified with distance inland as a TIBL forms (Figure 3.3).

The horizontal extent of the sea-breeze circulation is typically 10 to 100 km in the absence of topographic relief. Later in the day, when the horizontal extent of the circulation is largest, the Coriolis effect causes the direction to veer (back) in the northern (southern) hemisphere. Thus, if the circulation persists throughout the day, the onshore flow direction crosses the coast at less than right angles. The speed of a fully developed sea breeze is typically 2 to 5 m s^{-1}, but it can reach up to 10 m s^{-1}, in a layer roughly 500 to 1,000 m deep. The land breeze is usually weaker and shallower. Land-sea breezes develop primarily when synoptic flow is weak, and, though they are not strong, they can be the dominant ventilation pattern in coastal areas and have a fundamental influence on regional climate. Advection of cool marine air moderates summer temperatures along the coast. Convective uplift over land pumps surface moisture to higher levels, leading to frequent altocumulus cloud. Pollutants may be recirculated in the cell and can thus accumulate over several days.

Because of strong summer land-water temperature contrasts, Canada's extensive coastal zones are frequently subjected to land-sea or land-lake breeze circulations. Pure land-lake breeze circulations without the complications of relief occur around the shores of the Great Lakes. Munn (1963) reports an observational study of micrometeorology at Douglas Point on the east shore of Lake Huron. The study employed a meteorological tower and a tethered balloon, which carried an instrument package up to elevations of 500 m above lake level. Vertical profiles characteristic of lake breezes were present even in August, when land-lake temperature contrasts are no more than 3.6 Celsius degrees. In the Nanticoke study on the shore of Lake Erie discussed above, Ogawa et al. (1986) observed the lake breeze to penetrate landward with an "upward rolling motion" associated with uplift near the lake breeze front. Mukammal (1965) found "spikes" of ozone to be correlated with onshore surges of the lake breeze. The complete land/lake breeze circulation of the Lake Ontario region has been simulated using a mesoscale flow model (Comer and McKendry, 1993). It shows that the spatial pattern is sensitive to the shape of the lake, the influence of adjacent bodies of water, and the synoptic wind direction.

SHELTER EFFECTS, SLOPE BREEZES, AND MOUNTAIN/VALLEY WINDS

Combined Mean Flow and Turbulent Response to Obstacles and Topographic Relief

Both the turbulent and the mean-flow fields are significantly perturbed by obstructions to flow. An example is the flow around windbreaks (lines of trees, hedges or fences placed to modify winds). In principle, the flow has to respond to a windbreak in the same way it does to a terrain obstacle. However, in the case of terrain, the effect on the mean flow is usually much larger than on turbulence. In contrast, modification of turbulence and changes in mean flow are of similar importance with a windbreak. Indeed, windbreaks are often constructed for both of two reasons – to reduce wind speed behind the shelter and thus reduce wind loading and abrasion on structures or plants, and to ameliorate the microclimate by decreasing turbulence and exchanges of heat, water vapour, or dust. The aim is to modify the local energy and mass balances in favour of heat, water, and soil conservation for the benefit of plants, animals, and buildings.

The flow across a solid barrier is commonly divided into four zones (Figure 3.10a, b). In the cavity, or quiet zone, behind medium- or low-density windbreaks, mean wind speed and turbulence are reduced and small eddy sizes prevail. Further downwind is an extended wake region of increased turbulence where eddy sizes and wind speeds return to their undisturbed, open-field values (McNaughton, 1988). Wilson, Swaters, and Ustina (1990) show that maximum reduction of wind speed depends on a parameter that describes the effect of the porosity of the barrier. High-density barriers (with low porosity, such as walls and thick hedgerows) cause large reductions in wind speed, but shelter is confined to a short distance (only five to ten times barrier height – h – downstream; Figure 3.10c). More porous barriers, such as trees and open fences, provide less shelter, but shelter extends over a greater distance (about fifteen to twenty barrier heights downstream). There is also evidence that turbulent transport of heat and mass, such as water vapour, carbon dioxide, dust, and snow, is reduced in the quiet zone and enhanced in the wake (McNaughton, 1988).

On the Canadian Prairies, natural and artificial windbreaks are employed in windy areas to reduce soil erosion on fallow land, collect snowdrifts, and thus accumulate moisture for the growing season. They prevent snow from drifting onto highways, provide shelter for livestock, protect orchards, and reduce heating use by farm buildings. There are costs associated with providing this shelter (materials, shading, land, and possibly water use), so the question arises as to what design is best. Unfortunately definitive answers are not available, though the general pattern of microclimatic and airflow response to shelterbelts is well known (Oke, 1987) and they have been used by farmers for a long time.

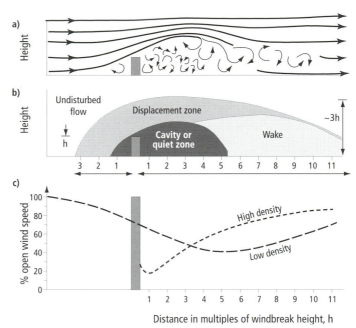

Figure 3.10
Air flow response to windbreaks: (a) streamlines; (b) flow zones over an individual windbreak; and (c) reduction in wind speed around individual windbreaks of two densities.
Sources: (a) and (c) Oke (1987); (b) McNaughton (1988).

   In Canada the economic value of shelter is especially important for climate-sensitive crops such as grain, tobacco, sugar beets, tomatoes, potatoes, and melons (Baldwin, 1988; Kort, 1988). The effect of these microclimatic modifications near shelterbelts is illustrated by the pattern of average crop yields behind a windbreak at Ridgetown, Ont. (see Figure 12.15). Use of snow fences to avoid accumulation of snowdrifts across highways and railway lines (Figure 3.11) exploits reduction of wind speed and turbulence in the cavity zone to enhance deposition about five to ten *h* downstream from the fence. This removal of snow causes a deficit over the target region further downwind and thus a reduced danger of snowdrift building up. Detailed knowledge is critical to this application. A snow fence too close to the road, or too high, may increase rather than reduce the danger of snow accumulation.
   In the absence of large-scale flow, thermal breezes generated by topography can dominate local or regional climates. Daytime surface heating, particularly on south-facing slopes, leads to a sharp vertical gradient of air temperature in a shallow layer close, and parallel to, the ground. As a result, temperatures near the ground may be considerably higher than the air at the same elevation but further away from the slope. Because of the inverse relationship between deviations of temperature and density (and thus pressure), the isobaric surfaces are lifted towards the slope, and a horizontal pressure gradient is created. The pressure gradient force, together with the buoyancy of the heated air, results in a shallow up-slope (anabatic) flow parallel to the surface. This thermal breeze is maintained as long as the rate of surface heating is sufficient to sustain the horizontal density gradients.

Figure 3.11
Snow fences in southern Alberta, showing a substantial amount of snow trapped in the wake region and a well-developed, snow-free scour zone immediately in the lee of the fence. Wind direction during drifting is from right to left.

When radiation cooling of the surface dominates, the situation is reversed, leading to down-slope (katabatic) drainage flows of cool air. In general, the stably stratified air in katabatic winds confines this flow phenomenon to a shallower layer with higher wind speeds than its anabatic counterpart. In both cases, the depth of the flow and its intensity depend on the rate of heating (or cooling), the slope-angle, and the temporal and spatial scale over which the flow is allowed to develop.

At the scale of a whole valley system, daily patterns of heating and cooling often result in circulations which reverse from day to night. The cell is oriented along the long-axis of the valley, with return flow aloft, similar to the sea breeze. The along- and across-valley (slope-breeze) components combine to produce a complex, linked flow system. By day these circulations may result in spatial patterns of cloud tied to the up-slope thermals and concomitant clearing in the sinking air over valleys (Figure 3.12). At night, cold air accumulation in valley bottoms, due to katabatic drainage, produces thermal stratification, which favours frost or fog at the lowest elevations (see Oke, 1987, for a review).

COMBINED LAND-SEA AND
MOUNTAIN-VALLEY CIRCULATIONS

Synthesis of Surface Forcings

Atmospheric flow which is forced by topographic relief produces wind fields of such spatial complexity that they cannot be adequately represented by observations from a network of fixed stations. Details of the spatial and temporal variability of wind fields can then be investigated only by use of numerical flow models operating at appropriate resolution – usually the mesoscale. Construction of a mesoscale numerical

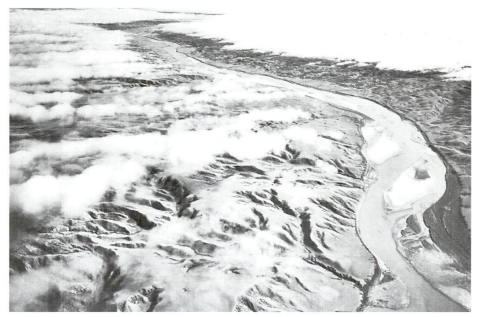

Figure 3.12
Cloud patterns associated with surface controls on convection, near Swift Current, Sask. Note the cloudless conditions associated with organized sinking over the moist and cooler river valley and the alignment of cumulus clouds along the ridgelines, where up-slope winds converge.

model cannot be fully explained here, but, we can learn from the results that it can generate.

In order to understand spatial distribution of photochemical pollutants in the Lower Fraser Valley of British Columbia, Steyn and McKendry (1988) and Miao and Steyn (1995) applied numerical models to explain the mesoscale winds in this coastal valley that are responsible for advection of pollutants on scales from tens to hundreds of kilometres. The synoptic conditions conducive to photochemical production of pollutants are the same sunny, anticyclonic weather that generates land-sea breeze circulations in the valley. Moreover, because of strong topographic forcing by the valley sides, sea breezes in this region are superposed on valley, mountain, and slope winds and probably on roughness effects and thermal breezes caused by the roughness and warmth of the Vancouver urban region (chapter 13). Patterns of wind flow are thus a result of forcing due to the combination of topographic relief and surface properties. The result is a complex regional wind system operating on many scales (Steyn and Faulkner, 1987).

Figure 3.13a is an example of the near-surface wind field from one of the model simulations. It suggests the existence of considerable spatial structure in the wind field, much of which could not be detected by measurements made at surface observing stations because of the impossibility of setting up a sufficiently dense network. The modelled wind field shows a strong onshore component driven by the sea-breeze system, which is channelled by the upper parts of the valley. There is also an indication of up-slope flow on the northern (south-facing) slopes of the valley. Comparison of the measured and modelled winds shows that the superposition of the different circulations is well represented by the numerical model results (Figure 3.13a).

a)

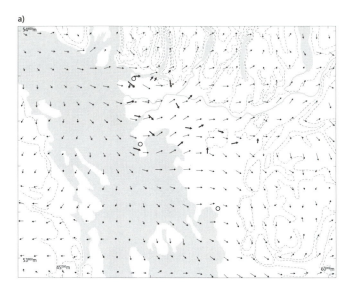

b)

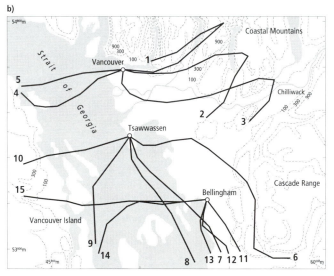

Figure 3.13
(a) Vector arrows depicting the modelled and observed wind field at about
50 m above ground at 1300 h on 23 August 1985, during sea-breeze flow
in the Lower Fraser Valley, BC. Bold arrows are the observed winds.
The distance between grid points (the position of the vector tails) is equivalent to a wind speed of 7.5 m s^{-1}. (b) Plan view of three-dimensional air
parcel trajectories from three coastal locations, with "releases" at five
different heights, for the same day as (a). (Trajectories 1 to 5 from Vancouver
are at release heights 176, 328, 670, 1,103, and 1,673 m above mean sea
level (m.s.l.), respectively; paths 6 to 10 from Tsawwassen are at 130,
335, 660, 1,094 and 1,668 m above m.s.l.; and paths labelled 11 to 15 from
Bellingham are at 206, 407, 733, 1,174 and 1,783 m above m.s.l.). Each
trajectory starts at 0800 h and lasts for fourteen hours. The dots on the
trajectories give the hourly position.
Source: Miao and Steyn (1995).

Miao and Steyn (1995) used computed wind fields from the mesoscale model to derive the trajectories of hypothetical precursor photochemical pollutants emitted at 0800 h from three coastal sites. The time coincides with the morning commuter rush. The trajectories follow "releases" from five different heights at each site. The paths that they take for the next fourteen hours (i.e., until 2200 h) are strikingly different because of the complex interactions of the sea- and valley-breeze circulations and the overlying synoptic flow (Figure 3.13b). In general, the upper two levels (above 1,100 m) are caught up in the easterly/northeasterly synoptic winds and travel across the Strait of Georgia. In contrast, the trajectories of the three lowest levels (below 750 m) are influenced by local and mesoscale flows and have very different histories and implications. The Vancouver trajectories are initially inland with the sea-breeze inflow, but in slightly varying directions at each height. Later in the afternoon and evening they are up the slopes of the north side of the Lower Fraser Valley and then seaward aloft in the sea-breeze return flow. Hence they remain in the vicinity of the valley. The trajectories for the three lowest levels at Tsawwassen, BC, and Bellingham, Wash., do not appear to be drawn into the valley. They are either drawn up the slopes of the south side of the valley or travel south around the western end of the Cascade Range and down the narrow coastal plain. If these patterns characterize flows during fine weather, when photochemical processes are most active, the location of Vancouver is clearly problematic. The city's emissions and their secondary pollutants are likely to be recirculated within the valley over the course of a day and hence may contribute to day-to-day accumulation. Prior to the modelling study, this behaviour could only be surmised.

CONCLUDING REMARKS

As the rest of this book shows, observational and theoretical studies of surface climates are dominated by research conducted over homogeneous surfaces. One reason for this tendency is the enormous simplification afforded by horizontal homogeneity. If homogeneity prevails, observational studies need only one measurement site to be representative of the whole surface, and numerical models can ignore advection of heat, moisture, or momentum from one surface type to another. By contrast, an observational study designed to account for horizontal inhomogeneity must include measurements of horizontal and vertical variations of all relevant meteorological quantities. This comprehensiveness greatly expands the instrumental and logistical requirements. Any theoretical study of spatially variable surface climates that explicitly accounts for advection must include two or three dimensions. This need greatly increases the complexity of the model and the computing power necessary.

Nevertheless, this is one of the directions that surface climatology must follow. Already there is pressing need to answer questions about homogeneity, representativeness, and what constitutes areally representative values of fluxes over patchy terrain in connection with global climate models (chapter 14). Full dimensional capability will greatly enrich our understanding of surface climate systems in their full geographical context.

The phenomena and processes described in this chapter do not constitute surface climates in themselves. Rather, they serve as interfaces or transitions between the surface climates presented in the remaining chapters.

REFERENCES

Baldwin, C.S. 1988. "The Influence of Field Windbreaks on Vegetable and Specialty Crops." *Agriculture, Ecosystem and Environment* 22 and 23: 191–203.

Briggs, G.A. 1988. "Surface Inhomogeneity Effects on Convective Diffusion." *Boundary-Layer Meteorology* 45: 117–36.

Comer, N.T., and McKendry, I.G. 1993. "Observations and Numerical Modelling of Lake Ontario Breezes." *Atmosphere-Ocean* 31: 481–99.

den Hartog, G., Smith, S.D., Anderson, R.J., Topham, D.R., and Perkin, R.G. 1983. "An Investigation of a Polynya in the Canadian Archipelago, III, Surface Heat Flux." *Journal of Geophysical Research* 88: 2911–16.

Garratt, J.R. 1990. "The Internal Boundary Layer – a Review." *Boundary-Layer Meteorology* 50: 171–203.

Holmes, R.M. 1969. "Airborne Measurements of Thermal Discontinuities in the Lowest Layers of the Atmosphere." *Conference on Agricultural Meteorology.* Boston, Mass.: American Meteorological Society.

Hunt, J.C.R., and Simpson, J.E. 1982. "Atmospheric Boundary Layers over Non-homogeneous Terrain." In E.J. Plate, ed., *Engineering Meteorology*, Amsterdam: Elsevier, 269–314.

Kerman, B.R., Mickle, R.E., Portelli, R.V., and Trivett, N.B.A. 1982. "The Nanticoke Shoreline Diffusion Experiment, June 1978 – II. Internal Boundary Layer Structure." *Atmospheric Environment* 16: 423–37.

Kort, J. 1988. "Benefits of Windbreaks to Field and Forage Crops." *Agriculture, Ecosystem and Environment* 22 and 23: 165–90.

Leclerc, M.Y., and Thurtell, G.W. 1990. "Footprint Prediction of Scalar Fluxes and Concentration Profiles Using a Markovian Stochastic Approach." *Boundary-Layer Meteorology* 52: 247–58.

McNaughton, K.G. 1988. "Effects of Windbreaks on Turbulent Transport and Microclimate." *Agriculture, Ecosystem and Environment* 22 and 23: 17–39.

Miao, Y., and Steyn, D.G. 1995: "Mesometeorological Modelling and Trajectory Studies during an Air Pollution Episode in the Lower Fraser Valley, B.C., Canada." In J.G. Watson, J.C. Chow, and P.A. Solomon, eds., *Regional Photochemical Measurement and Modelling Studies*, Pittsburgh: Air and Waste Management Association, 249–81.

Munn, R.E. 1963. "Micrometeorology of Douglas Point." *Canadian Meteorological Memoirs* No. 12. Toronto: Meteorological Branch.

Mukammal, E.I. 1965. "Ozone as a Cause of Tobacco Injury." *Agricultural Meteorology* 2: 145–65.

Munro, D.S., and Oke, T.R. 1975. "Aerodynamic Boundary-Layer Adjustment over a Crop in Neutral Stability." *Boundary-Layer Meteorology* 9: 53–61.

Ogawa, Y.T., Ohara, S., Wakamatsu, P., Diosey, G., and Uno, I. 1986. "Observation of Lake Breeze Penetration and Subsequent Development of the Thermal Internal Boundary Layer for the Nanticoke II Shoreline Diffusion Experiment." *Boundary-Layer Meteorology* 35: 207–30.

Oke, T.R. 1987. *Boundary Layer Climates.* 2nd edn. London: Routledge.

Portelli, R.V. 1982. "The Nanticoke Shoreline Diffusion Experiment, June 1978 – I. Experimental Design and Program Overview." *Atmospheric Environment* 16: 413–21.

Raupach, M.R., and Thom, A.S. 1981. "Turbulence in and above Plant Canopies." *Annals Review of Fluid Mechanics* 13: 97–129.

Sakiyama, S.K. 1990. "Drainage Flow and Inversion Breakup in Two Alberta Mountain Valleys." *Journal of Applied Meteorology* 29: 1015–30.

Schmid, H.-P. 1993. "Source Areas for Scalars and Scalar Fluxes." *Boundary-Layer Meteorology* 67: 293–318.

Schmid, H.-P., and Oke, T.R. 1990. "A Model to Estimate the Source Area Contributing to Turbulent Exchange in the Surface Layer over Patchy Terrain." *Quarterly Journal of the Royal Meteorological Society* 116: 965–88.

Schmid, H.-P., Cleugh, H.A., Grimmond, C.S.B., and Oke, T.R. 1991. "Spatial Variability of Energy Fluxes in Suburban Terrain." *Boundary-Layer Meteorology* 54: 249–76.

Schuepp, P.H., Leclerc, M.Y., McPherson, J.I., and Desjardin, R.L. 1990. "Footprint Prediction of Scalar Fluxes from Analytical Solutions of the Diffusion Equation." *Boundary-Layer Meteorology* 50: 355–74.

Shuttleworth, W.J. 1988. "Macrohydrology – the New Challenge for Process Hydrology." *Journal of Hydrology* 100: 31–56.

Smith, M.O., Simpson, J.R., and Fritschen, L.J. 1985. "Spatial and Temporal Variation of Eddy Flux Measures of Heat and Momentum in the Roughness Sublayer above a 30-m Douglas Fir Forest." In B.A. Hutchison and B.B. Hicks, eds., *The Forest–Atmosphere Interaction*, Dordrecht: Reidel, 563–82.

Smith, S.D., Anderson, R.J., den Hartog, G., Topham, D.R., and Perkin, R.G. 1983. "An Investigation of a Polynya in the Canadian Archipelago, I. Structure of Turbulence and Sensible Heat Flux." *Journal of Geophysical Research* 88: 2900–10.

Steyn, D.G., and Faulkner, D.A. 1987. "The Climatology of Sea Breezes in the Lower Fraser Valley, B.C." *Climatological Bulletin* 20: 21–39.

Steyn, D.G., and McKendry, I.G. 1988. "Quantitative and Qualitative Evaluation of a Three Dimensional Mesoscale Numerical Model of a Sea Breeze in Complex Terrain." *Monthly Weather Review* 116: 1914–26.

Steyn, D.G., and Oke, T.R. 1982. "The Depth of the Daytime Mixed Layer at Two Coastal Sites: A Model and Its Validation." *Boundary-Layer Meteorology* 24: 161–80.

Steyn, D.G., Oke, T.R., Hay, J.E., and Knox, J.L. 1981. "On Scales in Meteorology and Climatology." *McGill Climatological Bulletin* 30: 1–8.

Taylor, P.A., and Lee, R.J. 1984. "Simple Guidelines for Estimating Wind Speed Variations Due to Small Scale Topographic Features." *Climatological Bulletin* 18: 3–32.

Walmsley, J.L., Taylor, P.A., and Salmon, J.R. 1989. "Simple Guidelines for Estimating Wind Speed Variations Due to Small-Scale Topographic Features – an Update." *Climatological Bulletin* 23: 3–14.

Wilson, J.D., and Swaters, G.E. 1991. "The Source Area Influencing a Measurement in the Planetary Boundary Layer: The Footprint and the Distribution of Contact Distance." *Boundary-Layer Meteorology* 55: 25–46.

Wilson, J.D., Swaters, G.E., and Ustina, F. 1990. "A Perturbation Analysis of Turbulent Flow through a Porous Barrier." *Quarterly Journal of the Royal Meteorological Society* 116: 989–1004.

CHAPTER FOUR

Winter and Snow

JOHN W. POMEROY AND
BARRY E. GOODISON

INTRODUCTION

The presence of a such an unequivocal season as the Canadian winter distinguishes our climate from that of neighbours in the Americas, is deeply embedded in the national psyche, and remains a key to the worldwide perception of Canada. Comparative climatological analyses define December, January, and February as the winter period. Table 4.1 summarizes the average annual cold, snowy weather conditions in that period for the climatic regions of Canada. As expected, a strong north–south temperature gradient prevails in the interior (Resolute, NWT; Churchill, Man.; Thompson, Man.; Saskatoon, Sask.; and Toronto, Ont.), while the Atlantic and Pacific coasts have the warmest winter conditions. There is extreme variation in winter's length and severity across the country. Vancouver, BC, has only four days with a maximum temperature less than 0°C, while Resolute has over nine months like that. Resolute has a much colder winter period than Vancouver as well; the mean daily maximum from December to February is −27.8°C in Resolute and 5.9°C in Vancouver.

Snow is the most essential element typifying winter – in human perception, in plant and animal ecology, in global atmospheric dynamics, and in local climatic properties. Over most of Canada, winter is a time of increase of snow stored on the ground; a decrease in depth and eventual disappearance of snowcover mark the transition to spring. Abundant snow is the basis for an active winter recreation industry, and lack of snow translates into hardship for this sector. In some regions of Canada, snow is a short-term impediment to transportation, while in others it is used to create the only roads available. Variation in snowfall amounts and timings from expected conditions cause human hardships and costs for snow removal and traffic disruption can interfere with budgets of local governments. Severe snow storms occasionally halt transportation and isolate regions for several days, even requiring emergency-measures responses from governments. Snowmelt water provides over 80 per cent of annual runoff in the semi-arid Prairies and Arctic and hence provides a critical hydro-chemical replenishment to soils, lakes, and streams. Snowcover plays a key role in the ecology of much of Canada, especially at high altitudes and high latitudes, where its physical properties and chemistry have a dominating influence on ecosystem structure and productivity.

Table 4.1
Winter conditions for selected stations in Canada

Element	Resolute (1947–90)	Vancouver (1937–90)	Halifax (1953–90)	Churchill (1943–90)	Thompson (1967–90)	Saskatoon (1892–90)	Toronto (1937–90)
Total annual snowfall (mm water equivalent)	97.3	54.9	261.4	200.1	200.9	105.4	124.2
Percentage of annual precipitation as snow	64	4	17	43	34	27	15
Percentage of precipitation days as snow	86	8	37	65	55	53	33
Mean daily max. temp. (°C)	−13.5	13.5	10.7	−3.0	2.0	8.0	12.3
Mean daily max. temp. Dec–Feb. (°C)	−27.8	5.9	−0.6	−16.1	−22.7	−15.4	−1.2
Days with max. temp. < 0°C	278	4	60	196	155	115	61
Coldest month	Feb.	Jan.	Feb.	Jan.	Jan.	Jan.	Jan.
Months with mean temp. < 0°C	Sep. to June	None	Dec.–March	Oct.–May	Oct.–April	Nov.–March	Dec.–March

Roughly one-third of annual precipitation in Canada falls as snow and typically over half of the days with precipitation have snowfall. Over most of Canada, snowcover first forms by mid-November and finally melts in mid-April. The time of maximum accumulation is January–February in southern Canada and March–April in northern Canada. There are important deviations from these average conditions across the country. Only 8 per cent of the days with precipitation in Vancouver have snowfall, but in Resolute 86 per cent do (Table 4.1). On average snowcover disappears in early February along the temperate BC coast but stays until late June in the Arctic's Queen Elizabeth Islands. Just as winter is the season that helps define Canada, so snow is the quintessential element of the Canadian environment because of its persistence and unique climatic properties.

Snow accumulation is what remains after falling snow has been modified by concurrent melt, interception in vegetation canopies, wind transport, sublimation, and redistribution. It is incorrect to assume that the short-term increase in snow on the ground is equivalent to snowfall or that the snow accumulation measured at one location can be easily extrapolated to another location without appropriate regard for the processes of redistribution and phase change, resulting from melt or sublimation. The complex processes of accumulation and relocation of snowfall that result in snowcover formation are discussed in detail by Pomeroy and Gray (1995).

Snowcover is notable for causing distinct surface conditions to develop, irrespective of the underlying properties of the surface. In the sense that it "masks" underlying surface conditions with its own unique properties, the presence or absence of snowcover is therefore a critical factor in the winter climate. Another factor is the rapid transformation of surface properties associated with formation and melt of snowcover. A single storm over one day can transform an autumn landscape of soil, grass, rocks, and bushes into one of continuous snowcover and snow drifts. Even coniferous forests develop snow-covered canopies with 0.2–0.3 m depth of snow in their branches. Rapid melting, often associated with chinooks in the west and rain-on-snow in the east, can return snow-covered landscapes to their former conditions in a matter of days.

Snowcovers have unique surface properties because of the special physical properties of snow as compared to other solids on the earth's surface.

First, at 0°C water may exist as a solid, liquid, or vapour on the earth's surface. Below this temperature, it is primarily ice (snow) and vapour, supplemented by thin, liquid-like layers on the edge of snow crystals. Above this temperature water exists as liquid or vapour. Because of diurnal and annual temperature variations, most snow melts (to liquid) or sublimates (to vapour) on an annual basis under Canadian climatic regimes (the exceptions are glaciers) and interacts with vapour and liquid-like phases of water at temperatures below zero. The liquid water content of a snowcover is extremely variable even over time-scales of hours, because it depends on the temperature and rate of melt within the snowcover, the rate of rainfall to the snow-covered surface, the liquid-retention capacity of snow, and the rate of water drainage from the pack.

Second, the latent heat of vaporization is extremely large, being approximately 2.83 MJ kg^{-1} of snow. The energy required to sublimate one kilogram of snow is therefore equivalent to that required to raise the temperature of ten kilograms of liquid water by 67 Celsius degrees. Vaporization is reversible, as this energy is released to the environment on recrystallization of vapour to ice.

Third, the latent heat of fusion is large – about 333 kJ kg^{-1} of snow. The energy required to melt one kilogram of snow (already at 0°C) is therefore equivalent to that required to raise the temperature of one kilogram of water 79 Celsius degrees. Fusion is also reversible, because energy is released to the environment when water freezes.

Fourth, the thermal conductivity of a snowcover is low compared to that of soil and varies with density and liquid-water content of snowcover. A typical thermal conductivity for dry snow with a density of 100 kg m^{-3} is 0.045 W m^{-1} °C^{-1}, over six times less than that for soil. Thus snow can insulate over six times more effectively than soil.

Fifth, the albedo of a snowcover is high compared with soil and vegetation and varies over the winter. A fresh, continuous snowcover in open areas has an albedo of 0.8–0.9. As a snowcover ages and becomes patchy and wet, the albedo can drop to 0.2. Bare soil and vegetation therefore absorb up to eight times as much solar radiation as a fresh snowcover. Albedo in snow-covered forest areas can be as low as 0.2.

Sixth, snowcovers are aerodynamically smooth compared with most land surfaces. Snow surfaces have aerodynamic roughness lengths (z_o) of 0.01 to 0.7 mm, except in drifting snow, when z_o increases substantially. Land surfaces typically have aerodynamic roughness lengths several orders of magnitude greater than this. Thus for a given synoptic-scale air flow, the wind speed is usually greater over snowcover than over bare surfaces.

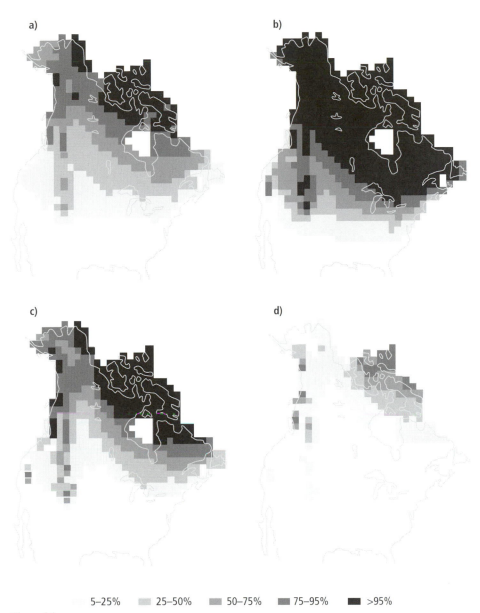

Figure 4.1
North American snow climatology on a mean annual basis from 1973 to 1991. (a) October and November; (b) December to March; (c) April and May; (d) June to September.
Source: Karl et al. (1993).

The unique physical properties of snow, the decrease in sensible and radiative energy at the earth's surface with increasing latitude, and global circulation patterns cause the circumpolar distribution of snowcover on the earth and the sensitivity of this distribution to seasonal and longer-term changes in climate. Karl et al. (1993) calculated the frequency of snowcover occurrence in North America as a function of season from 1972 to 1991. Their work shows that some areas almost always have a

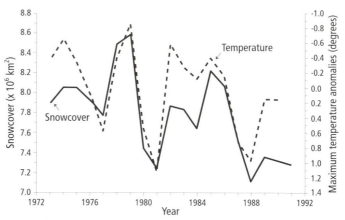

Figure 4.2
Variations in annual September–October snow-covered area over North
America and annual departures from the mean maximum temperature for the
same area.
Source: Karl et al. (1993).

snowcover in a particular season and some areas almost always do not (Figure 4.1).
The transition between these cases is a climatically sensitive zone where there is a
25–75 per cent probability of snowcover occurrence. This zone migrates across Canada from north to south in the autumn and retreats northward in spring and summer,
always occupying a notable portion of the landmass. In cold, wet years this transition
zone is snow-covered, whereas in dry, warm years it is snow-free. The sensitivity of
snow-covered area in North America to temperature variations is particularly strong
during the period of snowcover formation in the autumn. Figure 4.2 shows the yearly
September–October snow-covered area of North America and the yearly departure in
maximum temperature from the period average for September–October (Karl et al.,
1993). A drop in maximum temperature of 0.8 Celsius degrees results in an increase
in snow-covered area of 6×10^5 km^2, while a rise in maximum temperature of 1.1
Celsius degrees results in a decrease in snow-covered area of 8×10^5 km^2. It may be
presumed that a long-term shift in climate would result in a change in snow-covered
area of similar magnitude.

SNOWFALL

Snow forms in the atmosphere when the temperature is less than 0°C and supercooled
water and suitable aerosols are present. Ice crystals form around small aerosol nuclei
and rapidly grow through aggregation of small ice crystals and riming from water
droplets into the familiar snowflake form. For snowfall to occur, there must be sufficient depth of cloud to permit the growth of snow crystals, and enough moisture and
aerosol nuclei to replace those lost from the cloud as falling snowflakes (Schemenauer,
Berry, and Maxwell, 1981). Snowfall in meteorological records is the depth of fresh
snow that falls to the ground during a given period. In Canada this is expressed as a
depth of snow on the ground in centimetres, assuming a density of 100 kg m^{-3}, or as a
depth of equivalent water in millimetres.

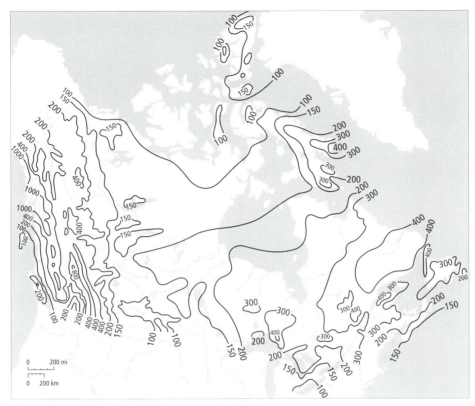

Figure 4.3
Macroscale distribution of mean annual snowfall in Canada from 1951 to 1980. Units are snow water equivalent (*SWE*) depths in millimetres.
Source: Environment Canada (1986).

Distribution of Snowfall

The distribution of snowfall is usefully considered at two scales – macro- and meso-. Local-scale effects are more strongly associated with redistribution after snowfall. Macroscale (distances of 100 to 1,000 km) snowfall distribution varies with latitude, physiography, and proximity to large bodies of water. At this scale, dynamic meteorological effects such as flow deviations caused by the Coriolis effect, standing waves in the atmosphere, flow around mountain ranges, latitudinal temperature changes, and regional moisture sources such as oceans are important. Mesoscale snowfall distribution includes distances of 1 to 100 km, over which changes occur in orographic cooling or warming, deviation of flow due to sharp topographic change, or convective precipitation in response to addition of heat and water vapour from lakes.

Figure 4.3 shows the macroscale distribution of mean annual snowfall in Canada as snow water equivalent (*SWE*). Central and northern Canada receive the smallest snowfall amounts, and the western and eastern regions receive from three to four times as much snowfall as the dry central provinces of Saskatchewan and Manitoba. The low snowfall amounts found in the Prairies and Arctic are associated with the blocking effects of the western Cordillera on the air streams from the Pacific, the lack of frontal

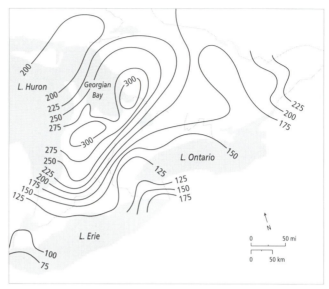

Figure 4.4
Mesoscale distribution of mean annual snowfall in southern Ontario.
Units are *SWE* depths in millimetres.
Source: Hare and Thomas (1979).

systems, and remoteness from moisture sources to the east and north. The greatest snowfall amounts are associated with northwestern Pacific coast locations, high elevations in British Columbia, Ontario northeast of Lake Superior, the highlands of southern Quebec and Labrador, and high elevations in Newfoundland. These regions are all associated with a regional winter source of moisture and some orographic cooling with respect to the elevation of the moisture source.

Figure 4.4 provides a more detailed example of the mesoscale distribution of snowfall in southern Ontario. As topographic influences are small, the major controls are the tracks of winter storms over the Great Lakes and the decrease in temperature with latitude. Intense local convection occurs over relatively warm lake water, producing cloud and snowfall, which moves inland with the prevailing wind. This causes an increase in annual snowfall, which ranges from 100 mm *SWE* depth in southwestern Ontario, where relatively few storms arrive from the Great Lakes, to 320 mm *SWE* just east of Lake Huron, where a northwesterly flow over the lake dominates after passage of a cold front. A similar degree of increase in snowfall occurs in British Columbia as one moves from the low-elevation Okanagan valley to the adjacent Monashee and Selkirk mountain ranges. However, in central British Columbia, the difference results not from lake effects but from orographic cooling over an approximate 2-km change in elevation.

Interpretation of Snowfall Records

Snowfall is more difficult to measure than rainfall, because weather radar and satellites are not able to supplement the point measurements made at meteorological stations. Canada had a network of 2500 meteorological stations in recent years that measured snowfall, however, large areas of the north and high elevations are devoid of stations. The ability of this network to determine areal snowfall depends on the accuracy of the point measurements and the density of the observing network. Most Canadian stations use manual observations of snowfall occurrence and measure the depth of freshly fallen snow, converting this to snowfall by assuming a density of fresh snow of 100 kg m^{-3}. The Canadian Nipher-shielded snow gauge was introduced to the network in 1960, and about three hundred and fifty stations now use it to measure the water equivalent of snowfall on a six-hourly basis. This gauge is the Canadian standard and has been shown to be among the most accurate in the world. As well, more than two hundred automatic meteorological stations measure snowfall using a recording weighing gauge or automatic snow depth sensor.

All snowfall measurements are subject to error, which varies considerably with method and environment. In general, it is more difficult to measure snowfall in a windy environment. There are three main types of errors.

First, there are manual errors in measurement of surface depth. Fresh-snow density often varies from its assumed value of 100 kg m^{-3}, as shown by Goodison (1981). He found seasonal fresh-snow densities vary from 70 to 165 kg m^{-3} across Canada. Densities from individual snowstorms vary even more widely. Wind transport of snow can erode fresh snow before it can be measured or mix fresh and old snow in deposits such that they cannot be distinguished. Because of these errors due to variability in density and, more important, wind transport, manual surface measurements of fresh snowfall are quite unreliable in Prairie and arctic environments.

Second, there can be manual errors in measurement with Nipher-shielded gauges. Though the Nipher shield is 95 per cent or more efficient at wind speeds of less than 5 m s^{-1}, at greater speeds its efficiency drops rapidly, to less than 80 per cent at 7 m s^{-1}. Its catch efficiency is dramatically better than that of unshielded gauges or gauges equipped with Alter-type shields, whose efficiency varies from 10 to 45 per cent at 7 m s^{-1} (Goodison, 1978). Another error for Nipher-shielded gauges is "wetting loss" – water left in the gauge canister when it is poured out for measurement. Goodison and Metcalfe (1989) found this systematic loss is 0.15 mm *SWE* per observation. "Trace amounts" of snow are another concern. Officially, a trace of snow in the canister is given a value of zero, while the actual value lies between 0 and 0.15 mm *SWE*. In northern Canada the large number of trace snowfall events (sometimes two hundred or more per winter) calls into question the published annual snowfall record for such areas (Woo et al., 1983).

Third, measurement with automatic Nipher-shielded gauges can produce errors. The automated gauge is subject to roughly the same undercatch in windy conditions as the manual version. Because snow is allowed to accumulate over long periods, wetting loss and trace-amount loss are not significant for the automatic gauges. Other errors may be important in environments with very low wind speeds. Heavy snow-loads may build up on the Nipher shield itself and block snow from entering the canister. Because the automatic gauge is usually not attended, snow build-up and icing are unlikely to be rectified in a timely manner, and the quality of snowfall data may be suspect in the absence of close inspection.

Corrections applied to address the errors noted above can be critical in assessing the annual snowfall for much of Canada. At Regina, Sask., manual Nipher-gauge measurements were corrected for wetting loss and trace-amount loss for three years.

The average correction was 24 mm *SWE*, which results in a 31 per cent increase in precipitation recorded as snowfall. At Resolute, NWT, manual Nipher-gauge measurements were corrected for trace amounts, and gauge inefficiency at high wind speeds, for six years, using Goodison's (1978) technique. The increase to recorded snowfall precipitation varied from 64 per cent to 161 per cent per year.

INTERCEPTION

Interception of snowfall by vegetation plays a major role in the snow hydrology and surface climatology of coniferous forests. The amount intercepted and stored is controlled by accumulation of falling snow in the canopy. However, the fate of this snow is affected by sublimation, melt, unloading by canopy branches, and wind redistribution. Intercepted snow acts as both a source and a sink of water (Figure 4.5). As a sink, interception receives snow primarily from snowfall and unloading from upper branches and, less important, drip from melting snow from upper branches and vapour deposition during supersaturated atmospheric conditions. As a source, intercepted snow sublimates to water vapour, melts and evaporates to water vapour, becomes suspended by atmospheric turbulence (and either sublimates or redeposits), melts and drips to the surface snowpack, or is unloaded from a branch and released to the surface snowpack. The significance of these specific processes for interception varies with climatic region, local weather pattern, tree species, and canopy density.

Interception Processes and Capacities

The interception efficiency of a canopy is the ratio of snowfall intercepted to total snowfall – a synthesis of the collection efficiencies for individual branches that compose the canopy. The collection efficiency is the ratio of snow retained by the branch to that incident on the horizontal area of the branch. Early in a storm, snowflakes fall through the spaces between branches and needles, lodging in the smallest spaces until small bridges form at narrow openings. These snow bridges increase the collection area and hence the efficiency with which a branch accumulates snow. Further snow is retained on the bridges by cohesion. Cohesion of snow crystals results from the formation of microscale ice-bonds between snow crystals shortly after contact. Ice-bonds form in response to movement of the thin, liquid-like layer surrounding the crystals or reformation of the crystals resulting from small-scale vaporization and condensation (Langham, 1981). Cohesion increases rapidly shortly after initial contact, with the rate of increase in cohesion growing as temperatures approach freezing. Hence the collection efficiency of a branch caused by cohesion of snow to snow rises with increasing temperature, other factors being equal.

Three factors limit the collection efficiency of branches. First, snow crystals falling onto branch elements and onto snow held by the branch create elastic rebound. Rebound is most pronounced below temperatures of −3°C, declining rapidly as temperature rises from −3 to 0°C (Schmidt and Gluns, 1991). Rebound occurs most effectively near the branch's edge. Hence large branches (with a small ratio of perimeter to horizontal area) lose proportionately less snow to this factor than do small ones (large ratio of perimeter to horizontal area).

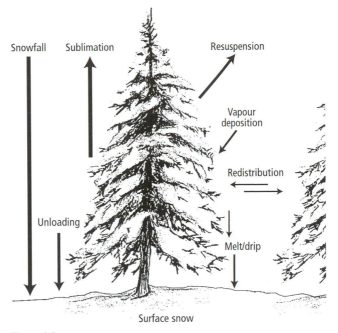

Figure 4.5
Disposition of winter snowfall in a forest environment.
Source: Pomeroy and Schmidt (1993).

Second, branches bend under a load of snow. Bending decreases the horizontal area of the branch and increases the vertical slope, thereby increasing the probability that falling snow crystals will bounce off. The degree to which a branch bends under a given load increases with the branch's elasticity. At sub-freezing temperatures branch elasticity is related to the ice-crystal content of the branch, and it increases linearly with temperature (Schmidt and Pomeroy, 1990).

Third, strength of snow structure affects branches' collection efficiency. As snow accumulates on a branch, the degree to which it holds together and adheres to the branch is related to the strength of interlocking snow crystals. As temperature goes up, so does the rate at which the snow structure "simplifies" (reduces number of bonds) because of metamorphism. Hence snow strength decreases with rising temperature, and when accumulations are large, this reduction may lead to lowered interception efficiency (Gubler and Rychetnik, 1991).

If meteorological conditions are constant, interception is most efficient during the bridging of the smallest branches. As accumulation continues, branch bending lessens the horizontal area of the branch and decreases the ability to intercept snow more rapidly than bridging can increase it. As interception approaches a maximum, sharp vertical angles of the snow surface promote crystal rebound and erosion, rather than continued accumulation. Low temperatures promote more efficient rebound of snow particles (greatest rate of change with temperature from −3 to 0°C), less bending of branches (greatest rate of change with temperature from −12 to 0°C) and stronger snow, with conflicting results for the interception efficiency.

It has been observed that snowfall above $-3°C$ has relatively high interception efficiency because of cohesion. Such highly cohesive snow is intercepted by tree trunks as well as branches for short periods. As the temperature drops below $-3°C$ (non-cohesive), it seems likely that reduced branch bending at lower temperatures and greater strength will increase interception efficiency, particularly for large accumulations that require stiff branches and strong snow to remain in the canopy. Hoover and Leaf (1967) and Gubler and Rychetnik (1991) note increased interception efficiency at lower temperatures. In assessing the effect of temperature on interception efficiency, one should carefully separate out other effects correlated with temperature, such as snow crystal habit and snow density.

High winds may induce snow redistribution from conifer branches during snowfall, reducing apparent interception efficiency during a storm. The lowest interception efficiencies in the measurements of Schmidt and Gluns (1991) occurred during the highest wind speeds. More effective snow particle rebound during high winds may be the reason for reduced interception efficiency in these cases, though release of accumulating snow, triggered by branch vibrations in the wind, may play a role as well.

One may conclude that interception efficiency generally rises with increasing size of falling snow crystals, decreasing temperature, decreasing wind speed, and decreasing density of intercepted snow. Satterlund and Haupt (1970) and Schmidt and Gluns (1991) show that these meteorological conditions are more important than is conifer species in determining interception efficiency for single branches or single trees.

Integrating the above observations from single branches or single trees up to the scale of a forest is difficult because within-canopy snow interactions have not been specifically measured in field experiments. In addition, bulk canopy characteristics may override many of the factors important to interception on single branches or trees. McNay, Peterson, and Nyberg (1988) compared snow-depth surveys in forested and open areas after individual snowfall events in upland Douglas fir and western hemlock forests on Vancouver Island, BC. They found a linear increase of new snow depth under the forest canopy with new snow depth in open areas, with the rate of increase controlled by the completeness of the canopy crown. While their model is directed towards predicting snow depth, it may be modified for predicting interception snow-water equivalent by assuming a density of fresh snow, $\rho_s = 0.15$ kg m^{-3}. Interception is now found to be a function of snowstorm snow–water equivalent and crown completeness (Figure 4.6). The model probably overestimates interception for high snowfalls, because it presumes a constant interception efficiency.

Sublimation of Intercepted Snow

Sublimation of snow consumes a considerable amount of energy, especially in relation to that available in a winter forest environment. However, the well-distributed nature of intercepted snow through the lower 10 to 20 m of the atmosphere in a canopy results in a large volume of air being exposed to snow. Forest canopies maintain a relatively low albedo in spite of their low snowload. This, in addition to the large ratio of surface area to mass in intercepted snow, results in the potential for notable rates of turbulent transfer of atmospheric heat to the intercepted snow and concomitant removal of water vapour produced at the snow surface. Pomeroy and Schmidt (1993)

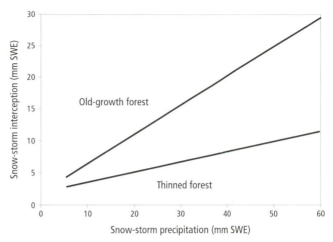

Figure 4.6
Interception by an old-growth coniferous forest and a commercially
thinned coniferous forest expressed in millimetres of *SWE* per square
metre of forest floor and as a function of size of snowstorm.

provide data which suggest that, for a unit mass of snow, the surface area of snow
intercepted in boreal-forest spruce canopies is 60 to 1,800 times greater than that of
snow on the ground. This high surface area, and the long period of exposure of snow-
covered canopies during northern winters, provide ample opportunity for sublimation, even when cold temperatures suppress the rate. Measurements in Prince Albert
National Park, Sask., during December 1992 and January 1993 show that 10 kg of
snow accumulated on a single black spruce tree in a cold period (−25 to −45°C). Over
several warmer weeks in late January (−20 to −2°C), about 6 kg of this snow sublimated. The mass of the tree and its snow during January is shown in Figure 4.7.
Landscape-based snow surveys over that winter showed that 31 per cent of the annual
snowfall sublimated from intercepted snow in spruce and pine canopies. A review of
intercepted-snow sublimation losses (Schmidt and Troendle, 1992) suggests that approximately one-third of annual snowfall is lost to sublimation from intercepted snow
in dense coniferous canopies in this region. Extrapolating to the boreal forest of western Canada (from northwestern Ontario to Great Slave Lake) and correcting for spatial distributions of coniferous canopy density, we find an average of 46 mm *SWE* in a
region that receives from 100 to 200 mm *SWE* of annual snowfall.

WIND TRANSPORT

The effects of wind on the evolution and distribution of a snowcover are most evident
in open environments. Wind hardens and increases the density of snow. When snow is
transported by wind, the crystals undergo changes to their shape, size, and other physical properties and, on redeposition, form drifts and banks denser than the parent material.

   The transport of snow particles by wind involves three modes of movement. Creep
is the movement of particles too heavy to be lifted by the prevailing wind by rolling

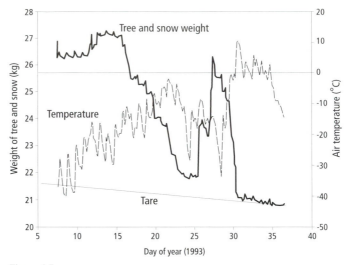

Figure 4.7
Combined weight of a 12-m black spruce and its intercepted snowload and the
ambient air temperature 5 January–9 February 1993 in a black-spruce stand in
Prince Albert National Park, Sask. The tare is the weight of the tree itself,
which was dessicating and losing needles during the experiment.

along the surface. Saltation is the movement of snow particles by a skipping or jump-
ing action along the snow surface. Turbulent diffusion (suspension) is the movement
of snow particles by suspension in the flow.

Saltation and Turbulent Diffusion

In the first few millimetres above the snow surface, particles move in saltation, and
the mass flux appears to decrease exponentially with height. Erosion of snow particles occurs when the shear stress exerted on the snow surface by the wind exceeds the
shear strength of the snow surface and any shear force exerted on non-erodible parts
of the surface, such as vegetation, ice, and rocks. This partitioning of shear stress
forms the basis of descriptions of saltation transport. The shear stress itself is proportional to the square of the atmospheric friction velocity.

The aerodynamic roughness of the surface during blowing snow differs from that
for non-transport conditions because roughness is increased by saltating snow particles. The greater the rate of saltation transport, the greater the aerodynamic roughness
length. Pomeroy, Gray, and Landine (1993) found that this increase in z_0 could be expressed as a function of friction velocity and the silhouette area produced by the vegetation units protruding through the snow surface.

Suspended snow is that supported by turbulent diffusion and carried along by moving air at a velocity approximately equal to that of the mean horizontal wind. Instead
of bouncing along the snow surface, suspended snow is lifted by turbulent eddies in
the atmosphere, often to well above the surface. Suspended snow exists from a several
millimetres above the snow surface to hundreds of metres above the ground. Its maximum height is determined by the duration of the blowing snowstorm, the intensity of

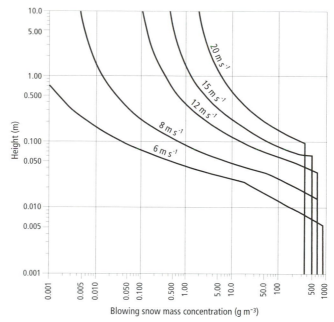

Figure 4.8
Vertical profiles of blowing snow mass as a function of wind speed, given
a threshold wind speed of 5 m s^{-1} at the 10-m height. The mass concentration in the saltating layer is considered constant, and concentration
drops off rapidly with height in the suspended layer.

turbulence, and the upwind distance over open, snow-covered terrain. Saltating snow
particles, rather than snow surface crystals, are generally considered the source of particles suspended by turbulence. As shown in Figure 4.8, the concentration of suspended snow above Prairie snowcovers reaches a maximum just above the saltation
layer and decreases with height at a rate that depends on the wind speed. Pomeroy and
Male (1992) show that 77 per cent and 40 per cent of this total transport would occur
below a height of 1 m, when the 10-m wind speed was 10 m s^{-1} and 30 m s^{-1}, respectively. Though mass concentration of suspended snow is usually smaller than that
of saltating snow, the total mass of snow moving in suspension may be large because
of the great vertical thickness of the layer. Experimental work by Tabler, Pomeroy,
and Santana (1990) indicates that the suspended-snow transport rate over complete,
unobstructed snowcovers can be expressed with some accuracy as an empirical function of wind speed.

Total snow transport can be estimated by summing the saltation and suspension
components. Figure 4.9 shows the relative contribution of saltation and suspension
transport for a threshold friction velocity corresponding to a wind speed at the 10-m
height of 5.5 m s^{-1} (dry snow of moderate hardness) over an unvegetated, level, and
extensive snowfield. Suspension dominates over saltation transport, especially at high
wind speeds. Empirical expressions are available that may be used to estimate the total
snow transport, directly from wind speed (Tabler, Pomeroy, and Santana, 1990, and
Pomeroy, Gray, and Landine, 1991).

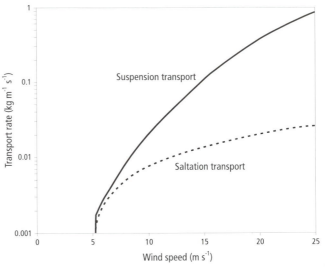

Figure 4.9
Transport rate of blowing snow by saltation (near surface) and suspension
as a function of the wind speed at 10 m.
Source: Pomeroy and Male (1992).

Sublimation of Blowing Snow

Sublimation from blowing snow is a significant loss to surface water-supply and a notable gain to atmospheric water vapour supplies during dry winter weather. The theory of sublimation of blowing snow is described by Schmidt (1972), Male (1980), and Pomeroy, Gray, and Landine (1993). Sublimation of a snow particle entrained in the atmosphere is governed by the laws of conservation of mass and energy. Sublimation of blowing snow particles occurs at high rates compared to the surface of a snow-cover because of the higher turbulent exchange in the air and the much greater ratio of surface area to mass for a snow particle removed from the pack. For instance, a unit mass of blowing snow has about three thousand times more surface area than does an equivalent mass of snow on the ground. The sublimation rate can be calculated using an involved equation for heat and mass transfer developed by Schmidt (1972) and implemented for standard meteorological measurements by Pomeroy (1989). The magnitude of sublimation loss depends on blowing snow mass concentration, wind speed, air temperature, relative humidity, and – less important – solar radiation.

In much of Canada, variations in temperature or humidity through their normal range cause at least an order-of-magnitude change in the sublimation rate. Variations in daily radiation input affect this rate very little. For example, the sublimation rate increases twenty-five-fold when air temperature rises from $-35°C$ to $-1°C$. In a severe, blowing snowstorm, with 10-m height wind speeds of 20 m s^{-1}, temperature of $-15°C$, and relative humidity of 70 per cent, the sublimation loss from blowing snow particles is equivalent to about 1 mm of *SWE* per horn. In the Prairies or the Arctic, where annual snowfalls of 100 mm *SWE* and frequent blizzards are common, blowing snow sublimation can substantially deplete winter snowcover.

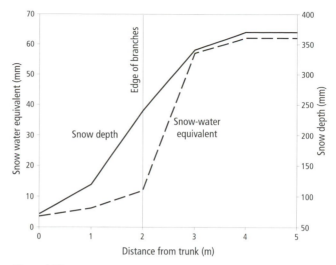

Figure 4.10
Variation in snow depth and *SWE* near a 10-m–tall white spruce, in a stand of trembling aspen, in Prince Albert National Park, Sask., December 1991.

ACCUMULATION

Because of variations in snowfall, snow interception, and snow transport over land-scapes, topographic factors such as elevation, slope, and aspect are important to snow accumulation, along with vegetation type, density, and height. Elevation alone is not a causative factor in snowcover distribution, because differences in snow accumulation with elevation are often substantially affected by altitudinal changes in vegetation cover and slope angle. Orographic snowfall is related more to slope and wind flow than to elevation. However, with vegetation and micro-relief held constant, the depth of snowcover in southern Canada increases with elevation, because of the greater number of snowfall events and the decrease in evaporation and melt. In northern Canada this relation often does not apply, because there are few midwinter melts at any elevation, the length of the snowfall season does not vary notably between high and low elevations, and the greater wind speeds at high elevations can erode more of the snowcover than takes place at low elevations.

Forested Environment

Within a forest canopy, *SWE* varies in relation to the distance to trees and between stands of different tree species. Snow accumulation decreases as a coniferous tree trunk is approached (Figure 4.10). Conversely, snow accumulation increases slightly closer to a deciduous tree trunk. Deciduous forests have higher *SWE*, and least accumulations occur under coniferous trees (Sturm, 1992; Pomeroy and Gray, 1995).

The influence of leaf area index on snow accumulation is shown in Figure 4.11, as measured in midwinter in the southern boreal forest of Saskatchewan. Leaf area index

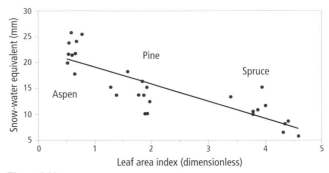

Figure 4.11
Relationship between leaf area index (dimensionless) measured in winter and *SWE* in a southern boreal forest, Prince Albert National Park, Sask., December 1992.

was measured for the canopy above each snow measurement point. There is a roughly linear decrease in *SWE* with increasing leaf area, though individual tree species have noticeable effects and there is considerable variation within a species type. An examination of the winter evolution of snowcover in the same boreal forest by stand species type is made in Figure 4.12. It is evident that throughout winter the aspen stand and small clearing accumulated more snow than did the jack pine or black spruce stands. A warming trend at the end of January removed all intercepted snow from the canopy. The differences in *SWE* at the end of that month represented differential losses due to sublimation, of about 32 per cent of the snowfall accumulated to that date.

Much of the research on snowcover distribution in forested environments has concentrated on the differences between the amounts of snow collected in forests and in openings and has demonstrated larger amounts of snow in clearings. In southeastern British Columbia, Toews and Gluns (1986) report that snow accumulation in clear-cuts ranged from 4 to 118 per cent more than that in adjacent coniferous forests, with a mean difference of 37 per cent. In the foothills of southern Alberta, Golding and Swanson (1986) report that snow accumulation increased from 20 to 45 per cent from forest to clearing.

Kuz'min (1960) suggests that the *SWE* in a forest can be related to that in a clearing and to the density of the forest canopy irrespective of clearing size. The average depth of snow in a clearing varies with the size of the opening. Snow depths decrease as the size of the opening increases. Swanson (1988) notes for forests in Alberta that an opening two to three times the height of the surrounding forest canopy accumulates the maximum amount of snow. Beyond diameters twelve times the average tree height, the clearing retains less snow than the adjacent coniferous forest. The reduction in snowcover in large clearings is caused by the growing likelihood that wind transport will erode snow from the clearing. This phenomenon has implications for the effects of clear-cut blocks on snow accumulation.

Open Environments

In open environments, vegetation and terrain features can produce remarkable variations in snow accumulation. The effect of a hedge-row in the centre of a field and of

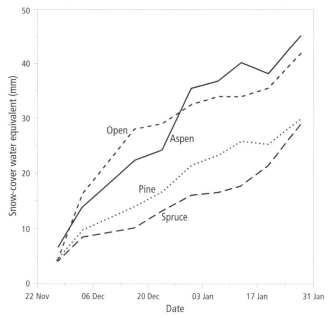

Figure 4.12
Winter evolution of *SWE* under various forest stands and in a small clearing in the southern boreal forest, Prince Albert National Park.
Source: Pomeroy and Gray (1995).

different crop heights is shown in Figure 4.13. Snow accumulation near the hedge is over ten times greater than that in adjacent fields, and snow accumulation in the fields depends on the height of the wheat stubble. The effect of a sharp topographic change is shown in Figure 4.14, which displays an *SWE* transect at Trail Valley Creek, 60 km north of Inuvik, NWT. There is a sharp drop from the upland tundra plateau to the valley bottom, which induced formation of a major side drift. Slightly higher vegetation on the valley bottom holds more snow than the short plants on the plateau. Drifts such as these formed near hedges and valley sides persist much later than the adjacent snowcover during melt and provide a major source of meltwater to the Arctic and the Prairies. The landscape-wide effects of the small-scale snowcover response to vegetation and terrain are particularly evident in Canada's Prairie landscapes. A long and windy winter season, with a mix of open, short grass terrain, farm yards, shelterbelts, plateaux, and gullies, produces distinctive snow accumulations in various landscape types (Figure 3.11). Table 4.2 lists the relative amounts of snow water retained by various landscapes in a Prairie environment in Canada near the time of maximum accumulation. It is evident that different terrain and vegetation combinations can create distinctive patterns of snow accumulation across the landscape.

Pomeroy, Gray, and Landine (1993) have calculated the annual losses of snow to wind transport (to field edges, hedgerows, gullies, and so on) and to sublimation during blowing snow. The average annual blowing-snow losses on one kilometre fetches of stubble (stalk height = 0.25 m) and fallow (surface roughness equivalent to grain stalk height = 0.01 m) at four locations in Saskatchewan show that on average at Prince Albert, Swift Current, and Yorkton at least 8 per cent (range of 8–11 per cent)

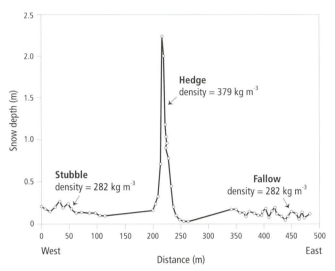

Figure 4.13
Transects of snow depth and density across a summer-fallowed field
and a wheat-stubble field separated by a hedge east of Saskatoon,
March 1989.
Source: Pomeroy, Gray, and Landine (1993).

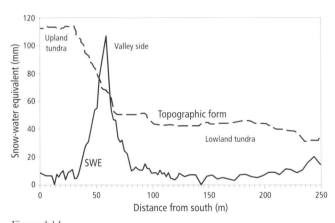

Figure 4.14
Transect showing the relationship of *SWE* and topography at Trail Valley
Creek, 50 km north of Inuvik, NWT, May 1991. The transect goes from
an upland tundra plateau down a valley side to a lowland tundra valley
bottom.

of annual snowfall is removed from a 1-km fetch of 0.25-m–tall stubble. At Regina
snow losses are about 19 per cent. The blowing-snow transport and sublimation
losses are greater on fallow than on stubble. The largest increases due to a change in
surface condition occur at Regina and Swift Current, where the snowcover losses due
to transport at least double and the sublimation losses increase by 7 per cent. The percentage of annual snowfall lost to sublimation from a 1-km fetch is equal to or

Table 4.2
Relative snow water retention on various landscape types in an open grassland environment

Landscape type	Variant	Relative SWE accumulation[a]
Level plains	Fallow	1.00
	Stubble	1.15
	Pasture (grazed)	0.60
Gradual hill and valley slopes	Fallow, stubble, hayland	1.0–1.10
	Pasture (ungrazed)	1.25
Steep hill and valley slopes	Pasture (ungrazed)	2.85
	Brush	4.20
Ridge and hilltops	Fallow, ungrazed pasture	0.40–0.50
	Stubble	0.75
Small, shallow drainageways	Fallow, stubble, pasture (ungrazed)	2.0–2.15
Wide valley bottoms	Pasture (grazed)	1.30
Farm yards	Mixed trees	2.40

[a] Normalized to SWE monitored on level plains under fallow.

greater than (ranging to 2.6 times) the amount of snow transported. Over the Prairies, blowing-snow losses generally tend to decrease with increasing latitude as one progresses from the grassland to boreal forest because of climatic differences and changes in vegetation. Lower wind speeds and air temperatures and vapour pressure deficits prevail at the northerly stations, and the vegetation changes from open grassland to mixed grassland and deciduous and boreal forest. In arctic regions, blowing-snow fluxes may also be large, despite low temperatures and humidity, because the open exposure, high winds, and reduced bonding of snow crystals result in a high frequency of blowing snow.

Variations in the mean annual fluxes of blowing snow with fetch distance are demonstrated in data for Prince Albert, Regina, Swift Current, and Yorkton, all in Saskatchewan. Figure 4.15a shows that snow transport on stubble at Prince Albert and Yorkton remains reasonably constant at between 9 and 10 Mg m^{-1}, independent of fetch distance. At Regina and Swift Current, the flux decreases slowly with increasing fetch at distances greater than approximately 1 km. These trends are due to sublimation becoming the dominant process governing snow disposition for most combinations of wind speed, temperature, and relative humidity. Figure 4.15b shows sublimation increasing with increasing fetch at Regina. At a fetch distance of 4 km the mean annual value is about 254 Mg m^{-1} (approximately 64 mm SWE). In contrast, the loss over the same distance at Yorkton is only 112 Mg m^{-1} (approximately 28 mm SWE). The higher sublimation rates at Regina and Swift Current are primarily the result of the extreme winds at these stations. Each year there are several occasions when high wind speeds scour snow from previously filled roughness elements and produce large sublimation losses due to high rates of particle ventilation. As a consequence, winds following these events, which would normally cause transport, do not do so, because of the increased surface roughness produced by scouring.

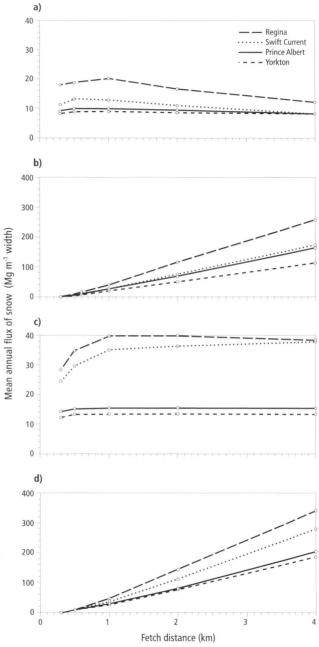

Figure 4.15
Variation in mean annual mass of snow (a) transported in saltation and
suspension and (b) sublimated as blowing snow across stubble fields
at four locations in the Prairies; (c) and (d): as for (a) and (b), respectively,
but for fallow fields. Annual snow masses are expressed per unit width
for the length of the fetch.
Source: Pomeroy, Gray, and Landine (1993).

The variation in blowing-snow fluxes with fetch distance on fallow land is illustrated in Figure 4.15c, d. At Regina and Swift Current as fetch distances increase from 300 m to 1 km, snow transport increases appreciably. At Regina the maximum rate of snow transport occurs with a fetch of 1 km, whereas at Swift Current it is still increasing at 4 km. The rapid increase in transport rate with distance for fetches less than 1 km is attributed to the combined effects of high winds and lack of snowcover.

Comparing the data in Figures 4.15b and 4.15d, notice that sublimation from a 4-km fetch of fallow at Regina is about 40 per cent greater than the loss from the same length of stubble. At Yorkton, it is about 60 per cent greater.

Pomeroy and Gray (1994) have also elucidated the role of other weather variables on the climatology of blowing snow in the Prairie provinces.

Snowcover Measurement

Different techniques are used to measure snowcover. Point snowcover depths are measured most easily with snow rulers. The ruler is pushed through the snowpack to the ground surface, and depth is read off. For automatic measurements, sonar is used. An ultrasonic pulse is emitted from a fixed, downward-facing horn, placed from 0.5 to 10 m above the snow surface, and the timing of its echo is recorded by a microprocessor. The distance from horn to snowcover is calculated from this timing (Goodison et al., 1988).

For snow density, a pit is dug to the bottom of the snowpack, and individual snow strata (layers of snow having distinctive properties, such as density, hardness, and grain size) are identified and their thicknesses measured. A small scoop (100 to 250 ml) is used to remove samples of snow from each layer. These are weighed to determine layer density.

Two techniques are routinely employed for point measurements of snow-water equivalent. The first, gravimetric, uses a graduated, hollow tube with a cutter fitted to the end to obtain a vertical core of snow. The water equivalent is determined by melting or weighing the snow core (Goodison, Ferguson, and McKay, 1981). The second, for automatic measurements, uses a snow pillow, which is a mattress-type apparatus filled with fluid (antifreeze) and installed below the snowpack. The snow pillow monitors *SWE* by responding to the weight of snow accumulated on its surface (Goodison, Ferguson, and McKay, 1981).

Snow surveys are made at regular intervals throughout winter at designated stations along a permanently marked traverse (snow course) to determine depth, vertically integrated density, and *SWE*. The length of a snow course and the distance between sampling points vary, depending on site conditions and uniformity of snowcover. In hilly terrain, a snow course is generally 120 m to 270 m in length, and observations are taken at 30-m intervals. In open environments, it may be longer, with density measurements taken 100 to 500 m apart and depth measurements made at about five equally spaced points between the density locations (Goodison, Ferguson, and McKay, 1981).

Stratified sampling lessens the data collection requirements of snow surveys by classifying the snowcover into landscape classes and sampling based on a landscape stratification. Random samples of density and depth are taken from a specific class, and these are used to calculate the average snow–water equivalent for that class. Mean depth of water on larger scales is obtained by weighting the values for various classes according to area. The number of depth and density measurements taken on a given landscape class should attempt to define the mean snow depth with a relative error of about 5 per cent and the standard deviation with a relative error of about 10 per cent. Sampling of snow density around a central depth (mean, median, or mode) is recommended. Where coefficients of variation for snow depth and density are available, these should be used to calculate the number of samples required for the specified basin-wide accuracy (Steppuhn and Dyck, 1974; Woo et al., 1983).

Areal distribution measurements of snowcover are time-consuming if conducted manually. Hence remote sensing methods complement the traditional snow surveys. Remote sensing of *SWE* in mountainous regions employs relations between the elevation above which there is continuous snowcover (snow-line) and the mean *SWE* above this line. These relations are based on the sequence of areal depletion of snowcover during melt. The snow-covered area/snow-line must be followed from pre-melt through melt if one is to calculate the pre-melt *SWE*. Visible or near-infrared imagery can be used to determine snow-covered area and the elevation of the snow-line in these regions. When a series of clear images is available during melt, a bulk *SWE* for the snow-covered area can be calculated (Martinec and Rango, 1991). For the Prairies, the natural gamma radiation emitted by the earth's surface can be detected by a spectrometer mounted in an aircraft. *SWE* is calculated from gamma particle counts and data on soil moisture. Initially, a background count (no snowcover) is established over a prescribed flight line (typically 16 km long by 300 m wide). Flights are made along the same flight line(s) during winter, and, as snow accumulates, the attenuation of gamma radiation emitted by the earth's surface increases and the count rate decreases. This method can be used in snowpacks with water equivalents up to 300 mm (Goodison, Ferguson, and McKay, 1981). A newer technique employs passive microwave radiation. The earth emits microwave radiation as a near blackbody, in that its radiant intensity depends largely on the temperature of the ground. Snow, a porous medium, scatters radiation and has a lower emissivity than the earth. The microwave emissivity of snow increases with increasing snow-surface roughness, temperature, underlying frozen soil, and snow wetness and decreasing snow particle radius and snow depth. Nevertheless, successful estimates of *SWE* have been made for cold, dry snowpacks over open agricultural land with frozen soils. Passive microwave satellite sensors (SMMR, SSM/I) detect vertically and horizontally polarized microwave radiation with a 25 km resolution and daily coverage over Canada. Goodison and Walker (1994) review the use of this technique.

ENERGETICS OF SNOW SURFACES

Snow surfaces have unique energetics because of their physical properties and because of the distinctive climatic conditions during winter. The surface energy balance is the framework for considering the energetics of snow. The major difference is that the snow surface itself undergoes phase changes through sublimation, melt, evaporation, refreezing, and condensation and mass changes through wind erosion, evaporation, infiltration to soils, and runoff. Thus consideration of the energetics of a snowcover is more complex than initially evident for a relatively uniform surface.

Energetics at the Soil Boundary

The ground heat flux (Q_G) to snow on a daily basis in southern Canada is typically considered to be a small component of the energy balance. However, because it is persistent, it can have a major cumulative effect over a season and retard or accelerate the time of melt. The ground heat flux under melting snow is not completely the result of the pre-melt temperature gradient and thermal conductivity of soils. Infiltration of meltwater into soils also involves a significant energy flow associated with the accompanying latent heat of fusion. Infiltration may warm upper soil layers, increase apparent thermal conductivity, and further melt frozen soil water or refreeze meltwater in the frozen soil. The refreezing of meltwater at the soil–snow interface forms a basal ice layer and releases a significant quantity of latent heat, which is a distinctive characteristic of melting arctic snowcovers (Marsh and Woo, 1984).

These phenomena change Q_G by transferring sensible heat to the soil and absorbing or releasing latent heat. Rouse (1990) divides Q_G into the storage of sensible heat and the storage of the latent heat of fusion in ground ice. However, relatively few values for these components are available. Male and Granger (1979) measured net daily Q_G totals of 0.45 to −0.38 MJ m^{-2} d^{-1} over three years in central Saskatchewan during melt, with strong variation from year to year. Granger and Male (1978) note that the soil heat flux on the Prairies shows a diurnal cycle that lags the radiation cycle and is of the opposite sign.

As shown in Figure 4.16, when snowmelt begins, this diurnal cycle is almost completely eliminated. In warmer locations, the soil heat flux is larger (Gold, 1957). However, in the Arctic, the flux is negative in late winter – for example, −0.9 MJ m^{-2} d^{-1} per day during snowmelt near Inuvik, and Marsh and Woo (1987) report similarly elevated levels near Resolute. The large negative ground heat flux delays the snowmelt season in northern Canada, despite high incident atmospheric and radiation heat fluxes.

Energetics at the Atmospheric Boundary

The turbulent fluxes of sensible and latent heat affect snow evaporation and the energy balance governing snowmelt. Chapter 2 describes the derivation of these fluxes from gradients of temperature, humidity, and wind speed and the appropriate exchange coefficients (the eddy diffusivities for heat, water vapour, and momentum (K_H, K_V, and K_M, respectively). The behaviour of the K_S over snow is unique. For the stable conditions that are common over snow, and with very light winds ($z/L > 0.1$), Granger and Male (1978) show that K_H/K_M decreases with increasing stability as:

$$K_H/K_M = (1 + 7\,(z/L))^{-0.1}, \tag{4.1}$$

where z/L is a measure of atmospheric stability and L is the Obukov length. In unstable conditions, K_H/K_M increased with decreasing stability as:

$$K_H/K_M = (1 - 58\,(z/L))^{0.25}, \tag{4.2}$$

for z/L from −0.003 to −0.2. K_V/K_M increases from 0.5 to 0.8 as z/L increases from −0.002 to −0.02 in the unstable range. However, it remains at about 0.5 over the stable range of z/L from 0.002 to 0.1.

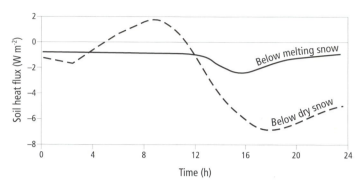

Figure 4.16
Soil heat flux below a shallow snowpack near Bad Lake, Sask., for dry snow and
wet (melting) snow conditions.
Source: Granger and Male (1978).

Often wind, temperature, and humidity gradients are not available, and bulk transfer equations must be used to estimate Q_H and Q_E. In general, Male (1980) reports that bulk transfer formulae are poor estimators of latent heat flux over snow, because the vapour pressure at one level above the snow surface is insufficient to define accurately a gradient and therefore the vapour flux. Male and Granger (1979) do report correlations between daily totals of sensible heat estimated by profile and bulk transfer approaches. However, significant differences can exist on any single day. Bulk transfer equations take the form:

$$Q_H = -\bar{D}_H \bar{u}_z (\bar{T}_a - \bar{T}_{io}),\tag{4.3a}$$

and

$$Q_E = -\bar{D}_E \bar{u}_z (\bar{e}_a - \bar{e}_{io}),\tag{4.3b}$$

where D_H and D_E are the convective heat bulk transfer coefficients for sensible-latent heat, respectively, and the subscripts a and i_o denote air and snow surface, respectively. For values of the D_s over snow consult Gold and Williams (1961) and Male and Granger (1979). Considerable variation of the D_s can be expected, depending on stability, fetch distance, and uniformity of terrain.

For continuous snowcovers, the latent heat flux corresponds to the radiation heat flux, with cycles of condensation at night and evaporation during the day. Male and Granger (1979) report daily net evaporation rates (less condensation) in central Saskatchewan ranging from 0.02 to 0.3 mm d^{-1} SWE, with a mean of 0.1 mm d^{-1} SWE. These fluxes accounted for 14 to 22 per cent of the incoming energy to the snowcover.

For discontinuous snowcovers, local advection from bare ground to snow patches becomes a major component of the snowpack energy balance. Gray and O'Neill (1974) report that sensible heat transfer supplied 44 per cent of the incoming energy to an isolated snow patch. When the patch was surrounded by continuous snowcover, this value drops to 7 per cent.

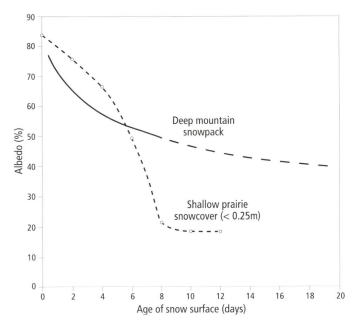

Figure 4.17
Variation in albedo for melting snowpacks in the mountains (deep snowpack)
and Prairies (shallow snowpack).
Source: O'Neill and Gray (1973).

Solar radiation is a major component of the energy flux to snow, so time of year and latitude strongly influence the snow energy balance. Much of the solar radiation incident on a snow surface is reflected, because albedos are as high as 0.9 for compact, dry, clean snow, dropping to 0.5–0.6 for wet snow and 0.3–0.4 for porous, dirty snow mixed with visible sea ice. Given this large variation, it is important to estimate the albedo correctly in calculations of snowmelt and surface energy balance. Male (1980) notes that albedo is a function of the roughness of the surface, snow grain size and density, and whether the surface is illuminated by diffuse radiation or by direct beam radiation. In general, the spectral reflectance of fresh snow is high, and it decreases once snow has thawed. Refreezing of snowcover causes no notable change in albedo, though contaminants such as soil dust can lower albedo. Albedo decays with time since the last snowfall, with very different decay rates for shallow and deep snowpacks (Figure 4.17). As snowcover less than 0.30 m deep melts, the underlying surface begins to influence the albedo, as shown by O'Neill and Gray (1973) for Canadian Prairie snowpacks. Over snow surfaces, outgoing longwave radiation is generally greater than incoming longwave, so the net longwave radiative energy is usually negative.

The net radiation balance over snow shows wide variations because of the influences of cloud cover, forest canopy, and topography. The simple situation of melting snow over an open area of low relief is shown in Figure 4.18 (Male and Granger, 1979). The first two days (11 and 12 April) are cloudless and are followed by continuous cloud on the third day (13 April). Fog occurs for a period on the first day. The

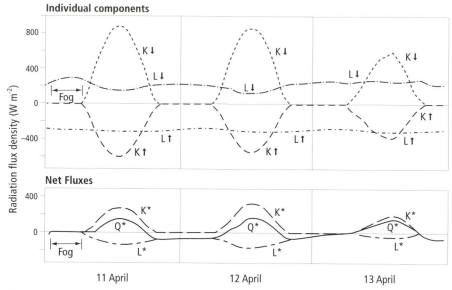

Figure 4.18
Fluxes comprising radiation balance of melting Prairie snowpacks during three days near Bad Lake,
Sask. Conditions: 11 April, fog, then clear; 12 April, clear; 13 April, cloud. $K\downarrow$ is global solar,
$K\uparrow$ reflected solar, K^* net solar, $L\downarrow$ incoming longwave, $L\uparrow$ outgoing longwave, L^* net longwave, and
Q^* net allwave radiation.
Source: Granger and Male (1978).

high snow albedo causes the reflected solar radiation to be nearly the same magnitude
as the incoming global radiation. Incoming longwave radiation and outgoing longwave radiation are also of similar magnitude; outgoing remains nearly constant because during melt the surface temperature stays close to 0°C.

The example shown in Figure 4.18 contrasts with non-melting snow, where longwave cooling at night and a low thermal conductivity of the snowpack cause the snow
surface temperature and hence outgoing longwave to drop sharply. Under coniferous
forests, the net solar radiation input to the snow decreases sharply with increasing canopy density. Similarly, the net radiation decreases with increasing canopy up to densities of about 60 per cent, after which the flux rises. Net radiation can be significantly
greater in a deciduous forest than in an open area.

SNOWMELT

The rate of snowmelt is controlled by the energy balance near the upper surface, where
melt normally occurs. The type of mathematical model required to simulate this energy balance depends on the depth of snow. Male (1980) indicates that, for deep snow,
only the vertical components of the energy and mass fluxes at the snow surface need
to be considered. But for shallow snow, the snowpack should be considered a "box" to
and from which energy is transferred by radiation, convection, and conduction and
across whose boundaries occur mass fluxes of solid, liquid, and gaseous water. This
box model includes the energy flux due to radiation, Q^*, sensible, Q_H, and latent, Q_E,
heat transfer by convection, and heat conducted from the ground, Q_G. It may also be

necessary to include Q_P, the energy flux resulting from precipitation Q_M, the energy available for melt (which drains from the base of the pack), and Q_A, the net energy advection (usually from bare ground to the snow). The rate of change in internal energy of the pack (dU/dt) is then given as:

$$dU/dt = Q^* + Q_H + Q_E + Q_G + Q_P + Q_M + Q_A. \tag{4.4}$$

The change in internal energy of the snowpack can be a significant term in the energy balance of melting shallow snowpacks. Meltwater is released from the pack in a diurnal cycle in response to the cycling of energy inputs. The nighttime energy deficit must be compensated for the next day before the pack can return to 0°C and release water. The internal energy, U (in J m^{-2}), is comprised of solid, liquid, and vapour components:

$$U = (C_i + C_w + C_v) \, T_i \, z_i, \tag{4.5}$$

where C is heat capacity, and the subscripts i, w, and v refer to snow and ice, liquid water, and vapour, respectively; T_i is the snow temperature, and z_i is the depth of snow. Before melt occurs, the w component is zero and U may be found by using internal snowpack water-vapour diffusion calculations. During melt, the flow and refreezing of liquid water are extremely difficult to model and must be measured in the field. Dielectric devices can quickly determine the liquid water content of snowpacks, and passive microwave remote sensing holds promise for mapping areas of wet snow.

Expressing the energy balance for melting snow in terms of Q_M – the energy available for melting a unit volume of snow – equation (4.4) shows a correspondence with most of the terms in the general energy balance expressed in chapter 2. The relative importance of some of these terms during the melt of a low-arctic snowpack is shown in Figure 4.19, which gives hourly and daily heat fluxes near Inuvik, NWT (Marsh and Pomeroy, 1996). The advected term in Figure 4.19 presumes that all positive net energy of bare ground is advected onto remaining snow patches. Hence this is an overestimate when snow-covered area is less than 50 per cent.

The amount of meltwater can be calculated from the expression

$$SWE_M = Q_M/(\rho_w L_f B_i), \tag{4.6}$$

where SWE_M is the snowmelt water expressed in mm, ρ_W the density of water, L_f the latent heat of fusion, and B_i the thermal quality of snow or the fraction of ice in unit mass of wet snow. B_i is usually in the range of 0.95 to 0.97.

When advection is negligible, net radiation and sensible heat govern the melt of shallow Prairie snowpacks (Male and Gray, 1981). Generally, at the beginning of melt, radiation is the dominant flux, with sensible heat growing in contribution through the melt. Temperature index models do not predict snowmelt well for open, continuous snowcovers. However, considerable success has been achieved in relating melt to air temperature in mountainous and forested regions, and in areas of patchy snowcover, where advection becomes a dominant term. Shook, Gray, and Pomeroy (1993) have shown that the potential for advection increases dramatically as basin snowcover falls from 70 per cent to 40 per cent and remains relatively high as basin

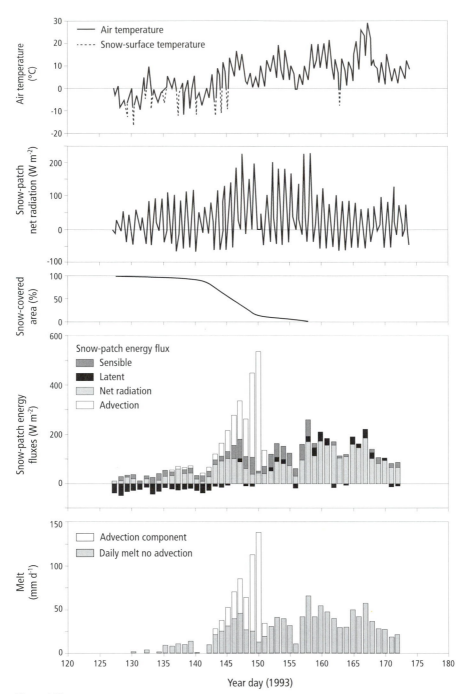

Figure 4.19
Energy and melt fluxes for a melting snow patch in the Arctic. Air and snow temperature, snow patch net radiation, snow-covered area, daily energy fluxes and melt rate calculated with and without maximum possible advected energy from bare ground.
Source: Marsh and Pomeroy (1996).

snowcover further decreases as shown from Days 145 to 152 in Figure 4.19. Temperature-index calculations employing a melt factor M_f have the form

$$SWE_M = M_f (T_a - T_b), \tag{4.7}$$

where T_a is the mean air temperature over a given time period and T_b is a base temperature below which melt does not occur (usually taken as 0°C). Granger and Male (1978) found the melt factor to vary from 6 to 28 mm °C^{-1} d^{-1}, for snowmelt on the Prairies. Exceptionally high values of M_f (50 mm °C^{-1} d^{-1}) were occasionally measured in April, during times of high radiant input. Melt factors should be used cautiously in calculating snowmelt because of the considerable variation in their value from year to year and with location.

CONCLUSIONS

Snow, as one of the lowest-density materials, is subject to large spatial variations in snowfall and accumulation. This variation is sensitive to meteorological factors such as regional wind speed, location of open water, topography, and local aerodynamic roughness. Over large areas, open water and topography dominate snowfall patterns. At smaller scales, factors such as vegetation, land-use, and local relief strongly influence the accumulation patterns of snow. Within forests, areas with less vegetation accumulate the greatest snowcover, while in open areas, sites with greater vegetation and topographic roughness accumulate the most. Sublimation and melt reduce snow depths. They in turn are sensitive to wind, aerodynamic roughness, and net radiation. Sublimation of blowing snow can ablate over 50 per cent of annual snowfall in the windswept southern Canadian Prairies. Sublimation of intercepted snow can ablate over 30 per cent of annual snowfall from conifer stands in the boreal forest.

Snow presents special measurement difficulties, and much care is necessary before data archives of snowfall and snow accumulation are employed in climatological analyses. In general, they cannot be used without correction. Snow-accumulation surveys are best designed to represent categories of land cover.

The energetics of snowmelt and evaporation are unique because of the high albedo and small roughness of snowfields. The high albedo complicates estimation of net radiation available for melting snow. The small roughness produces large errors in estimation of sensible and latent heat fluxes to the snowpack.

Researchers in Canada have recently done much to increase our understanding of the climate and hydrology of snow. However, much remains to be learned about this type of precipitation, which is so important in Canada's environment.

REFERENCES

Environment Canada. 1986. *Climatic Atlas – Canada. Map Series 2 – Precipitation*. Downsview, Ont.: Environment Canada.

Gold, L.W. 1957. "The Influence of Snowcover on Heat Flow from the Ground." In International Association of Scientific Hydrology Publication 46, Wallingford, England, 13–21.

Gold, L.W., and Williams, G.P. 1961. "Energy Balance during the Snowmelt Period at an Ottawa Site." In International Association of Scientific Hydrology Publication 54, Wallingford, England, 288–94.

Golding, D.L., and Swanson, R.H. 1986. "Snow Distribution Patterns in Clearings and Adjacent Forest." *Water Resources Research* 22: 1931–40.

Goodison, B.E. 1978. "Accuracy of Canadian Snow Gauge Measurements." *Journal of Applied Meteorology* 27: 1542–8.

– 1981. "Compatibility of Canadian Snowfall and Snow Cover Data." *Water Resources Research* 17: 893–900.

Goodison, B.E., Ferguson, H.L., and McKay, G.A. 1981. "Measurement and Data Analysis." In D.M. Gray and D.H. Male, eds., *Handbook of Snow, Principles, Processes, Management and Use*, Toronto: Pergamon Press, 191–274.

Goodison, B.E., and Metcalfe, J.R. 1989. "Automation of Winter Precipitation Measurement in Canada." In *WMO/IAHS/ETH Workshop on Precipitation Measurement*, St Moritz, Switzerland, 35–40.

Goodison, B.E., Metcalfe, J.R., Wilson, R.A., and Jones, K. 1988. "The Canadian Automatic Snow Depth Sensor: A Performance Update." In *Proceedings of the Western Snow Conference*, 178–181.

Goodison, B.E., and Walker, A.E. 1994. "Canadian Development and Use of Snow Cover Information from Passive Microwave Satellite Data." In B.J. Choudhury, Y.H. Kerr, E.G. Njoku and P. Pampaloni, eds., *ESA/NASA* International Workshop on Passive Microwave Remote Sensing, VSP Publications, 245–62.

Granger, R.J., and Male, D.H. 1978. "Melting of a Prairie Snowpack." *Journal of Applied Meteorology* 17: 1833–42.

Gray, D.M., and O'Neill, A.D.J. 1974. "Application of the Energy Budget for Predicting Snowmelt Runoff." In H.S. Santeford and J.L. Smith, eds., *Advanced Concepts in the Study of Snow and Ice Resources*, Washington, DC: National Academy of Sciences, 108–18.

Gubler, H., and Rychetnik, J. 1991. "Effects of Forests near the Timberline on Avalanche Formation." In *Snow, Hydrology and Forests in High Alpine Areas*, International Association of Scientific Hydrology Publication 205, Wallingford, England: IAHS Press, 19–38.

Hare, F.K., and Thomas, M.K. 1979. *Climate Canada*. 2nd edn. Toronto: John Wiley and Sons Canada Ltd.

Hoover, M.D., and Leaf, C.F. 1967. "Process and Significance of Interception in Colorado Subalpine Forest." In W.E. Sopper and H.W. Lull, eds., *Forest Hydrology*, New York: Pergamon Press, 213–24.

Karl, T.R., Groisman, P.Ya., Knight, R.W., and Heim, R.R. 1993. "Recent Variations of Snow Cover and Snowfall in North America and Their Relation to Precipitation and Temperature Variations." *Journal of Climate* 6: 1327–44.

Kuz'min, P.P. 1960. "Snowcover and Snow Reserves." *Gidrometeorologicheskoe, Izdatelsko* (Leningrad). Trans. U.S. National Science Foundation, Washington, DC, 1963.

Langham, E.J. 1981. "Physics and Properties of Snowcover." In D.M. Gray and D.H. Male, eds., *Handbook of Snow, Principles, Processes, Management and Use*, Toronto: Pergamon Press, 275–337.

McNay, R.S., Peterson, L.D., and Nyberg, J.B. 1988. "The Influence of Forest Stand Characteristics on Snow Interception in the Coastal Forests of British Columbia." *Canadian Journal of Forest Research* 18: 566–73.

Male, D.H. 1980. "The Seasonal Snowcover." In S. Colbeck, ed., *Dynamics of Snow and Ice Masses*, Toronto: Academic Press, 305–95.

Male, D.H., and Granger, R.J. 1979. "Energy and Mass Fluxes at the Snow Surface in a Prairie Environment." In S. Colbeck and M. Ray, eds., *Proceedings of the Modelling of Snow Cover Runoff*, Hanover, NH: U.S. Army Cold Regions Research and Engineering Laboratory, 101–24.

Male, D.H., and Gray, D.M. 1981. "Snowcover Ablation and Runoff." In D.M. Gray and D.H. Male, eds., *Handbook of Snow, Principles, Processes, Management and Use*, Toronto: Pergamon Press, 360–436.

Marsh, P., and Pomeroy, J.W. 1996. "Meltwater Fluxes at an Arctic Forest-Tundra Site." *Hydrological Processes* 10: 1383–1400.

Marsh, P., and Woo, M-K. 1984. "Wetting Front Advance and Freezing of Meltwater within a Snowcover: Observations in the Canadian Arctic." *Water Resources Research* 20: 1853–4.

– 1987. "Soil Heat Flux, Wetting Front Advance and Ice Layer Growth in Cold, Dry Snow Covers." In *Snow Property Measurement Workshop*, Technical Memorandum No. 140, Snow and Ice Subcommittee, Associate Committee on Geotechnical Research, National Research Council of Canada, Ottawa, 497–524.

Martinec, J., and Rango, A. 1991. "Indirect Evaluation of Snow Reserves in Mountain Basins." In *Snow, Hydrology and Forests in High Alpine Areas*, International Association of Scientific Hydrology Publication 205, Wallingford, England, 111–19.

O'Neill, A.D.J., and Gray, D.M. 1973. "Spatial and Temporal Variations of the Albedo of a Prairie Snowpack." In *The Role of Snow and Ice in Hydrology*, Proceedings of the Banff Symposium 1: 176–86.

Pomeroy, J.W. 1989. "A Process-Based Model of Snow Drifting." *Annals of Glaciology* 13: 237–40.

Pomeroy, J.W., and Gray, D.M. 1994. "Sensitivity of Snow Relocation and Sublimation to Climate and Surface Vegetation." In *Snow and Ice Covers: Interactions with Climate and Ecosystems*, Wallingford, England: International Association of Scientific Hydrology Press, 213–26.

– 1995. *Snowcover Accumulation, Relocation and Management*. NHRI Science Report No. 7, Saskatoon: Environment Canada, National Hydrology Research Institute.

Pomeroy, J.W., Gray, D.M., and Landine, P.G. 1991. "Modelling the Transport and Sublimation of Blowing Snow on the Prairies." *Proceedings of the 48th Eastern Snow Conference*, 175–188.

– 1993. "The Prairie Blowing Snow Model: Characteristics, Validation, Operation." *Journal of Hydrology* 144: 165–92.

Pomeroy, J.W., and Male, D.H. 1992. "Steady-state Suspension of Snow: A Model." *Journal of Hydrology* 136: 275–301.

Pomeroy, J.W., and Schmidt, R.A. 1993. "The Use of Fractal Geometry in Modelling Intercepted Snow Accumulation and Sublimation." *Proceedings of the 50th Annual Eastern Snow Conference*: 1–10.

Rouse, W.R. 1990. "The Regional Energy Balance." In T. Prowse and S. Ommanney, eds., *Northern Hydrology, Canadian Perspectives*, Saskatoon: Supply and Services Canada, 187–206.

Satterlund, D.R., and Haupt, H.F. 1970. "The Disposition of Snow Caught by Conifer Crowns." *Water Resources Research* 6: 649–52.

Schemenauer, R.S., Berry, M.O., and Maxwell, J.B. 1981. "Snowfall Formation." In D.M. Gray and D.H. Male, eds., *Handbook of Snow, Principles, Processes, Management and Use*, Toronto: Pergamon Press, 129–52.

Schmidt, R.A. 1972. "Sublimation of Wind-transported Snow: A Model." In USDA Forest Service Research Paper RM-90, Fort Collins, Col.: Rocky Mountain Forest and Range Experiment Station, 304–13.

– 1991. "Sublimation of Snow Intercepted by an Artificial Conifer." *Agricultural and Forest Meteorology* 54: 1–27.

Schmidt, R.A., and Gluns, D.R. 1991. "Snowfall Interception on Branches of Three Conifer Species." *Canadian Journal of Forest Research* 21: 1262–9.

Schmidt, R.A., and Pomeroy, J.W. 1990. "Bending of a Conifer Branch at Subfreezing Temperatures: Implications for Snow Interception." *Canadian Journal of Forest Research* 20: 1250–3.

Schmidt, R.A., and Troendle, C.A. 1992. "Sublimation of Intercepted Snow as a Global Source of Water Vapour." In *Proceedings of the 60th Western Snow Conference*, 1–9.

Shook, K., Gray, D.M., and Pomeroy, J.W. 1993. "Temporal Variations in Snowcover Area during Melt in Prairie and Alpine Environments." *Nordic Hydrology* 24: 183–98.

Steppuhn, H., and Dyck, G.E. 1974. "Estimating True Basin Snowcover." In *Advanced Concepts and Techniques in the Study of Snow and Ice Resources*, Washington, DC: National Academy of Sciences.

Sturm, M. 1992. "Snow Distribution and Heat Flow in the Taiga." *Arctic and Alpine Research* 24: 145–52.

Swanson, R.H. 1988. "The Effect of *in Situ* Evaporation on Perceived Snow Distribution in Partially Clear-cut Forests." In *Proceedings of the 56th Western Snow Conference*, 87–92.

Tabler, R.D., Pomeroy, J.W., and Santana, B.W. 1990. "Drifting Snow." In W.L. Ryan and R.D. Crissman, eds., *Cold Regions Hydrology and Hydraulics*, New York: American Society of Civil Engineers, 95–146.

Tabler, R.D., Benson, C.S., Santana, B.W., and Ganguly, P. 1990. "Estimating Snow Transport from Wind Speed Records: Estimates Versus Measurements at Prudhoe Bay, Alaska." In *Proceedings of the 58th Western Snow Conference*.

Toews, D.A., and Gluns, D.R. 1986. "Snow Accumulation and Ablation on Adjacent Forested and Clearcut Sites in Southeastern British Columbia." In *Proceedings of the 54th Western Snow Conference*, 101–11.

Woo, M-K., Heron, R., Marsh, P., and Steer, P. 1983. "Comparison of Weather Station Snowfall with Winter Snow Accumulation in High Arctic Basins." *Atmosphere-Ocean* 21: 312–25.

CHAPTER FIVE

Oceans and the Coastal Zone

OWEN HERTZMAN

INTRODUCTION

What are the coastal zones of Canada? In British Columbia we might include the region west of and including the slopes of the Coast Range, plus all of Vancouver Island and the Queen Charlotte Islands but not the precipitation shadow behind the Coast Range. In northern and Atlantic Canada zone boundaries are harder to draw. Areas within about 200 km of the coast are affected by the adjacent ocean and should be included. However, placing a boundary on the coastal zones in these regions is very difficult. For our purposes, we bring in all of Nova Scotia, New Brunswick, Prince Edward Island, and Newfoundland in the Atlantic coastal zone, along with the coastal settlements of Labrador. The Baffin Island, Hudson Bay, and Mackenzie delta regions are in the Arctic Ocean coastal region. The high Arctic is dealt with in chapter 8.

The Three Coasts

The three oceans that border Canada (Pacific, Atlantic, and Arctic) strongly influence the surface climates of the adjoining regions (British Columbia, Atlantic Canada, and the Mackenzie Valley and Arctic coast, respectively). In addition, they help determine the overall climate of the country through their effects on both the large-scale (continental and global) atmospheric circulation and on the cyclonic storms that dominate the country's precipitation distribution. Each of the three oceans acts to moderate the coastal climate in certain seasons and to delay the end of the cold and warm seasons. How this moderation and delay are accomplished differs widely among the three oceans. The primary differences among them are the duration and extent of sea ice cover and the annual range of sea surface temperature (SST).

Oceans as Thermal Capacitors

Oceans are significant to the climate of Canada's coastal zone primarily because of their thermal properties, including latent heat transfer to (and from) the overlying air. Oceans store heat in seasonal mixed layers, which are typically tens of metres deep,

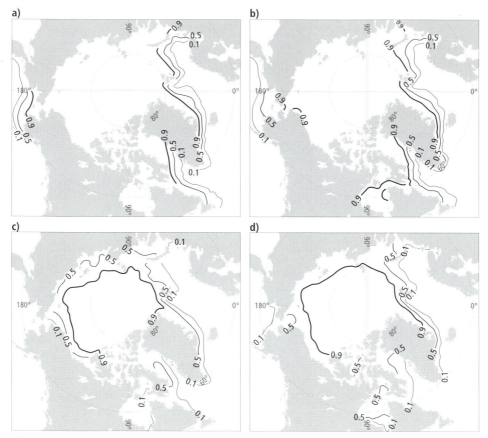

Figure 5.1
Seasonal climatologies of sea-ice extent in the Arctic, 1953–84: (a) winter; (b) spring; (c) summer;
(d) autumn. Contours shown are 0.1, 0.5, and 0.9 ice coverage.
Source: Mysak and Manak (1989).

much deeper than the annual thermal zone of any type of soil, rock, or vegetation. During the cold season the mixed layer typically deepens in response to stronger wind forcing, while in the lighter winds of the warm season the ocean typically becomes stratified, with layers of lighter warm water lying above denser cold water. Hence, one expects ssTs to vary less, annually and diurnally, than air temperatures over adjacent land areas. Data from Ocean Weather Station Papa in the North Pacific (Hertzman, Miyake, and Pond, 1974) and from Ocean Station Bravo in the Labrador Sea (Lazier, 1980) confirm this tendency. So, for the coastal zone, oceans in general act as moderators both diurnally (through sea–land and land–sea breeze circulations) and annually (through advection towards the land of warmer air in the cold season and cooler air in the hot season).

When sea ice forms, an ocean's moderating role is reduced, and in some cases reversed (for a useful review see Barry, 1994). Sea ice in the Gulf of St Lawrence and off Newfoundland in the period from March to May (Figure 5.1) prolongs the cold

Figure 5.2
Aerial view along the north coast of Ellesmere Island (83°00′N, 71°50′W) in late June, showing ribbed ice in the foreground and the ice shelf of the central Arctic Ocean in the background.
Source: Photograph from the National Air Photo Library, Ottawa.

season in the adjacent land areas. Similar behaviour is observed in and near Hudson Bay (Prinsenberg and Danard, 1985; Rouse, Hardill, and Silis, 1989; Silis, Rouse, and Hardill, 1989). Open water "leads" in the ice affect the winter and spring coastal meteorology of the area. Additional snowfall results when cold air off the sea ice combines with water vapour from the leads. In the Arctic Ocean, sea ice is present in all seasons (Figure 5.2). The climate of the adjacent coastal zone is strongly influenced by the sea ice and the frozen ground or cold meltwater in the active layer. Monthly mean daily minimum air temperatures for even the warmest months rarely exceed 8°C in these regions.

Air–Sea Coupling

The coupling between the atmosphere and the underlying ocean consists essentially of the latent and sensible heat fluxes between the boundary layers of the two fluids, and the wind stress imposed by the atmosphere on the ocean. Winds over ocean regions tend to be stronger than those over adjacent land areas because of reduced drag, and vapour pressure generally decreases upwards, away from the ocean surface. Thus continental air advected over coastal ocean water often experiences large, upward fluxes of latent heat (Hopkins and SethuRaman, 1987). These fluxes help initiate and develop storms in these areas. Air advected from marine regions to adjacent coastal land areas often has dewpoint temperatures much different from those of the local air and land surface, thereby increasing the probability of fog formation at and near the coast.

The Deep Ocean

The primary agents controlling oceanic effects on the climates of coastal regions are the coastal oceans. However, the deep oceans also play a role, because they contain circulation systems that produce overturning on time-scales of decades to millennia. The ocean climate that we observe today may be very different from the one present several decades or centuries ago. Many scientists (for example, Broecker and Peng, 1982; Schmitz, 1995) believe that the entire group of deep ocean basins functions as a slow conveyor belt, with surface waters convecting downward in certain preferred regions (such as the Greenland and Labrador seas) and outcropping (emerging) in other regions (such as the North Pacific). Because of the slow speeds of ocean currents (typically much less than 0.5 m s^{-1}), oceanic conditions along the Atlantic or BC coasts may be influenced by tropical and/or arctic meteorological events that occurred months and years earlier. Thus some portion of the variability in SST distribution on any of Canada's coasts is not controlled by local radiation and energy balances or by recent local meteorology. This fact has notable implications for postulated future climate change.

LARGE-SCALE BOUNDARY CONDITIONS

Ocean Basin Currents

The main currents that affect Canada's coastal oceans are shown in Figure 5.3. They are the Western Arctic (or Beaufort Sea) gyre, the Kuroshio/North Pacific Current, the Labrador Current, and the Gulf Stream/North Atlantic Current.

The cold waters of the Labrador Current and the warm waters of the Gulf Stream (Figure 5.3a) are two of the major controls on the climate of Atlantic Canada. Because the Gulf Stream separates from the North American coast just north of Cape Hatteras, its warm waters (SSTs > 20°C) are usually several hundred kilometres south or southeast of Nova Scotia and Newfoundland. Occasionally, warm core rings, including modified Gulf Stream water, approach the coast more closely. On the continental shelf, the water is a mixture of St Lawrence River water (in the Gulf of St Lawrence and on the Scotian Shelf) and Labrador Current water. The latter, which represents a portion of the outflow of the Arctic Ocean through Fram Strait east of Greenland and through Hudson Strait and the Canadian Arctic Islands, separates into onshore and offshore branches east of Labrador. The offshore branch interacts with the extension of the Gulf Stream east of Newfoundland. The onshore branch provides most of the coastal water for Newfoundland and some of the water for the Scotian Shelf.

a)
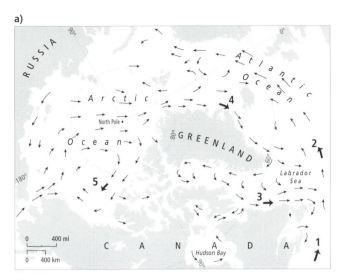

b)
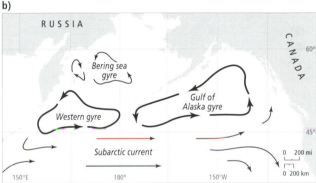

Figure 5.3
Ocean surface currents for (a) the Arctic and Atlantic oceans and (b) the
North Pacific. Currents are solid arrows. In (a) the thick arrows indicate
(1) the Gulf Stream, (2) the North Atlantic Current, (3) the Labrador
Current, (4) the Arctic outflow east of Greenland, and (5) the Western
Arctic or Beaufort Gyre. In (b) note the two subarctic gyres north of the
main west-to-east subarctic current (also known as the North Pacific
Current, or the Kuroshio extension). The Kuroshio current itself is in the
lower left corner.
Sources: (a) Pickard and Emery (1982); (b) Miller et al. (1991).

Because the Atlantic region lacks high mountains and is downstream of the centre
of the continent, the effects of the dominant ocean currents on mean air temperatures
and precipitation are quite different in summer and winter.

The Pacific Ocean, primarily via the Kuroshio/North Pacific Current, has a substan-
tial effect on BC air temperatures, but smaller than similar effects downstream of the
North Atlantic because the Pacific is larger than the Atlantic in both volume and area.
The Kuroshio and its extension feed into two anti-clockwise gyres as they cross the
North Pacific (Figure 5.3b). The eastern Gulf of Alaskan gyre covers the area north of

45°N in winter and north of 50°N in summer. Thus much of the low-level air in the marine cyclones that arrive on the BC coast has been in thermal contact with water from either this gyre or the Kuroshio current. Since this water has a usual temperature of 8 to 12°C, it keeps BC coastal water and air temperatures warmer in the winter than those of lower-latitude locations in Atlantic Canada. These warmer SSTs, the greater water depths, larger tides, and the general absence of cold air over near-shore waters because of the coastal mountains, combine to impede formation of sea ice. The lack of sea ice in the region in turn influences subsequent meteorology and accounts for the earlier start of spring than at lower latitudes on the Atlantic coast. The water temperatures in the Bay of Fundy also remain high enough during the winter to impede formation of sea ice. The effects of this narrow region of warm open water are felt primarily within a few tens of kilometres.

In the Arctic Ocean, the ocean circulation is dominated by the clockwise Beaufort Sea gyre (Figure 5.3a). This anticyclonic feature, seen primarily in ice-drift and drifting-buoy measurements (McLaren et al., 1987), is driven by a combination of the transpolar current and the mean annual anticyclonic atmospheric circulation in the region. The seasonal ice climatology for northern Canada (Figure 5.1) clearly shows that, in the Beaufort Sea, the only time with substantial open water is late summer, unlike the region near Hudson Bay and Baffin Island, where there is a substantial low-ice period in summer and autumn. Ledrew, Johnson, and Maslanik (1991) explore the atmospheric forcing of the late-summer change (interruption) of the mean circulation in the Beaufort Sea. They conclude that the hemispheric atmospheric change from warm- to cold-season regimes is accompanied by large amounts of very stable stratospheric air moving closer to the surface. This in turn helps lock in for a few weeks a low-pressure system over the Beaufort Sea, which then forces cyclonic (counterclockwise) ice circulation and thinning and break-up of the ice-field near the Canadian coast. The open water thus generated contributes to the formation of cyclonic storms during autumn.

Upwelling

From oceanic Ekman-layer theory, we can discover when and where in Canada's coastal zone subsurface ocean water directly affects meteorology and climate. The Ekman theory states that in the northern hemisphere the mean current in the upper oceanic layers is rotated 90 degrees clockwise from the mean surface wind blowing over the water. Thus if the mean wind over a period of days is approximately along the coast, with land to the left (when one is looking downstream) in the northern hemisphere, then the mean current will be directed away from the coast. The equation of mass continuity implies that water from the layer(s) below must upwell and replace the original surface water. In British Columbia this occurs weakly during the summer, but in Nova Scotia during the same season the climatological mean wind is from the appropriate direction (southwest). The strong resulting upwelling is a major cause of water temperatures' being colder in summer than those expected from purely atmospherically driven energy considerations (Figure 5.4a). This cold water increases the incidence of fog in spring and summer by lowering the air temperature of warm, moist air from the south and southwest below its dewpoint.

Storm Tracks

The influence of both coastlines and sst distributions on preferred cyclo
tracks is an actively debated research topic. In Canada, it is clear that the tracks
in each of the three oceanic coastal regions relate rather differently to their adjacent
oceans.

The Pacific Ocean surface plays a rather minor role in storm tracks affecting British
Columbia. Most of the major cyclones that pass through the coastal region are formed
well upstream (Gyakum et al., 1989). The mainly zonal track along which most
cyclones are steered is mainly parallel to sst isotherms. The track typically migrates
from the Oregon/northern California coast in winter to the northern bc/southern
Alaska coast in summer. The very warm water near Hawaii is too far away to affect
coastal weather often, though it can influence warm-sector air in exceptional periods.

In the Arctic, the presence of open water in summer enhances storm development
and redevelopment in the region (Ledrew, 1992). The mean winter and annual anticyclone in the high Arctic can change to a seasonal mean cyclone in summer and
autumn mainly because of enhanced cyclogenesis resulting from large upward fluxes
of sensible and latent heat from open-water leads. Because these leads appear most
often close to the coast, there is some speculation that coastal cyclone tracks are
preferred.

In the Atlantic there appear to be two main storm tracks for winter cyclones. One
is along the New England and Nova Scotia coast and across Newfoundland and
the Strait of Belle Isle. The other is along the northern edge of the Gulf Stream (Figure 5.4c and d). On both tracks, storms often intensify downstream of the main
cyclone-generation zone in the hemisphere, which is east of Cape Hatteras (Sanders
and Gyakum, 1980). Both tracks have been deduced from both upper-air and surface
meteorological data by workers from Canada, the United States, and Russia. Both
are close to zones of high January surface-temperature gradients (Figure 5.4b),
and the mean air temperature over coastal land areas in January is −5 to −7°C (Figure 5.8c, below).

Cold Air Outbreaks

The primary continental effect on the coastal regions in winter is the appearance of
very dry, very cold air associated with anticyclones. On the west coast this air usually
appears as a density current travelling down a river valley or fjord. Its effects are often local (for example, there may be rain in Vancouver and snow in the central bc
Fraser Valley at the same time), and the climatological importance is rather small.
Occasionally, a substantial amount of cold air flows out over the Gulf of Alaska,
causing large upward heat fluxes and significant baroclinicity, both of which enhance
cyclone development.

Major outbreaks of cold air in the Hudson Bay/Baffin Island region accompany
substantial cyclogenesis in the autumn in this region (Mailhot et al., 1996). Further
south and east, there are outflows of cold air (sometimes as cold as −35°C) from the
land and sea-ice regions over the open water of the Labrador Sea, the Grand Banks,

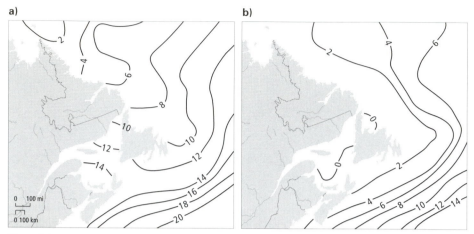

Figure 5.4
Atlantic coastal average sea-surface temperatures (ssts) in (a) July and (b) January. Average Atlantic
coastal and ocean weather in (c) August and (d) February.
Source: Phillips (1990).

the Gulf of St Lawrence, and the banks off Nova Scotia. These outflows contribute to
total upward fluxes of latent plus sensible heat greater than the solar constant during
otherwise rather routine synoptic conditions in autumn and winter. These fluxes in
turn contribute to local precipitation regimes, offshore cyclogenesis (including small,
short-lived polar lows), and higher-density surface ocean water (with lower temperatures and higher salinities than the water below), which may lead to deep oceanic convection in some of these areas (Lazier, 1980; Kelley, 1994).

COASTAL CLIMATOLOGY

Radiation Budget

The radiation budget of the ocean is dominated by that of the ocean surface. The basic
surface radiation budget is presented in chapter 2, above, in equation (2.1). Applying
this equation to an ocean region at 45°N with cloudless skies, we note that for periods
of from several hours to perhaps one to two days, the sst is roughly constant. Thus the
emitted longwave radiation, $L\uparrow = \varepsilon_o \sigma T_o^4$, is approximately constant, with T_o representing the sst and the emissivity $\varepsilon_o = 0.95$ (Oke, 1987). $L\uparrow$ can be quite readily measured from satellite sensors, and therefore with estimates of the emissivity, T_0, can be
derived. The magnitude of incoming solar radiation, $K\downarrow$, is a function of latitude, date,
time, and the attenuation effects of water vapour, aerosols, and clouds. Attenuation is
confined primarily to the lower half of the atmosphere, including the boundary layer,
and can be calculated by using bulk radiation transfer models. Because of the freely
available water, vapour pressures in the lower 10–15 kPa of the atmosphere tend to
be within 25 per cent of saturation, except during subsidence periods associated
with strong anticyclones. On cloudy days, observations from nearby land stations can

c)

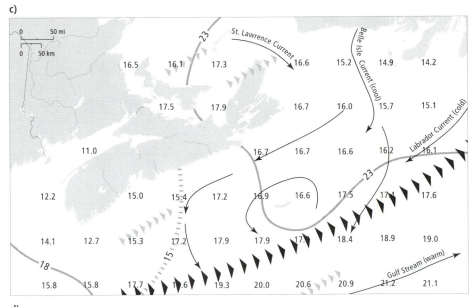

d)

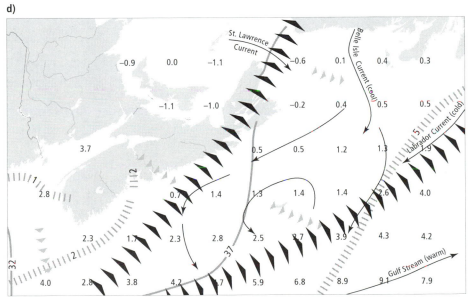

| ⁞15⁞ | Percentage of time wave heights less than 2 m | 6.8 | Sea surface temperature (°C) |

⁞⁞2⁞⁞ Percentage of time wave heights above 6 m

▷ ▷ ▷ Prevailing wind direction

—37— Average wind speed (km h⁻¹)

⟶ Ocean current

◣◣◣ Winter storm tracks

▸▸▸ Summer storm tracks

often be used to estimate $K\downarrow$ if the cloud field is continuous across the coast. This is often not the case; therefore measurements of offshore radiation are required to obtain detailed descriptions of coastal radiation regimes.

The reflected solar radiation, $K\uparrow = \alpha\ K\downarrow$, is a simple proportion of global solar radiation. The albedo for smooth water at low zenith angles is 0.02–0.03 (Pickard and Emery, 1982). During most of the late spring and summer daytime hours in southern Canada, the solar zenith angle is less than 60 degrees. For these angles, the albedo is less than 0.10 (Nunez, Davies, and Robinson, 1972). For other areas and seasons, on cloudless days, the albedo is a strong function of zenith angle, reaching close to 0.5 near 80 degrees. For overcast days, albedos near 0.1 apply for angles up to 80 degrees.

Incoming longwave radiation, $L\downarrow$, is the most difficult term to estimate. If upper-air soundings of temperature and humidity are available, then simple, one-dimensional emission models can be used to compute this flux density. More usually, an effective air emissivity is used, together with the surface or screen-height air temperature, to represent the water vapour assumed to exist aloft. These emissivities can range from about 0.5 in the dry, cold winter to values near 0.9 in moist air in summer. If cloud is present, the base of the lowest cloud is considered to radiate as a near blackbody at the temperature of the cloud base. If the surface temperature and the height of the cloud base are known, this temperature can be derived from standard thermodynamic diagrams. In some daytime, cloudless cases (for example, Oke, 1987), net longwave radiation over water surfaces can be neglected, because it is insignificant with respect to net solar radiation.

If one considers a slab of water of depth z in the upper ocean, a portion of the absorbed radiation (K^*), will pass through to the ocean beneath. For a well-mixed layer of 20 m depth, this proportion ranges between about 0.03 and 0.54 of K^*, depending on the turbidity of the water. If the well-mixed surface layer is deeper, this proportion is reduced. Values of $0.01\ K^*$ are reached at depths between 25 and 150 m (Lewis, Cullen, and Platt, 1983; Lewis, Kuring, and Yentsch, 1988). Thus for clear days in coastal regions, ocean surface waters heat up much more slowly due to absorption of radiation on summer days, and cool much more slowly on summer nights, than adjacent land areas (except for a thin strip of shallow coastal water heated from below by sand and rocks). In regions where net radiation dominates the energy budget and resulting SST, the radiation budget becomes a strong local meteorological forcing term.

Figure 5.5 shows some of the radiation budget's components for Sable Island, an ocean station in the Atlantic off Nova Scotia. The annual cycle of monthly mean daily total solar radiation (Figure 5.5a) is rather ordinary for a mid-latitude station. (Dobson and Smith, 1989, indicate that the values of monthly maxima of daily solar radiation are rather like those at Swift Current, Sask.) The standard deviation of daily values about the monthly means is greatest from April to June, indicating the strong tendency for winter-like meteorological conditions (with increased thick cloud and decreased $K\downarrow$ on some days) to persist well into the "warm" season. Cloudless days in these months have high $K\downarrow$ because of low solar zenith angles. The data for hourly solar radiation (Figure 5.5b) indicate the usual seasonal variation for a mid-latitude station. The thermal lag in the region is emphasized when one considers that October temperatures are higher than those in April (Figure 5.7b, below).

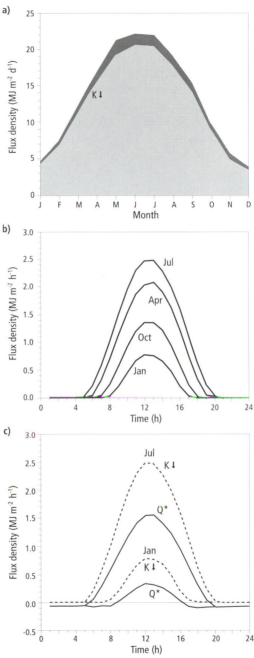

Figure 5.5
Radiation data for Sable Island, NS: (a) Monthly mean daily global solar radiation (grey) and standard deviation (black); (b) monthly mean hourly global solar radiation for January, April, July, and October; and (c) monthly mean hourly global solar radiation and net radiation for January (lower curves) and July (upper curves).

Net radiation values are compared to global solar radiation in Figure 5.5c. The nocturnal values of $Q*$ give a relatively accurate indication of the net longwave balance for daytime as well as for nighttime, since SST is typically fairly constant over twenty-four–hour periods (changes of less than 1 to 2 Celsius degrees). Nocturnal $Q*$ values are about -0.1 MJ m^{-2} h^{-1}, with diurnal changes of less than 0.40 MJ m^{-2} h^{-1}. Thus the majority of the decrease between $K\downarrow$ and $Q*$ occurs because of reflected and transmitted solar radiation at the surface. Because the fractional reduction in July between $K\downarrow$ and $Q*$ varies between about 0.4 and 0.5 for different hours of the day, up to one-third of net solar radiation can be involved in heating the water column at that time of year. In January, with higher solar zenith angles, the majority of incident solar radiation is reflected.

ENERGY BUDGET

The energy budget equation for a surface slab of ocean of depth z is

$$Q* = Q_H + Q_E + Q_Z + Q_A + Q_S, \tag{5.1}$$

where the last three terms represent heat fluxes through the bottom of the layer, net heat flux advected out of the volume horizontally, and net heat stored in the volume, respectively. Q_S is of the form $C_o(\partial \overline{T}/\partial t)$, with C_o representing the oceanic heat capacity, which is about 4×10^6 J m^{-3} °C^{-1}. For a layer depth of 50 m, and $(Q* - Q_H - Q_E) = 20$ W m^{-2}, with no advection or loss through the bottom, a temperature change of about 1 Celsius degree per day is indicated for the layer. The fluxes of sensible and latent heat are driven by surface winds generating turbulent eddies, which exchange heat and water molecules between the ocean and the air above. These exchanges depend on wind speed, the vertical gradient of the quantity being exchanged, and an exchange coefficient. The last depends on the stability of the air, wind speed, and the wave field at the surface. For any but the very lowest wind speeds, transfer of heat (both sensible and latent) upward and of momentum downward ensures a more rapid change in the properties of the overlying air than in those of the underlying water.

The horizontal advective term Q_A is important in regions where there is a substantial warm or cold current (velocity > 0.1 m s^{-1}) entering the region. Assuming 10^5 s d^{-1}, the effects of the advective term are felt 10 km into the control volume after one day. Current speeds of approximately 1 m s^{-1} have been observed near the Gulf Stream. However, since Q_A is for net advection, it becomes significant only when downstream temperature gradients of a few Celsius degrees per 100 km are observed. The advection on the air side of the interface must also be included – typically, by using the upstream SST distribution and a back trajectory to compute the thermal properties of the air entering and exiting the boundary layer above the ocean volume being considered. For small volumes and short periods these terms will be small. However, near strong SST gradients, such as on the north wall of the Gulf Stream, the effects can be significant.

The vertical heat-loss term Q_Z is important with shallow, mixed layers, since, when winds generate substantial upward surface fluxes, they are usually accompanied by downward entrainment of surface water and upward entrainment of deep water. This term is the one where the thermal effects of coastal upwelling are represented. In ocean mixed-layer studies, this term is often expressed as a function of the surface's turbulent fluxes.

When comparing the energy balance of oceans to those of coastal land regions, we see that they can be quite different on the same day, and so coastal regions commonly display thermal contrasts. The effects of the coastlines of Atlantic Canada on fronts and precipitation within cyclones are being studied as part of the Canadian Atlantic Storms Program (CASP). In particular, mesoscale networks of meteorological stations on mainland Nova Scotia, Sable Island (Taylor, Salmon, and Stewart, 1993), and the Avalon Peninsula of Newfoundland have yielded useful data on these subjects.

There are few complete energy budgets done for Canadian ocean stations. The energy budget of east coast Ocean Weather Station Bravo was examined by Smith and Dobson (1984). The station is in the Labrador Sea (56°N, 51°W), seaward of the mean position of the Labrador Current. Long-term heat loss from the ocean to the atmosphere was found to be 28 W m^{-2}, compared to 98 W m^{-2} from earlier work. The differences occurred in computations of incoming solar radiation, sensible heat flux, and latent heat flux. The heat flux coefficients used were updated to reflect less dependence on wind speed than most earlier estimates. The monthly radiation budgets were totally dominated by the net solar flux (with a range from near 0 to 200 W m^{-2}), while the net longwave flux was about 20 to 40 W m^{-2} upward in each month over the ten years examined (1965–74). The sensible and latent heat fluxes are about equal in some years, with the latter larger by up to a factor of two in other years. Maximum upward fluxes occur in winter, and peak monthly magnitudes of each flux are in the range 50 to 150 W m^{-2}, with marked interannual variability. No month in the entire record had net downward heat flux, though one month in 1966 had a negative value for sensible flux. The time series of monthly anomalies of the total surface flux from the long-time average indicates values of up to +150 and −200 W m^{-2}. Thus the interannual variability of the monthly energy budget in this region is a rather large fraction of the mean values. Seasonal heat storage in the water occurs from April to August, peaking at about 150 W m^{-2} in June and July, with loss to the atmosphere occurring in the other months, peaking at 200 W m^{-2} in January.

The above flux values can be compared with those for Ocean Weather Station Papa in the North Pacific (50°N, 145°W). Smith and Dobson (1984) report a net gain by the ocean of 32 W m^{-2} there, compared to a net loss of 21 W m^{-2} from earlier estimates. The turbulent flux is dominated by the latent heat flux (Bowen ratios of about 0.2 to 0.3), with the largest flux values in November–December and a secondary maximum in February–March (Hertzman, Miyake, and Pond, 1974). Seasonal heat storage in the water occurs from March to September, peaking at 150 W m^{-2}, with loss to the atmosphere taking place in the other months, peaking at about 80 W m^{-2} in January. The differences between these stations reflect both their latitudes and differences between the oceans. The major differences are caused by the length and strength of the fluxes during the winter season and the length of winter at each station.

Water Balance

It may seem strange to talk about a water balance in the ocean. However, one can think of the budget as one of fresh water, with the vast amount of salt water acting as a "liquid soil" or solvent. The basic water balance equation for the ocean is

$$P - E - R - D_W - z(1 - S_f) = 0. \tag{5.2}$$

The terms in equation (5.2) are the same as in equation (2.16) except for $z(1-S_f)$, in which z is water depth and S_f fractional salinity. We can simplify this balance by considering a well-mixed surface layer of depth z, which mixes only occasionally with lower layers, when forced by the wind. The "runoff" term can be equated to net horizontal advection by mean currents and treated like the advective term in the energy balance (equation 5.1). The "drainage" term can be positive or negative, depending on the salinity and density difference at the base of the mixed layer. The $z(1-S_f)$ term represents a change of salinity in the surface layer. $P-E$ (precipitation minus evaporation) can be described as hydrologic forcing. Areas of the ocean are characterized by their climatology of $P-E$. For example, the northwest Atlantic is an area of negative $P-E$ annually because of the strong net upward latent heat flux forced by the warm waters of the Gulf Stream.

We can examine the effects of forcing on the water storage. Since equilibrium salinities are about 3 per cent, mixed layers are typically 10 to 100 m in depth, and P and E are usually less than 100 mm d^{-1}, substantial day-to-day changes in the salinity of a mixed layer are very unlikely. However, small changes in salinity (less than 1 per cent) can in special cases critically affect the density of the mixed layer water, as compared to the density of the layers below. Many hours of intense evaporation (with latent heat fluxes up to 1,000 W m^{-2}, as in the strong outflows of cold air noted on pp. 107–8) is postulated as a triggering mechanism in some models of oceanic deep convection. The increased density of a shallow, mixed layer is posited to have enough negative buoyancy with respect to deeper layers to convect parcels of surface water rapidly downward by several hundred metres in a few hours (Seung, 1987). A few days of evaporation (at 5 mm d^{-1}) might have an equivalent effect, while substantial precipitation would act to oppose deep convection. In the case of entrainment of deeper water into a well-mixed surface layer, a surface layer (stratified by warming or freshening processes) may be mixed with cooler, saltier water to create a new mixed layer with intermediate properties.

Representative Stations

One way to describe the coastal climates of Canada is to look at climate data from representative stations. This has been done for several stations across the country by Phillips (1990). Here we summarize some of the data from the southern coastal zones, plus some from an arctic station (Atmospheric Environment Service, 1973).

Analysis of coastal zone climates on Canada's three coasts must address "continentality," or degree of marine control of the climate. In a recent work on the Gulf of Maine, Hertzman (1992) stated that "the Gulf of Maine should be classified as a 'cold-type' marine climate, rather than continental." Figure 5.6 shows that Yarmouth, NS, is clearly intermediate between the continental climate of Regina, Sask., and the marine climate of Vancouver, BC, in terms of annual range of mean monthly daily mean temperatures. Vancouver and Yarmouth show substantially more thermal delay (lag of temperature versus solar radiation) than does Regina. The primary climatic difference between Yarmouth and Vancouver is the number of months when mean temperature is at or below zero: four versus zero.

A broader view of the coastal zone thermal climate makes several points clear (Figure 5.7). The monthly means of daily maximum and minimum air temperatures

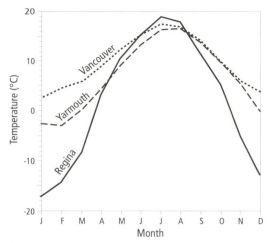

Figure 5.6
Climate comparison of Vancouver, BC, Yarmouth, NS, and
Regina, Sask. Mean monthly daily mean air temperature
(°C) is shown.

show that differences among the west, north, and east coasts are primarily present in
the winter. Indeed, mean monthly near-coastal SSTs (not shown) are quite similar in
Georgia Strait, BC, and near Halifax, NS, in both warm and cold seasons. The temperature differences between the east and west coasts in winter are due to the temperature of the air 500 km upstream and the ability of nearby ocean water to warm the
air that arrives over the coastal stations. In British Columbia, the initial air is warm
and the trajectories are mainly from the sea, while on the east coast the mean wind is
from the land (northwest) in the cold season. In winter St John's, Nfld, has warmer
average minimum temperatures than Halifax or Saint John, NB, because the latter stations are closer to the source of cold continental air (Ontario and Quebec). In extreme
years (such as early 1992), St John's does receive substantial amounts of cold, continental air.

In the spring (April, May, and June on the east coast), the Maritime stations
usually warm more quickly than St John's because they are further from the residual
sea ice, which can persist off Newfoundland until June or July (as in 1991). All the
stations in the Atlantic region are substantially colder than those in British Columbia
in this season because of lack of sea ice on the west coast. During summer, maximum
temperatures at Prince Rupert are up to 6 Celsius degrees colder on average than
those at Vancouver because the cyclonic storm track is closer to the northern station
in these months. Being on the track leads to increased relative amounts of precipitation and number of days with precipitation and fewer hours of sunshine (Phillips,
1990) at Prince Rupert, BC, compared to Vancouver. In summer all the east coast stations have temperatures similar to those of Vancouver. From June to August maximum temperatures at Inuvik, NWT, are close to those of the southern coastal regions,
but minima are noticeably colder. In autumn, east coast and west coast temperatures
are quite similar, with much colder temperatures in the north, despite some open
ocean water, because of the much lower solar radiative fluxes at northern latitudes.

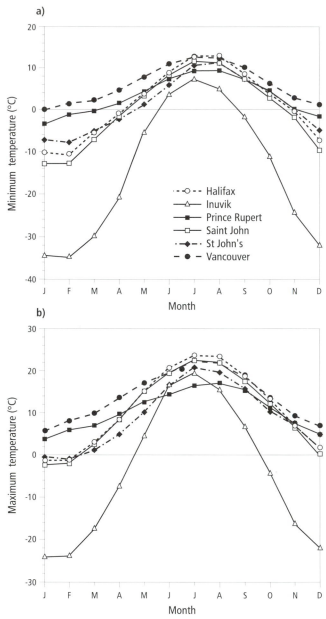

Figure 5.7
Climate comparison within the coastal zone, showing, for six stations,
mean monthly (a) daily minimum and (b) daily maximum air temperatures.

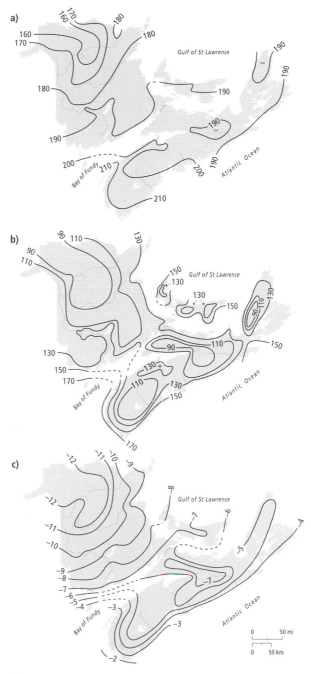

Figure 5.8
Climate variability within the coastal zone of the Maritimes:
(a) length of growing season (d); (b) average frost-free period (d);
and (c) mean January air temperature (°C).
Source: Dzikowski et al. (1984).

In the Maritimes coastal zone, growing seasons can vary by more than forty days (Figure 5.8a), frost-free periods by more than seventy days (Figure 5.8b), and average January temperatures by more than 7 Celsius degrees (Figure 5.8c). When the comparison is extended to Sable Island, the latter number rises to 10 Celsius degrees, with half of the variability present between Sable Island and stations directly on the Nova Scotia coast. Since this coastal zone is typical of ones with few mountains, we can infer that rather strong climate variability occurs even within the coastal zone. Why should this be so? Primarily because the contrast between marine and continental air, and between land and water, causes large differences in cloud amount, winter albedo, and summer surface temperature, all of which feed back into the local energy balance.

To summarize the thermal behaviour of the coastal zone, one can use growing degree days (*GDD*s), as defined by equation (12.4). Figure 5.9 shows substantial differences among the stations from April to September. It is clear that Vancouver has a far earlier start to growth than all other stations. The maximum spread appears in May and June, while maximum values are in July and August. Total annual *GDD*s in the Atlantic coastal region are between about 1,200 and 1,700, while on the west coast, Prince Rupert has about 1,150 and Vancouver 2,000.

Examination of the mean monthly precipitation at seven stations on the three coasts shows different moisture regimes in each area: the north is dry (less than 50 mm each month, with wetter summers than winters); the east is moist and relatively constant spatially and temporally (rates near 100 mm each month); and the west is more variable spatially and seasonally, with wet winters and quite dry summers in some places. These features are functions of latitude (poleward implies colder temperatures and less precipitation) and orography (more mountainous implies higher values and greater spatial variability). Within the west coast region, precipitation varies considerably, with particularly large differences in summer between Prince Rupert and Vancouver. On the east coast of Vancouver Island, rain-shadow effects combine with general synoptic forcing to encourage vegetation types that reflect water limitations.

COASTAL METEOROLOGY

In addition to the climate-related information offered above, there are some peculiarities about coastal meteorology in Canada that should be addressed.

Coastal Fog and Stratus

If you ask someone in Yarmouth, NS, or Sooke, BC, or Trepassey, Nfld, what is their most annoying weather fact of life, they are likely to respond: "Fog or low cloud when it wasn't forecast." Some of these fog events are associated with air flowing over the cold water carried to the surface by upwelling (described above and shown in Figure 5.10). During the spring and early summer, fog and low stratus cloud associated with weak, onshore, synoptic-scale flow or diurnal sea breezes account for daily maximum temperatures that are persistently lower at coastal stations than at inland stations in parts of Atlantic Canada. Similar effects occur on the west coast, primarily in northern California and less frequently on the BC coast. Differences of 5 to 10 Celsius degrees are not uncommon in Nova Scotia during late spring and summer. On

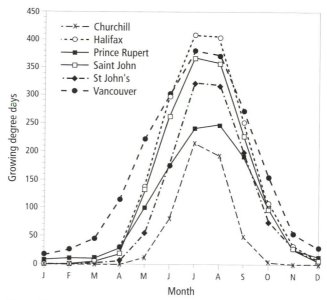

Figure 5.9

Mean monthly growing degree days (*GDD*s) above 5°C for six stations in the coastal zone.

a monthly average basis, differences of 2 to 5 Celsius degrees over 20 to 30 km inland from the Atlantic Ocean are well documented (Atmospheric Environment Service, 1981).

Sea and Land Breezes

Diurnal circulations of sea breezes have been studied in recent years in British Columbia (Steyn and Faulkner, 1986) and Newfoundland (Banfield, 1991).

Sea breezes are forced by differential heating of coastal land versus nearby ocean areas by radiatively driven differences in the energy balances between the two surfaces. They can act in concert with mountain–valley and other topographically forced circulations, as in British Columbia and Newfoundland, or they can be relatively free of these influences, as in Nova Scotia. Their effects can reach tens of kilometres inland, but more usually less than 20 km. They tend to form under weak or anticyclonic, synoptic-scale forcing. Hence, when they carry marine air over land areas, the changes in temperature and sky condition are often pronounced. Decreases in temperature of up to 10 Celsius degrees are not uncommon during sea breeze events in Nova Scotia and British Columbia.

Humid Events

In summer, when mean winds in the Atlantic coastal region are from the southwest, warm air with dewpoint temperatures above 20°C can dominate coastal Atlantic Canada. Some of this air comes from the Gulf of Mexico and the eastern United States,

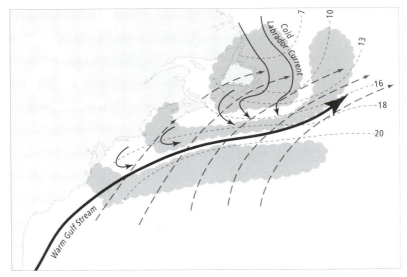

Figure 5.10
Depiction of fog events (shaded) over the western North Atlantic in summer. Dashed ssTs
are summer averages in °C. Solid arrows represent the cold Labrador Current and the
warm Gulf Stream; dashed arrows are prevailing surface air currents.
Source: Hsu (1988).

but a sizeable fraction passes over a considerable amount of water before reaching
the Maritimes or Newfoundland. For boundary-layer air (that with a back trajectory
below about 2 km), the time-weighted average of the ssT along such a path generally
determines dewpoint temperature. Since the 20°C or higher water temperature of the
Gulf Stream and its associated eddies dominates the ssT from Cape Hatteras to the
Tail of the Grand Banks and several hundred kilometres north and south of that line,
its role in the summer climate of the coastal region is considerable. Such humid-air
events occur a few times each summer.

On the west coast, occasional humid-air events are associated with storms on relatively southern tracks, which have incorporated moist air from near Hawaii and/or the
coastal areas of southern California.

Multiple Coast Anomalies

There are at least two places in eastern Canada that experience two simultaneous
coastal climates: the isthmus separating New Brunswick and Nova Scotia and that separating the Avalon Peninsula from the rest of Newfoundland. Both of these areas are
chronic bad-weather regions, with greater-than-normal numbers of high winds, snow
flurries, rain showers, and instances of freezing precipitation. These phenomena occur
primarily because the ocean on one coast can have quite a different temperature than
that on the other coast. In the Nova Scotia example, there can be continuous sea ice on
one side and 10°C sea water on the other. The local winds and precipitation processes
in these areas are complex, especially in the spring.

CURRENT AND FUTURE QUESTIONS

In the next several years, great improvements are expected in the routine satellite remote sensing of the earth's surface and atmosphere. Many of the planned satellite missions incorporate instruments that will give more accurate information for the sea than for the land (for example, microwave radiometers and scatterometers). Other instruments, such as those working in the thermal infrared, will be unable to distinguish fog from low cloud and snow. It will be a great challenge to use many of the new satellite sensors in the coastal regions, because those areas possess all of the most difficult elements for satellite remote sensing.

There is much ongoing interest in the development of climate models that couple the atmosphere and the ocean together. When these models are produced, those wishing to use them for diagnostic or prognostic studies of coastal regions will need to recognize inherent problems. The large gradients of properties, especially on the western edges of continents and ocean basins, constitute a modelling challenge that may be beyond our best efforts for a number of years to come. Thus clever analyses of simple data may still play a major role in coastal climate research.

REFERENCES

Atmospheric Environment Service (AES). 1973. *Climate Normals – Volume 1: Temperature, Volume 2: Precipitation, 1941–1970*. Downsview, Ont.: Environment Canada, Atmospheric Environment Service.

– 1981. *Climate Normals – Temperature and Precipitation: Atlantic Canada 1941–1970*. Downsview, Ont.: Environment Canada, Atmospheric Environment Service.

Banfield, C.E. 1991. "The Frequency and Surface Characteristics of Sea Breezes." *Climatological Bulletin* 25: 3–20.

Barry, R.G. 1994. "Canadian Cold Seas." In H.O. Slaymaker and H.M. French, eds., *Canada's Cold Environments*. Montreal: McGill-Queen's University Press, 29–61.

Broecker, W.S., and Peng, T.-H. 1982. *Tracers in the Sea*. New York: Eldigio Press.

Dobson, F.W., and Smith, S.D. 1989. "A Comparison of Incoming Solar Radiation at Marine and Continental Stations." *Quarterly Journal of the Royal Meteorological Society* 115: 353–64.

Dzikowski, P.A., Kirby, G., Read, C., and Richards, W.G. 1984. *The Climate for Agriculture in Atlantic Canada*. Report for Atlantic Advisory Committee on Agrometeorology. Bedford, NS: AES.

Gyakum, J.R., Anderson, J.R., Grumm, R.H., and Gruner, E.L. 1989. "North Pacific Cold-Season Surface Cyclone Activity: 1975–1983." *Monthly Weather Review* 117: 1141–55.

Hertzman, O. 1992. "Meteorology of the Gulf of Maine." In J. Wiggin and C.N.K. Mooers, eds., *Proceedings of the Gulf of Maine Workshop*, Woods Hole, Mass., 8–10 Jan. 1991, Boston, Mass.: Urban Harbors Institute, University of Massachusetts at Boston, 39–50.

Hertzman, O., Miyake, M., and Pond, S. 1974. "Ten Years of Meteorological Data at Ocean Weather Station Papa." Manuscript Report No. 29. Institute of Oceanography, University of British Columbia, Vancouver.

Hopkins, T.S., and SethuRaman, M. 1987. "Atmospheric Variables and Patterns." In R.H. Backus, ed., *George's Bank*, Cambridge, Mass.: MIT Press, 66–73.

Hsu, S.A. 1988. *Coastal Meteorology*. San Diego: Academic Press.

Kelley, D.A. 1994. "Temperature-Salinity Criterion for Inhibition of Deep Convection." *Journal of Physical Oceanography* 24: 2424–33.

Lazier, J.R.N. 1980. "Oceanographic Conditions at Ocean Weather Ship Bravo, 1964–1974." *Atmosphere-Ocean* 18: 227–38.

Ledrew, E.F. 1992. "Polar Regions' Influence on Climate Variability and Change." *Encyclopedia of Earth System Science*, 3: 647–59.

Ledrew, E.F., Johnson, D., and Maslanik, J.A. 1991. "An Examination of Atmospheric Mechanisms That May Be Responsible for the Annual Reversal of the Beaufort Sea Ice Field." *International Journal of Climatology* 11: 841–59.

Lewis, M.R., Cullen, J., and Platt, T. 1983. "Phytoplankton and Thermal Structure in the Upper Ocean: Consequences of Nonuniformity in Chlorophyll Profile." *Journal of Geophysical Research* 88: 2565–70.

Lewis, M.R., Kuring, N., and Yentsch, C. 1988. "Global Patterns of Ocean Transparency: Implications for the New Production of the Open Ocean." *Journal of Geophysical Research* 93: 6847–56.

McLaren, A.S., Serreze, M.C., and Barry, R.G. 1987. "Seasonal Variations of Sea Ice Motion in the Canada Basin and Their Implications." *Geophysical Research Letters* 14: 1123–6.

Mailhot, J., Hanley, D., Bilodeau, B., and Hertzman, O. 1996. "A Numerical Case Study of a Polar Low in the Labrador Sea." *Tellus* 48A: 383–402.

Miller, C.B., Frost, B.W., Wheeler, P.A., Landry, M.R., Welschmeyer, N., and Powell, T.M. 1991. "Ecological Dynamics in the Subarctic Pacific, a Possibly Iron-limited Ecosystem." *Limnology and Oceanography* 36: 1600–15.

Mysak, L.A., and Manak, D.K. 1989. "Arctic Sea-Ice Extent and Anomalies, 1953–1984." *Atmosphere-Ocean* 27: 376–405.

Nunez, M., Davies, J.A., and Robinson, P.J. 1972. "Surface Albedo at a Tower Site in Lake Ontario." *Boundary Layer Meteorology* 3: 77–86.

Oke, T.R. 1987. *Boundary Layer Climates*. 2nd edn. London: Routledge.

Phillips, D. 1990. *The Climates of Canada*. Ottawa: Canadian Government Publishing Centre, Supply and Services Canada.

Pickard, G.L., and Emery, W.J. 1982. *Descriptive Physical Oceanography: An Introduction*. 4th, revised edn. Toronto: Pergamon Press.

Prinsenberg, S.J., and Danard, M. 1985. "Variations in Momentum, Mass and Heat Fluxes Due to Changes in the Sea Surface Temperature." *Atmosphere-Ocean* 23: 228–37.

Rouse, W.R., Hardill, S., and Silis, A. 1989. "Energy Balance of the Intertidal Zone of Western Hudson Bay. II. Ice-Dominated Periods and Seasonal Patterns." *Atmosphere-Ocean* 27: 346–66.

Sanders, F., and Gyakum, J.R. 1980. "Synoptic-Dynamic Climatology of the 'Bomb.'" *Monthly Weather Review* 108: 1589–1606.

Schmitz, W.J. 1995. "On the Interbasin-Scale Thermohaline Circulation." *Review of Geophysics* 33: 151–74.

Seung, Y-H. 1987. "A Buoyancy Flux-Driven Cyclonic Gyre in the Labrador Sea." *Journal of Physical Oceanography* 17: 134–46.

Silis, A., Rouse, W.R., and Hardill, S. 1989. "Energy Balance of the Intertidal Zone of Western Hudson Bay. I. Ice-Free Period." *Atmosphere-Ocean* 27: 327–45.

Smith, S.D., and Dobson, F.W. 1984. "The Heat Budget at Ocean Weather Station Bravo." *Atmosphere-Ocean* 22: 1–22.

Steyn, D.G., and Faulkner, D.A. 1986. "The Climatology of Sea-breezes in the Lower Fraser Valley, B.C." *Climatological Bulletin* 20: 21–39.

Taylor, P.A., Salmon, J.R., and Stewart, R.E. 1993. "Mesoscale Observations of Surface Fronts and Low Centres in Canadian East Coast Winter Storms." *Boundary-Layer Meteorology* 64: 15–54.

CHAPTER SIX

Freshwater Lakes

WILLIAM M. SCHERTZER

INTRODUCTION

While oceans are vital to climate on a global scale, lakes can significantly affect local and regional climate. Lakes transport and store heat, respond to atmospheric forcing and feed-back to affect the overlying air. Air and water exchanges of heat and moisture have climatological implications not only for the water body but also on the climate of the overlying air and areas adjacent to the water mass. To understand the interactions and effects of lakes on climate we must examine the physical characteristics of water bodies and the temporal and spatial variation of the exchanges of momentum, heat, and water.

Much research on the physical climatology of lakes has focused on the variation of heat and moisture exchanges at the air–water interface with the aim of estimating lake evaporation. In Canada there are as many as two million freshwater lakes located within five major drainage basins, covering 7.6 per cent of the country's total area (Figure 6.1a). Nearly 14 per cent of the world's lakes having surface areas over 500 km^2 are located in Canada. The Great Lakes system constitutes the largest interconnected body of fresh water on earth, consisting of Lake Superior, Lake Huron, Lake Erie, and Lake Ontario, all bordering Ontario and the United States, and Lake Michigan, entirely within the United States. These lakes are the dominant geographical feature of the western part of the Great Lakes–St Lawrence River climatic region.

Though Canada's water-resource base is large, much of the physical climatology of its two million lakes is largely unexplored. Recently, concern over degradation of lake water quality by pollution, acid rain, and toxic inputs and issues such as climate change have increased research on water resources including lakes in the vast north (Prowse and Ommanney, 1990) and in basins with special characteristics. Much physical, biological, and chemical research has been conducted on the Great Lakes, which support large socio-economic-industrial concerns. Large-scale experiments have included work on the physical climatology of lakes – for example, the International Field Year for the Great Lakes (IFYGL), on Lake Ontario (IFYGL, 1981); the Upper Great Lakes Reference Study, which examined Lake Superior and Lake Huron (International Joint Commission [IJC], 1977); the Lake Erie Binational Study (Boyce, Moody,

a)

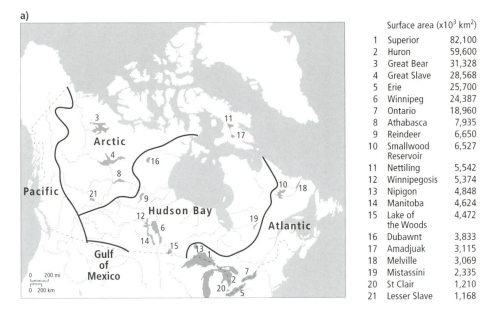

		Surface area ($\times 10^3$ km^2)
1	Superior	82,100
2	Huron	59,600
3	Great Bear	31,328
4	Great Slave	28,568
5	Erie	25,700
6	Winnipeg	24,387
7	Ontario	18,960
8	Athabasca	7,935
9	Reindeer	6,650
10	Smallwood Reservoir	6,527
11	Nettiling	5,542
12	Winnipegosis	5,374
13	Nipigon	4,848
14	Manitoba	4,624
15	Lake of the Woods	4,472
16	Dubawnt	3,833
17	Amadjuak	3,115
18	Melville	3,069
19	Mistassini	2,335
20	St Clair	1,210
21	Lesser Slave	1,168

b)

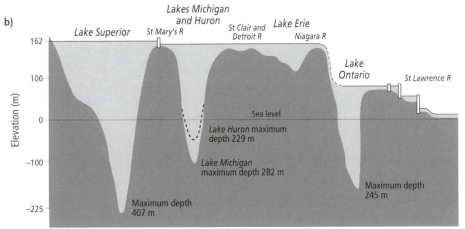

Figure 6.1
(a) Location of large lakes and major drainage basins in Canada, and their order according to surface area;
(b) elevation profile of the Great Lakes system.

and Killins, 1987); and the Upper Great Lakes Connecting Channel Study (Shimizu and Finch, 1988).

While it is not possible to cover here the physical climatology of all the sizes and types of lakes across Canada, this chapter provides an overview of the primary elements of this field. I give particular emphasis to the description of the momentum, radiation, energy and water budget, and evaporation of large lake systems, with the Great Lakes as the primary focus.

PHYSICAL CHARACTERISTICS
OF LAKES

The Canadian climate is diverse, and in large measure Canada's lakes reflect this diversity.

Formation and Type

Canada has tectonic (tectonic impacts and subsidence), barrier (volcanic), and excavated (glacial-action) lakes. Morphometric characteristics such as area, depth, and volume greatly influence the lakes' physical responses and air–water exchanges.

The diversity in the Great Lakes is shown in the depth profiles (Figure 6.1b). Water volume (V) is greatest in Lake Superior (11,600 km^3), compared to 3,580, 1,710, and 545 km^3 for Lakes Huron, Ontario, and Erie, respectively. The flushing time (F) of water $F = V/(R - E)$, where R is discharge and E evaporation, is longer for closed lakes, in which water leaves primarily through evaporation, than for open lakes in which loss also occurs through drainage and seepage. Flushing times in the Great Lakes (Schertzer, 1980) are for Lake Superior 165.0 years, for Lake Michigan 69.5 years, for main Lake Huron 10.6 years, for Georgian Bay 5.7 years, for Lake Erie 2.5 years, and for Lake Ontario 7.5 years.

Climate conditions are often reflected in the general thermal and circulation characteristics of lakes. Polar and mountain lakes with permanent ice cover are termed "amictic." Lakes that circulate fully only during summer are monomictic, while temperate lakes with complete turnover during both spring and autumn, such as the Great Lakes, are dimictic.

Physical Properties

Density of water is responsible for the temperature stratification in large lakes. Maximum density of fresh water occurs at about 4°C and decreases both at warmer temperatures and as it cools to its freezing point at 0°C. The temperature of maximum density is affected by pressure, decreasing by 0.2 Celsius degrees for every one hundred–metre depth of water.

Water is significantly different from other surfaces. It has one of the highest specific heat capacities of all substances (4.186 kJ kg^{-1} °C^{-1}), about twice that of ice (2.04 kJ kg^{-1} °C^{-1}). Thus it has a small diurnal temperature range. The thermal conductivity of still water at 25°C is only 0.57 W m^{-1} °C^{-1}. Since water is a fluid, most heat transport occurs not by molecular conduction but as a result of mixing processes which involve water movement. Solar radiation penetrates through the surface, and as a consequence energy absorption is distributed through a large volume and is further spread throughout the water volume by wind mixing, convection, and advection. Water is not compressible like air, but it can be deformed, resulting in surface waves. Dynamical processes such as internal waves, upwelling and downwelling, and seiches can redistribute heat within the water column. The other critical physical property is that open water surfaces have unlimited amounts of water available for evaporation.

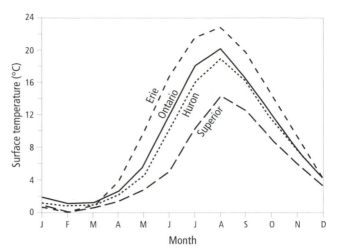

Figure 6.2
Monthly mean surface water temperatures of the Great Lakes.
Source: Phillips (1990).

Seasonal Thermal Cycles

Summer water-surface temperatures (Figure 6.2) are higher for shallow Lake Erie than for the other Great Lakes, while the thermal inertia of Lake Huron and Lake Ontario gives them higher surface temperatures in winter. Characteristics of the seasonal thermal cycles vary among lakes of different physical dimensions and latitudinal location.

Large dimictic lakes mix (overturn) twice yearly, in spring and the autumn, when the surface-water temperature reaches its maximum density. Increased wind speed and the flux of surface heat help to stir the lake (Schertzer, 1980). In winter, below the temperature of maximum density, lake temperatures are generally isothermal. In spring, shallow nearshore water warms more rapidly, forming a convergence zone (thermal bar). Its disappearance marks the onset of summer stratification, a process that in deep lakes such as the Great Lakes takes as long as six to eight weeks (Rodgers, 1987; Schertzer 1987). Thermal stratification proceeds at varying rates, forming a warm upper layer (epilimnion) and a cold lower layer (hypolimnion) separated by the thermocline, which inhibits heat and particle transfer (Schertzer et al., 1987). Maximum heat storage occurs towards the end of summer. Storms that exert high wind stress are responsible for very large energy transfer from the lake to the air and vertical mixing, which breaks down the thermocline, culminating in autumn overturn. Wind mixing and heat losses continue to cool the lake. By the end of winter, the entire water mass cools below 4°C, with the coldest water remaining close to shore.

Satellite imagery shows non-uniform spatial characteristics of surface temperature. Nearshore areas warm faster. Wind stresses may act to tilt the thermocline, resulting in upwelling of colder mesolimnion or hypolimnion waters on one shore and downwelling of warmer water on the opposite shoreline (Simons and Schertzer, 1987). Dynamical circulation processes such as internal waves and seiches (Boyce et al., 1989) result in increased variability of lake temperatures. In the case of a large lake such as

Ontario, variable wind (and other meteorological fields) complicate sampling and averaging of observations, making it difficult to develop a representative climatology in the short term.

Reduced thermal inertia in small, shallow lakes results in faster spring heating and autumn cooling and more uniform temperature distribution. Penetration of solar radiation to the lake bottom becomes an important factor in the lake's heat budget as the water column is heated both from above and from below. In very shallow lakes, bottom vegetation can also enhance warming. Further, net advection of heat from tributary streams may be significant, as can conduction of heat between water and the land. Shallow lakes often have complete ice cover, and, because of their size, small lakes generally do not modify the weather of the surrounding land to the extent observed for large lakes such as the Great Lakes.

Measurement Systems

Figure 6.3a depicts a standard meteorological buoy used on the Great Lakes. Typically it includes instrumentation to observe air temperature, surface water temperature, relative humidity, and wind speed and direction and can accommodate measurement of radiation fluxes. Other systems such as towers (Figure 6.3b) offer stable platforms from which specialized measurements can be conducted.

MOMENTUM TRANSFER

Surface wind stress (τ) is a critical component in lake studies. It is the process that drives lake circulation, lake set-up, surface waves, and turbulent transfer at the air–water interface.

Accurate estimates of the surface wind stress over lake surfaces require information on surface roughness characteristics and the drag coefficient. The roughness length (z_o) of a surface is related to the height, shape, and spacing of its roughness elements (chapter 2). For a water surface, z_o is complicated by the mobility of the roughness elements. Its magnitude ranges from 0.001 to 10 mm over smooth and rough water surfaces (Donelan, 1990).

The drag coefficient for momentum shows weak linear dependence on wind speed, and, in non-steady, inhomogeneous surface conditions, both it and z_o are required in order to account for the wave field. The effect of nonstationarity of the wind field on the value of the drag coefficient is significant. Over ultrasmooth to fully rough surfaces, drag coefficients, measured at a 10-m height above the water, may vary by an order of magnitude (3×10^{-4} to 3×10^{-3}), depending on wind speed, fetch, stability, and wave conditions (Donelan, 1990).

For large lakes the spatial and temporal variability of estimates of τ can be high, because detailed wind observations over the lake are often lacking. The presence of overlake fog, lake–land breeze circulations, and modifications of vertical profiles of meteorological variables by warm-air or cold-air advection (chapter 3) also contribute to difficulty in estimating the stress.

Surface wind stress in excess of 0.2 N m^{-2} on Lake Erie can result in significant vertical mixing of heat, especially in spring and autumn, under unstable thermal strati-

a
b

Figure 6.3
Examples of measurement systems deployed on large lakes: (a) meteorological buoy; and (b) tower platform.

fication (Schertzer et al., 1987). It also influences upwelling and downwelling events in large lakes, thereby affecting horizontal and vertical temperature distributions (Lam and Schertzer, 1987), currents, and heat transports (Simons and Schertzer, 1987).

RADIATION EXCHANGES

Important elements of physical climatology in lake studies include assessment of evaporation and of surface heat exchange (Figure 6.4), because they are source terms for lake characteristics, such as thermal stratification and biological productivity. Determining surface heat flux requires estimation of radiant and turbulent heat flux components, as well as the heat storage in the lake. We start with the radiation balance equation (2.1), which states the net amount of energy available to the lake surface resulting from radiative exchanges in the atmosphere and at the surface.

Surface Controls on the Radiation Balance

The reflectivity of a surface to solar radiation defines its albedo. Estimates of back-scattered light from subsurface layers range from 2 to 3 per cent for turbid water to 6 per cent for clear ocean water. Albedo over a water surface increases with increasing zenith angle and is consistently larger under rough conditions (wave heights > 0.3 m) (Nunez, Davies, and Robinson, 1971; 1972). Under both cloudless and scattered cloud conditions the albedo increases with zenith angle in a similar way (Figure 6.5a). The diurnal changes for broken cloud are considerably less, and under overcast conditions mean albedo averages only about 7.5 per cent. Daily values of albedo for Lake Ontario over the period from July to November (Figure 6.5b) increase from 7 per cent

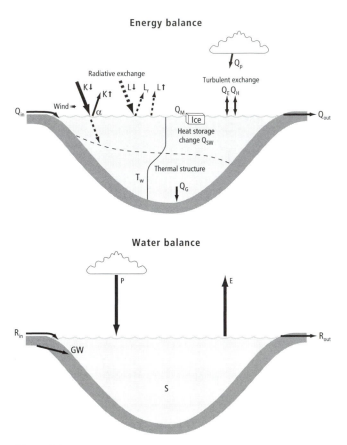

Figure 6.4
Energy and water balance components of lakes.

in early July to 11 per cent in mid-November, due to the combined effects of seasonal changes in zenith angle, cloud, and lake state.

When a lake is frozen, its albedo may be that of fresh or old snow (77–95 per cent and 40–70 per cent, respectively) or that of lake ice. The albedo of sea-ice ranges from 30 to 45 per cent (Oke, 1987). Higher reflectivity occurs at shorter wavelengths for all ice types; it is highest over densely consolidated brash, and lowest for thin skim ice (Leshkevich, 1984). Decay of the ice surface results in significant diurnal variations in albedo. However, Bolsenga (1977) suggests that an average mid-day albedo under average sky conditions suffices for computation of $K\uparrow$ in most studies of lake energy budgets. The daily mean value of albedo of Lake Ontario for winter months has been estimated to be 16 per cent (Davies, Schertzer, and Nunez, 1975).

The emissivity of water and ice ranges from 0.92 to 0.97; hence it is close to being a blackbody in the infrared. For Lake Ontario, Robinson, Davies, and Nunez (1972) determined $\varepsilon = 0.972$, with a standard deviation of 0.021, except that a thin film of oil can reduce emissivity by about 3 per cent.

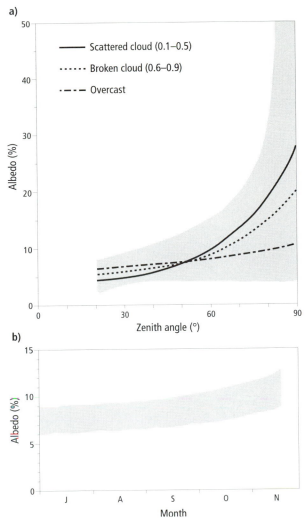

Figure 6.5
Lake surface albedo: (a) albedo in response to cloud and zenith angle; (b) range of observed mean daily surface albedo. The shading indicates the relative range of albedo observed over all types of sky conditions. *Source*: Nunez, Davies, and Robinson (1971).

Solar Radiation Penetration

Penetration of water by solar radiation is greatest at the shorter wavelengths 0.2 to 0.6 micrometres, while longer wavelengths are absorbed in the upper few centimetres. The energy that is absorbed is used for heating and contributes to biological productivity (Schertzer, 1978). The decrease in the intensity of irradiance between the surface and lower depths follows Beer's law.

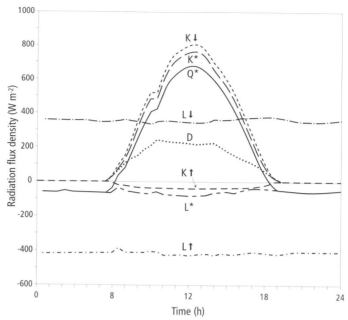

Figure 6.6
Diurnal variation of radiation balance components for Lake Ontario under
cloudless conditions on 28 August 1969.
Source: Davies, Robinson, and Nunez (1970).

The rate of light attenuation (a_v), defined as the mean vertical extinction coefficient, is estimated at 0.1 m^{-1} for pure water. Spatial and temporal variability in a_v over the Great Lakes ranges from 0.15 to 1.0 m^{-1} (Schertzer, Elder, and Jerome, 1978). Recent investigations of lake optics (Bukata, Jerome, and Bruton, 1988) have elucidated the detailed spectral characteristics of light attenuation in the Great Lakes.

Diurnal/Seasonal Variations of Radiation Components

Figure 6.6 shows the diurnal variation of radiation budget components for Lake Ontario under cloudless conditions on 28 August 1969 (Davies, Robinson, and Nunez, 1970). Under cloudy conditions solar radiation at the water's surface would be significantly less. The diffuse-beam portion of $K\downarrow$ ranges from 25 to 75 per cent and varies, depending on factors such as cloudiness and solar altitude (Oke, 1987). Since the albedo is low, so is the reflected solar radiation ($K\uparrow$), and as a result the net solar radiation (K^*) is only slightly less than $K\downarrow$ (Figure 6.6).

Incoming longwave radiation depends on atmospheric absorbers and emitters such as water vapour, clouds, and carbon dioxide. Under cloudless skies, $L\downarrow$ varies diurnally, with short-term variations of up to 20 per cent (Robinson, Davies, and Nunez, 1972). $L\downarrow$ increases in the presence of medium-to-low cloud, but high cloud has little influence. With constant emissivity outgoing longwave radiation ($L\uparrow$) depends on water surface temperature alone. The surface water temperature varies diurnally and is affected by lake state, surface sensible and latent heat exchange, mixing, and advection. Common sources of water temperature estimates are sensors at water intakes, airborne radiometric techniques (ARTs), satellite remote sensing, and ship surveillance. Comparison of ART and satellite temperature measurements with *in-situ* observations shows accuracies in the range of 1.5 Celsius degrees under controlled conditions (Robinson, Davies, and Nunez, 1972). Comparisons of $L\uparrow$ determined using surface water temperatures from float thermistors with those from infrared radiometer "skin temperature" techniques showed the latter to be more variable diurnally. Comparison of $L\uparrow$ determined at the surface and at radiometer height indicated significant divergence in summer and convergence in autumn because of temperature differences between the lake surface and the air. Flux errors of about 1 per cent may be introduced if water surface temperatures are not accurately observed. The reflectivity $(1 - \varepsilon_o)$ of a water surface for atmospheric radiation is approximately 3 per cent, and consequently reflected longwave radiation is a very small component of the surface radiation budget for lakes. As a result of small diurnal ranges in bulk-air and water-surface temperatures, both incoming and outgoing longwave radiation fluxes are relatively constant (Robinson, Davies, and Nunez, 1972). The net diurnal longwave flux $(L* = L\downarrow - Lr - L\uparrow)$ is negative (Figure 6.6).

Seasonal variation of radiation budget components for the Great Lakes is shown in Figure 6.7. As expected, reflected components of both solar and longwave radiation are small, relative to incoming fluxes. For all lakes, incoming solar radiation is minimal during winter and maximal during June and July. Monthly mean $K\downarrow$ during May, June, and July is greater on Lake Huron than on the other Great Lakes, implying less cloudiness. Overlake fog on Lake Ontario during May and June results in substantially lower $K\downarrow$ compared to other lakes. Incoming longwave radiation shows a similar seasonal pattern for the Great Lakes, ranging from 20 to 25 MJ m^{-2} d^{-1} during winter to 30 MJ m^{-2} d^{-1} in August. The lower $L\downarrow$ values for Lake Superior arise from lower air temperatures and humidities, compared to the lower Great Lakes. Losses of outgoing longwave radiation for the Great Lakes are strongly related to the seasonal pattern of surface water temperature. During winter, $L\uparrow$ is similar for all the lakes – approximately 27 MJ m^{-2} d^{-1}. During summer, $L\uparrow$ is lowest for cooler Lake Superior and highest for shallow Lake Erie. Monthly means of $K*$ for all the Great Lakes are similar to the seasonal cycle of $K\downarrow$. Monthly mean $L*$ is negative, with values ranging from -8 to -2 MJ m^{-2} d^{-1}.

Making direct measurements of $Q*$ over large lakes is costly, and researchers therefore depend on estimates of the individual components of the budget. Large errors are possible because of inaccuracies in estimating overlake meteorological conditions from land-station data (Phillips and Irbe, 1978). Figure 6.6 shows clearly that the diurnal net radiation budget is dominated by net solar radiation during the day and is equal to net longwave radiation at night. Robinson, Davies, and Nunez (1972) concluded that the influence of $L*$ on $Q*$ is variable but generally small.

Comparisons between $Q*$ as determined from measurements (Davies and Schertzer, 1974) and from numerical model calculations (Atwater and Ball, 1974) show that overlake fog can significantly affect overlake estimates. During spring, failure to account for overlake fog had the effect of reducing weekly averaged estimates of $Q*$ on Lake Ontario by as much as 40 per cent. Variation in cloud height (surface to 1200 m) affects incoming and outgoing fluxes by about 1 per cent at the surface. Changes in amount of

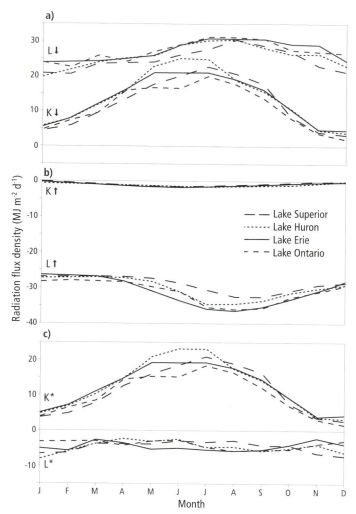

Figure 6.7
Seasonal variation of radiation balance components for Lake Ontario:
for (a) incoming; (b) outgoing; and (c) net radiation components.
Sources: (Lake Superior) Schertzer (1978); (Lake Huron) Bolsenga (1975);
(Lake Erie) Derecki (1975) and Schertzer (1987); (Lake Ontario) Atwater and
Ball (1974).

cloud (0 to 100 per cent) produced changes in $K\downarrow$ and $L\downarrow$ similar to the daily average net radiative flux (Pinsak and Rodgers, 1981).

The seasonal variation of Q^* for the Great Lakes (Figure 6.8a) shows lowest values during winter and peak values during summer. The lower Great Lakes demonstrate similar monthly mean Q^* for all months of the year, whereas the upper Great Lakes show a phase difference. Summertime Q^* for Lake Huron significantly exceeds values computed for the lower Great Lakes. This is attributable to the larger $K\downarrow$ and relatively lower $L\uparrow$ losses for Huron during these months. Q^* for Lake Superior shows a

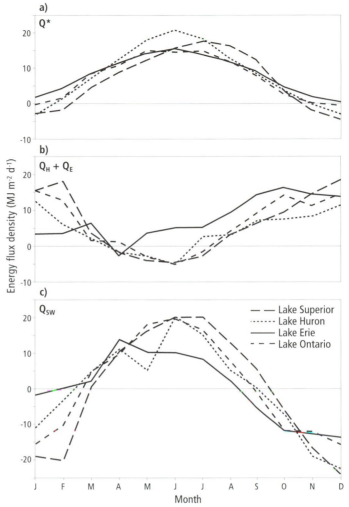

Figure 6.8
Long-term monthly mean for the Great Lakes: (a) net radiation; (b) total
turbulent exchange; and (c) storage heat flux into the water for the Great Lakes.
Sources: As for Figure 6.7.

substantial lag of approximately one month compared to the other Great Lakes, peaking in July rather than in June. $Q*$ is negative for Lake Superior from November
through February, while $Q*$ is negative primarily during December and January for
the lower Great Lakes.

ENERGY AND WATER BALANCES

Figure 6.4 provides a summary of the components forming a lake's energy and water
balances. In this section, I describe the variation of the primary energy budget components for the Great Lakes. The advection component of a lake's energy balance

includes heat input resulting from water inflows and outflows (precipitation, runoff, and tributary inflow and outflow).

A lake's energy balance can be expressed:

$$Q^* - Q_{Sw} + Q_A = Q_H + Q_E + Q_G, \qquad (6.1)$$

where Q^* is net radiation, Q_{Sw} net heat storage change in the lake volume expressed as a flux density through the lake surface, Q_A net advected heat, Q_H convection of sensible heat to the atmosphere, Q_E heat used by evaporation, and Q_G net heat conduction across the bottom of the lake.

Heat Storage

Because of the high specific heat of water, large lakes store great quantities of heat and react slowly to short-term changes in temperature (Schertzer and Sawchuk, 1990). Heat storage in lakes represents the integrated outcome of heating and cooling processes resulting from air–water interactions and hydrological balances. Annual heat storage change consists of two components – the summer heat income, which is the heat required to raise the lake temperature from 4°C to its maximum, and the winter heat income, which is required to raise the lake temperature from its minimum heat content up to 4°C. Long-term heat content varies widely among different lakes but is positively correlated with morphological characteristics such as lake area, depth, and volume (Gorham, 1964).

The largest annual heat budget for a lake is that of Lake Baikal, Russia (approximately 3.1 GJ m^{-2}). In comparison, Figure 6.9a shows annual and spring heat incomes for the Great Lakes. Analysis of the thermal regime of Lake Superior (Bennett, 1978; Schertzer, 1978) indicates that the average total spring heat income of 1.47 GJ m^{-2} is responsible for raising the mean lake temperature from a minimum of 1.4°C to the average temperature of maximum density, which is 3.82°C. The total summer heat income of 1.26 GJ m^{-2} produces a further increase to the maximum mean lake temperature of 5.9°C. Thus for Lake Superior, approximately 54 per cent of the annual heat income is used for spring warming of the water to the temperature of maximum density. In general, Figure 6.9a shows that as a lake's surface area, depth, or volume increases, there is an increase in heat uptake and storage. The large heat incomes for the Great Lakes are related to their dimictic nature, which allows the entire lake volume to be involved in heat exchange semi-annually.

The seasonal cycle of heat content for the Great Lakes is shown in Figure 6.9b, estimated from Bolsenga (1977), Boyce, Moody, and Killins (1977), and Schertzer (1978; 1987). On the Great Lakes, minimum heat storage occurs in late winter, and maxima in late summer/early autumn. Substantial differences in the timing of maximum and minimum heat content in each lake are related largely to differences in volume.

Advected Heat and Hydrological Balances

Net advected heat (Q_A) is due to the heat content of water entering the lake (Q_{in}) from tributaries, runoff, precipitation, and waste heat (Q_w) and of water leaving the lake

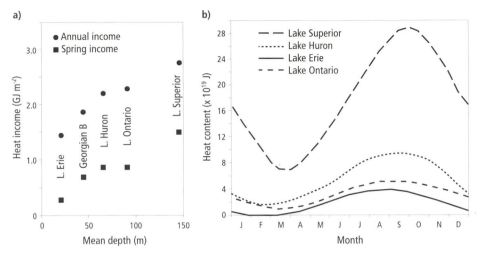

Figure 6.9
Heat storage in the Great Lakes: (a) heat income per unit of surface area versus mean depth; (b) Great Lakes seasonal heat-content cycle.
Sources: (a) Bennett (1978); (b) Bolsenga (1975); Boyce, Moody, and Killins (1977); Schertzer (1978; 1987).

(Q_{out}) through evaporation, outflow, and snowmelt heat loss (Q_M). In contrast to large, deep lakes, small, shallow lakes and ponds may have a significant heat budget associated with their bottom sediments, so that information is required on the structure and conduction characteristics of the sediments.

As already noted, water has low thermal conductivity, so molecular transport of heat is negligible, and it takes place almost entirely as a result of water movement. Whereas the transport of heat by water movement is negligibly small in a soil column, the net heat advection resulting from flows ($Q_{in} - Q_{out}$) can be significant, especially in smaller lakes. Channel flows connecting to the Great Lakes are large (5,000 to 7,000 m^3 s^{-1}); however, the net impact on the energy budget of a whole lake is reduced because of the similar magnitude of T_{in} and T_{out}, so ($Q_{in} - Q_{out}$) is generally small compared to the total heat content of the lake.

During summer, overlake precipitation adds heat to the lake. The opposite occurs during late autumn or winter; frozen precipitation on open water affects heat storage, because heat of fusion is extracted from the lake for melt. Heat gain or loss through formation or decay of ice also affects the lake's heat balance, and ice cover partially seals the surface and reduces heat losses (Schertzer, 1978; Pinsak and Rodgers, 1981). Including the heat exchange related to precipitation in the energy budget of Lake Ontario during IFYGL decreased evaporation estimates by an average of 1 per cent (April through October) and 5 per cent (November through March). Omitting the role of ice formation and decay changed evaporation from 1 to −2.4 mm per week (Pinsak and Rodgers, 1981).

Inputs of waste-heat from point sources into the Great Lakes can represent a significant heat input at the local scale (Boyce et al., 1993). Modifications to seasonal heat storage can also occur through artificial regulation of flows and flow changes resulting from such factors as introduction of ice booms.

For large lakes, the advective component represents a small contribution to the lake's energy balance, especially when averaged over the long term (Elder, Boyce, and Davies, 1974; Bolsenga, 1975; Derecki, 1975; Schertzer, 1978; 1987). For smaller bodies of water, net advection can become significant.

Turbulent Heat Fluxes

Observations and models are used to determine the Q^*, Q_{Sw}, Q_A, and Q_G terms in equation (6.1), leaving $(Q_H + Q_E)$ as the residual (Figure 6.8b). Direct observation of the turbulent heat fluxes (for example, using eddy correlation systems) is difficult and not practical over the spatial scales of large lakes. Bulk aerodynamic methods are commonly used to estimate turbulent transfer components (Derecki, 1975). In these approaches the bulk-transfer coefficients for sensible heat and latent heat fluxes are generally assumed to be equal under near-neutral conditions, averaging 1.5×10^{-3} (± 20 per cent) (Phillips, 1973). It is necessary to allow the coefficients to vary with friction velocity, stability, sea state, and so on (Henderson-Sellers, 1986).

Another approach is to partition the terms in the residual $(Q_H + Q_E)$ by application of the Bowen ratio (β). The latent heat flux can then be determined, based on independently derived overlake observations of temperature and vapour pressure. The Bowen ratio approach also assumes that the transfer processes for heat and water vapour are similar. Difficulties over large lakes are related to estimation of overlake T_a and T_o, and least confidence is held in values for periods when water temperature approaches air temperature (Quinn, 1978). The Bowen ratio is very sensitive to small changes in moisture and temperature, so daily estimates derived from hourly averaged ratios may vary significantly, compared to estimates derived from hourly averaged input data (Pinsak and Rodgers, 1981). For the temperature ranges encountered for large, dimictic lakes, a critical value of β occurs as it approaches the value -1; this results in unreasonably large computed values of Q_E. The standard deviation of β for Lake Ontario during IFYGL was found to be highest in the heating season (1.4 to 19), and smallest during the cooling season (< 0.6) (Pinsak and Rodgers, 1981). There is wide variation in the climatological monthly mean β for the Great Lakes in the spring months (Figure 6.10c). For Lake Ontario, 95 per cent of the evaporation occurs during the cooling period (August through February). While the spatial and temporal representativeness of the Bowen ratio is of concern for very large lakes, the technique is probably adequate for determining evaporation over longer time-scales, with least confidence for daily computations, especially during the heating season (Pinsak and Rodgers, 1981).

On an annual basis, Q_H is approximately 30–35 per cent of Q_E on Lake Ontario (Pinsak and Rodgers, 1981) and Lake Huron (Bolsenga, 1975) and ranges from a low of about 5–10 per cent for Lake Erie (Derecki, 1975; Schertzer, 1987) to 60 per cent on Lake Superior (Schertzer, 1978). The direction of Q_H should conform to the air–water temperature gradient. Negative values of Q_E observed in all the Great Lakes indicate episodes of condensation, primarily during spring (Figure 6.10b, c). Large errors in Q_H and Q_E occur in near-neutral conditions, when the vapour pressure of the water surface approaches that in the air and when β approaches infinity. During autumn, decreased storage of heat in the Lakes results primarily from increased heat losses through Q_H and Q_E and lower Q^*.

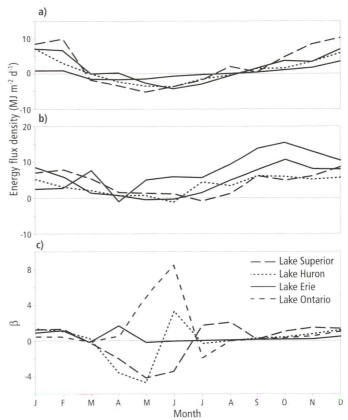

Figure 6.10
Great Lakes turbulent heat exchanges: (a) long-term monthly mean sensible heat flux (Q_H); (b) latent heat flux (Q_E); (c) computed Bowen ratio (β) (dimensionless).
Sources: As for Figure 6.7.

When changes in air-mass characteristics cause local advection, Phillips (1973) observed that a dry airmass enhances the evaporation rate (because the water–air humidity gradient is increased) and humid air suppresses it. Similarly, introduction of a relatively cold airmass over a water surface tends to increase convective instability, and the convective heat exchange, Q_H, whereas introduction of a warm airmass has a dampening effect. Phillips (1973) reports an example of enhanced Q_H over Lake Ontario in January when the lake water is considerably warmer than the cold, continental air traversing it. Q_H may then be as large as 20 MJ m^{-2} d^{-1}, and since $Q*$ is very small at this time, the energy output to the atmosphere must be derived from heat storage (Oke, 1987).

Figure 6.8b shows variation of the monthly mean total turbulent exchange ($Q_H + Q_E$) on the Great Lakes, in which positive values indicate net gain of heat to the atmosphere. In general, the total turbulent exchange is directed to the atmosphere during winter and autumn, because lake temperatures are higher than air temperature and evaporation is highest especially during autumn. Heat gains by the lake occur during spring and early summer, when the air temperature is warmer than that of the surface water, and heat gains also occur through condensation. The seasonal pattern of $(Q_H + Q_E)$ is similar for all the Great Lakes, with the notable exception of shallow Lake Erie during the heating season. For Lake Erie, turbulent exchange losses are much greater because of higher water-surface temperatures, which have a large effect on losses through Q_E. Maximum losses of heat through turbulent transfer occur when gains through net radiation are minimal. Examination of the spatial distribution of turbulent heat exchange on Lake Ontario (Phillips 1973) showed substantial differences across the lake, and, as expected, patterns resembled the spatial distribution of surface temperature.

Net Storage Heat Flux

The change in the amount of heat stored in a lake (Q_{Sw}) can be determined using differences of the vertical temperature structure between two surveys (Boyce, Moody, and Killins, 1977). The limiting factor is the availability of detailed temperature-profile measurements across the lake. The alternative is to observe or compute the other components of the lake's energy balance and to determine the storage heat flux as the residual.

Figure 6.8c shows long-term monthly mean storage heat flux, Q_{Sw}, for the Great Lakes. Positive changes in Q_{Sw} (lake heat gains) occur from February to September, primarily through heating by net radiation, because Q_E and Q_H are relatively small. During the lake's cooling period in autumn, negative changes in lake heat storage for the Great Lakes occurs primarily through the turbulent exchange components $(Q_H + Q_E)$, since net radiation is small during these months. Maximum heat loss takes place in December and January. Maximum heat storage change occurs in late August and early September (Figure 6.9b), coinciding with the period in which the Q_{Sw} turns negative (Figure 6.8c).

LAKE EVAPORATION

Water resources are crucial for such purposes as irrigation, navigation, industrial cooling water, and hydro-electric potential. Management of water resources (quantity and quality) is a major issue, and one in which the accurate assessment of water loss is a key concern. Evaporation is the common component linking energy and hydrological balance of a lake (Figure 6.4). Evaporation from large lakes is estimated primarily through the water budget, energy budget, mass transfer, or pan methods.

Both the water budget and energy budget approaches are applicable to determine long-term evaporation averages for whole lakes. Where inflows and outflows are very large and similar, small errors may significantly affect smaller components of the water balance, such as evaporation (de Cooke and Witherspoon, 1981). The energy budget approach is relatively accurate and practical for use on lakes of moderate size with small net advection. Empirical techniques, such as mass transfer and the evaporation pan, can be used for short periods, but they may require calibration for each lake, thereby necessitating independent evaporation estimates. The advantage of the mass transfer approach is that relatively accurate evaporation rates can be derived from a few simple measurements.

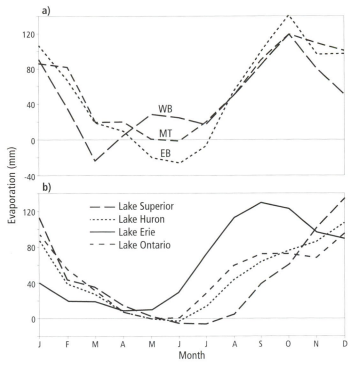

Figure 6.11
Monthly mean evaporation from the Great Lakes. (a) Comparison of evaporation estimates for Lake Ontario (IFYGL) from the water budget (WB), energy
budget (EB), and mass transfer (MT) approaches; (b) long-term monthly mean
evaporation from mass transfer estimates.
Sources: (a) Pinsak and Rodgers (1981); (b) AES (1992).

Comparison of figures on monthly mean evaporation on Lake Ontario during
IFYGL derived from three common methods shows similar seasonal patterns (Figure 6.11a), but for a few months differences in the evaporation estimate can be large.
Considering that the IFYGL represented one of the most intensive investigations of
a large lake, the results highlight the difficulty in determining evaporation for such
a body of water. The greatest differences occurred in spring and winter. In May, June,
and July, the energy budget method showed significant condensation; the aerodynam-
ical approach, slight condensation; and the water budget, significant evaporation.

Great Lakes Seasonal Evaporation

The seasonal distribution of evaporation for the Great Lakes shows highest evaporation
during late summer, autumn, and early winter, due to generally higher winds with large
vapour-pressure gradients (Figure 6.11b). Lowest evaporation occurs in late winter and
early spring, because ice cover suppresses vapour exchange and surface-water temperatures are low. Summer and early-autumn evaporation for Lake Superior deviates
significantly from that of the other lakes in the system, primarily because of different
surface-water temperature. On time-scales shorter than the month, lake evaporation

can be highly episodic. IFYGL buoy data indicates that 51 per cent of evaporation occurred during only 15 per cent of the days. Highest daily evaporation, 13 mm, was noted after passage of a cold front on 9 October 1972 (Quinn and den Hartog, 1981). Seventeen per cent of the evaporation took place on only eight days, when evaporation exceeded 7 mm d^{-1}.

Spatial variability of evaporation on very large, deep lakes can be very high and reflects variability in the seasonal temperature distribution. Analysis of spatial variability of evaporation using a dense buoy network revealed an increase in E from west to east, ranging from 289 mm to 488 mm (Quinn and den Hartog, 1981).

Evaporation from Small Lakes

Lakes have a wide range of physical dimensions and latitudinal locations, and these factors affect the seasonal energy and evaporation regime. In general, shallow-lake evaporation is controlled primarily by variations in the daily energy input, and consequently evaporation tends to correspond with the ice-free period from May through October. In large, deep lakes (for example, the Great Lakes), maximum evaporation occurs during autumn, spurred by energy derived from heat stored during summer.

Lake Diefenbaker, Sask., and Perch Lake, Ont., differ markedly from the physical dimensions of the Great Lakes. Lake Diefenbaker, created by damming the South Saskatchewan River, is 225 km long, averages 2.5 km in width, and has a mean depth of 24 m. In contrast, Perch Lake is small and circular (451,000 m^2), with a mean depth of 2 m and a maximum depth of 3.7 m. Extensive studies have been conducted to determine the energy and water balance and evaporation characteristics of both lakes.

Figure 6.12 shows the seasonal energy balance and evaporation for Lake Diefenbaker during 1973 (Cork, 1974). The lake's energy balance is considered typical of smaller lakes in the temperate and subarctic regions during the ice-free season (Rouse, 1979). Maximum $Q*$ occurs during June and July, but Q_E is maximal from August through October. Between May and August, Q_H is small and negative, which indicates heat transfer from the warmer air to the lake surface. After maximum heat storage in July and August, the warmer lake surface heats the overlying cooler air and Q_H becomes positive. Advection, while small, is important primarily during spring. Heat storage losses, especially after August, occur with decreases in $Q*$ and high losses through the turbulent components, mainly Q_E. The pattern of heat storage losses for Lake Diefenbaker is comparable, though lagged in time to that for shallow Lake Erie (Figure 6.8c), as the energy used for evaporation from August to October is derived partly from the heat storage. But the seasonal evaporation cycle is significantly shifted in Lake Diefenbaker, peaking earlier in the year than on the Great Lakes.

The seasonal energy balance and evaporation for Perch Lake in 1970 (Barry and Robertson, 1975) indicate that within two weeks after ice melt the lake heats rapidly (Figure 6.12), with mean surface-water temperatures exceeding those of the air. The mean water temperature of Perch Lake continues to exceed mean air temperature throughout summer and autumn. Vertical temperature profiles show generally weak stratification for all months except August, near the time of peak heat content. In such a small, shallow lake, heat storage not only increases rapidly in the spring and decreases rapidly in the autumn, it also tends to oscillate through a narrow range throughout sum-

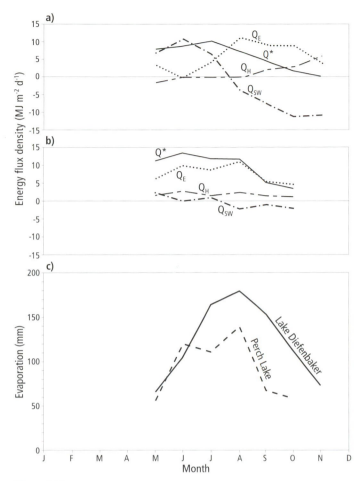

Figure 6.12
Seasonal energy balance of (a) Lake Diefenbaker, Sask., and (b) Perch Lake,
Ont.; (c) evaporation in both lakes.
Sources: Data abstracted from Cork (1974) and Barry and Robertson (1975).

mer and autumn in response to energy input. Q_H is positive throughout summer and
autumn as the warm water surface heats the overlying air. Q_E is the dominant turbulent
transfer component on Perch Lake, and Figure 6.12 shows that nearly all the net radiation is expended in evaporation and in heating air passing over the lake. Cumulative
curves (May through mid-October 1970) of the individual energy components (Barry
and Robertson, 1975) show that of the total energy input of $Q^* = 1.62$ GJ, 80 per cent
was used to evaporate water and 19 per cent to heat the air, and the remainder was
accounted for by heating of the sediments and advection losses. The seasonal evaporation (mass transfer approach), while not as high as that on Lake Diefenbaker, is seasonally displaced in a pattern similar to that for the larger Great Lakes. For a shallow body
of water such as Perch Lake, the rate of evaporation averaged over a week or longer is
nearly proportional to net radiation at the surface.

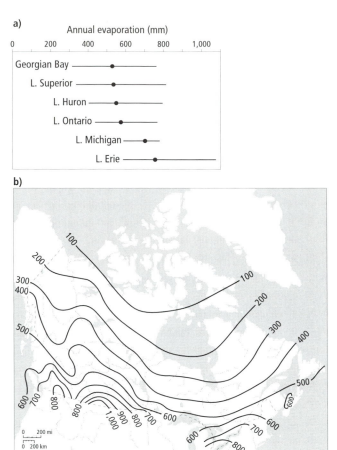

Figure 6.13
Average annual lake evaporation (mm) for (a) the Great Lakes, estimated
with use of the mass transfer technique; and (b) Canada, estimated with
use of pan evaporation.
Sources: (a) AES (1992); (b) Phillips (1990).

Annual Lake Evaporation

Long-term mean annual evaporation from the Great Lakes (Figure 6.13a) shows values ranging from 500 to 750 mm y^{-1}. Highest evaporation occurs in warmer and shallower Lake Erie, and lowest evaporation on lakes with cooler surfaces. The wide range in evaporation over each of the Great Lakes, shown by the bars, reflects weather variability.

Mean annual lake evaporation across Canada (Figure 6.13b) was summarized by Phillips (1990). Lake evaporation is greatest in the dry Prairies, where it can be as large as 1,000 mm a year. Least evaporation occurs in the extreme north, averaging around 100 mm. This is largely due to nine or more months with ice and snowcover, which effectively suppress evaporation from the lake surface. Prowse and Ommanney

(1990) reported that for much of the Canadian north, lake evaporation is equal to or greater than 50 per cent of precipitation.

The Pacific coast has the longest evaporation season, with annual losses averaging from 400 to 700 mm. At high elevations in the Cordillera, the season is short and lake evaporation is low because of the lower temperatures and longer ice season. July is the month of maximum evaporation in the south, and August is the peak month in the Arctic Islands. On the Prairies, evaporation losses are high in May. Across eastern Canada lake evaporation ranges from about 400 mm in the north to 800 mm in south-western Ontario. In the east there is a regular seasonal cycle, with a fairly well-marked maximum in July for smaller lakes.

PHYSICAL CLIMATOLOGY AND LAKES RESEARCH

Physical climatology of large lake systems (momentum transfer, and heat and moisture exchange) helps explain and assess questions relating to water quantity and quality. Models of lakes incorporate aspects of physical climatology in the assessment of eutrophication, the effects of acid rain, the fate and pathways of toxic contaminants, and, recently, the potential effects of climate change on water resources. It has been suggested that regional changes in climate may be detected through monitoring of the long-term physical climatology of large, deep lakes and the physical changes such as thermal stratification responses. Continuing research is required to specify surface-exchange coefficients for exchanges of momentum, heat, and moisture under variable surface conditions.

Spatial and temporal scales of analysis are often dictated by the adequacy of over-lake meteorological fields. Physical climatological research on the Great Lakes has often been at the large scale, providing essential input for models of circulation, temperature structure, or water quality. In such cases, fluxes are generally determined through occasional ship surveillance and data from stationary buoys or shore-line stations. A major problem with shoreline data is the requirement to apply transformations to derive overlake values. Current methods are generally lake-specific and require improvement. Recent use of satellite data for such variables as surface-water temperature has greatly improved flux computations in the Great Lakes.

REFERENCES

Atmospheric Environment Service (AES). 1992. *Monthly and Annual Evaporation from the Great Lakes*. Downsview, Ont.: Environment Canada, Atmospheric Environment Service.

Atwater, M.A., and Ball, J.T. 1974. *Cloud Cover and the Radiation Budget over Lake Ontario during IFYGL*. CEM Report No. 4130–513a, Vol. 1. Hartford, Conn.: Center for the Environment and Man, Inc.

Barry, P.J., and Robertson, E. 1975. "The Energy Budget of Perch Lake." In P.J. Barry, ed., *Hydrological Studies on a Small Basin on the Canadian Shield*, Chalk River, Ont.: Chalk River Nuclear Laboratories, AECL, 375–415.

Bennett, E.B. 1978. "Characteristics of the Thermal Regime of Lake Superior." *Journal of Great Lakes Research* 4: 310–19.

Bolsenga, S.J. 1975. "Estimating Energy Budget Components to Determine Lake Huron Evaporation." *Water Resources Research* 11: 661–6.

– 1977. "Preliminary Observations on the Daily Variation of Ice Albedo." *Journal of Glaciology* 18: 517–21.

Boyce, F.M., Donelan, M.A., Hamblin, P.F., Murthy, C.R., and Simons, T.J. 1989. "Thermal Structure and Circulation in the Great Lakes." *Atmosphere-Ocean* 27: 607–42.

Boyce, F.M., Hamblin, P.F., Harvey, L.D.D., Schertzer, W.M., and McCrimmon, R.C. 1993. "Response of the Thermal Structure of Lake Ontario to Deep Cooling Water Withdrawals and to Global Warming." *Journal of Great Lakes Research* 19: 603–16.

Boyce, F.M., Moody, W.J., and Killins, B.L. 1977. *Heat Content of Lake Ontario and Estimates of Average Surface Heat Fluxes during IFYGL.* Technical Bulletin No. 101, Canada Centre for Inland Waters, Burlington, Ont.

Bukata, R.P., Jerome, J.H., and Bruton, J.E. 1988. "Relationships among Secchi Disk Depth, Beam Attenuation Coefficient, and Irradiance Attenuation Coefficient for Great Lakes Waters." *Journal of Great Lakes Research* 14: 347–55.

Cork, H.E. 1974. *Lake Diefenbaker Energy Budget – Summer 1973.* Hydrometeorology Report 11, Atmospheric Environment Service, Regina Airport, Sask.

Davies, J.A., Robinson, P.J., and Nunez, M. 1970. *Radiation Measurements over Lake Ontario and the Determination of Emissivity.* First Report, Contract No. HO 81276, Department of Geography, McMaster University, Hamilton, Ont.

Davies, J.A., and Schertzer, W.M. 1974. *Canadian Radiation Measurements and Surface Radiation Balance Estimates for Lake Ontario during IFYGL.* Final Report, IFYGL Project Nos. 71EB and 80EB. Department of Geography, McMaster University, Hamilton, Ont.

Davies, J.A., Schertzer, W.M., and Nunez, M. 1975. "Estimating Global Solar Radiation." *Boundary-Layer Meteorology* 9: 33–52.

de Cooke, B.G., and Witherspoon, D.F. 1981. "Terrestrial Water Balance." In E.J. Aubert and T.L. Richards, eds., *IFYGL – the International Field Year for the Great Lakes*, Ann Arbor, Mich.: NOAA-Great Lakes Environmental Research Laboratory, 199–219.

Derecki, J.A. 1975. *Evaporation from Lake Erie.* NOAA Research Report, ERL 342-GLERL3. Ann Arbor, Mich.: Great Lakes Environmental Research Laboratory.

Donelan, M.A., 1990. "Air–Sea Interaction." In B. Lemehaute and D. Hanes, eds., *The Sea: Ocean Engineering Science*, New York: John Wiley and Sons Inc., 9: 239–92.

Elder, F.C., Boyce, F.M., and Davies, J.A. 1974. "Preliminary Energy Budget of Lake Ontario for the Period May through November 1972, (IFYGL)." *Proceedings of the 17th Conference of Great Lakes Research*, 713–24.

Gorham, E. 1964. "Morphometric Control of Annual Heat Budgets in Temperate Lakes." *Limnology and Oceanography* 9: 525–9.

Henderson-Sellers, B. 1986. "Calculating the Surface Energy Balance for Lake and Reservoir Modelling: A Review." *Review of Geophysics* 24: 625–49.

IFYGL. 1981. *IFYGL – the International Field Year for the Great Lakes.* By E.J. Aubert and T.L. Richards, eds., Ann Arbor, Mich.: NOAA-Great Lakes Environmental Research Laboratory.

International Joint Commission (IJC). 1977. *The Waters of Lake Huron and Lake Superior.* 3 vols. Windsor, Ont.: International Joint Commission.

Lam, D.C.L. and Schertzer, W.M. 1987. "Lake Erie Thermocline Model Results: Comparison with 1967–1982 Data and Relation to Anoxic Occurrences." *Journal of Great Lakes Research* 13: 757–69.

Leshkevich, G.A. 1984. "Airborne Measurements of Freshwater Ice Albedos." In *Proceedings 18th International Symposium on Remote Sensing of Environment*, Paris, France, 1–5 Oct. 1984, Ann Arbor, Mich.

Nunez, M., Davies, J.A., and Robinson, P.J. 1971. *Solar Radiation and Albedo at a Lake Ontario Tower Site. Third Report, Contract HO81276*, McMaster University, Hamilton, Ont.

– 1972. "Surface Albedo at a Tower Site in Lake Ontario." *Boundary-Layer Meteorology* 3: 77–86.

Oke, T. R. 1987. *Boundary Layer Climates*. 2nd edn. London: Routledge.

Phillips, D.W., 1973. "Contribution of the Monthly Turbulent Heat Flux to the Energy Balance of Lake Ontario." *Canadian Geographer* 17: 354–72.

– 1990. *The Climates of Canada*. Ottawa: Canadian Government Publishing Centre, Supply and Services Canada.

Phillips, D.W., and Irbe, G.J. 1978. "Lake to Land Comparison of Wind, Temperature and Humidity of Lake Ontario during International Field Year for the Great Lakes." CLI 2–77, Atmospheric Environment Service, Environment Canada, Downsview, Ont.

Pinsak, A.P., and Rodgers, G.K. 1981. "Energy Balance." In E.J. Aubert and T.L. Richards, eds., IFYGL – the International Field Year for the Great Lakes, Ann Arbor, Mich.: NOAA-Great Lakes Environmental Research Laboratory, 169–98.

Prowse, T.D., and Ommanney, C.S.L. 1990. *Northern Hydrology – Canadian Perspectives*. NHRI Science Report No. 1. Saskatoon: National Hydrology Research Institute, Inland Waters Directorate, Environment Canada.

Quinn, F.H. 1978. "An Improved Aerodynamic Evaporation Technique for Large Lakes with Application to the IFYGL." *Water Resources Research* 15: 935–40.

Quinn, F.H., and den Hartog, G. 1981. "Evaporation Synthesis." In E.J. Aubert and T.L. Richards, eds., IFYGL – the International Field Year for the Great Lakes, Ann Arbor, Mich.: NOAA-Great Lakes Environmental Research Laboratory, 241–6.

Robinson, P.J., Davies, J.A., and Nunez, M. 1972. *Longwave Radiation Exchanges over Lake Ontario*. Fourth Report, P.O. HO.81276. Department of Geography, McMaster University, Hamilton.

Rodgers, G.K. 1987. "Time of Onset of Full Thermal Stratification in Lake Ontario in Relation to Lake Temperatures in Winter." *Canadian Journal of Fish and Aquatic Science* 44: 2225–9.

Rouse, W.R. 1979. "Man-modified Climates." In K.J. Gregory and D.E. Walling, eds., *Man and Environmental Processes: A Physical Geography Perspective*, London: Dawson and Sons Ltd., 38–54.

Schertzer, W.M. 1978. "Energy Budget and Monthly Evaporation Estimates for Lake Superior, 1973." *Journal of Great Lakes Research* 4: 320–30.

– 1980. "How Great Lakes Water Moves." In *Decisions for the Great Lakes, Great Lakes Tomorrow*, Purdue University Press.

– 1987. "Heat Balance and Heat Storage Estimates for Lake Erie, 1967 to 1982." *Journal of Great Lakes Research* 13: 51–64.

Schertzer, W.M., Elder, F.C., and Jerome, J. 1978. "Water Transparency of Lake Superior in 1973." *Journal of Great Lakes Research* 4: 350–8.

Schertzer, W.M., and Sawchuk, A.M. 1990. "Thermal Structure of the Lower Great Lakes in a Warm Year: Implications for the Occurrence of Hypolimnion Anoxia." *Transactions American Fish Society* 119: 195–209.

Schertzer, W.M., Saylor, J.H., Boyce, F.M., Robertson, D.G., and Rosa, F. 1987. "Seasonal Thermal Cycle of Lake Erie." *Journal of Great Lakes Research* 13: 468–86.

Shimizu, R., and Finch, C. 1988. *Upper Great Lakes Connecting Channels Study.* 2 vols. Ottawa: Environment Canada.

Simons, T.J., and Schertzer, W.M. 1987. "Stratification, Currents and Upwelling in Lake Ontario, Summer 1982." *Canadian Journal of Fisheries and Aquatic Science* 44: 2047–58.

CHAPTER SEVEN

Wetlands

NIGEL T. ROULET, D. SCOTT MUNRO,
AND LINDA MORTSCH

INTRODUCTION

Wetland ecosystems can be found over the entire land surface of the earth in a wide range of climatic settings from the tropics to the polar regions. Wetlands are estimated to cover 5.3 million km^2 (Matthews and Fung, 1987), or a little less than 4 per cent of the earth's total land area. Over 50 per cent of all wetlands occur in the northern hemisphere, between 45°N and 70°N, in the latitudes that correspond to the cool temperate, boreal, and subarctic ecotones. At these latitudes, wetlands can cover 25 per cent of the landscape, and in Canada and Russia continuous wetlands cover 100 per cent of the land in some regions and are in excess of 300,000 km^2 in area.

Wetlands are now highly valued ecosytems, but this has not always been the case. Historically, wetlands were looked on very negatively. Colonel William Byrd, III (1674–1744), as quoted in Mitsch and Gosselink (1993: 8), described the Great Dismal Swamp of North Carolina as "A horrible desert, the foul damps ascend without ceasing, corrupting the air and render it unfit for respiration – Never was Rum, that cordial of Life, found more necessary than in this dirty place." Wetlands are now recognized as important. In some settings they can significantly reduce floods by temporarily detaining water. They can assist in short-term removal of waterborne nutrients, metals, and sediments from streams and rivers. In many northern countries, they supply peat for agriculture and fuel. In Finland, over 75 per cent of the commerical forests are situated on wetlands. The most notable function of wetlands is their value as habitats for water-fowl and wildlife. Prairie potholes, or sloughs, account for over half of the production of North American water-fowl (National Wetlands Working Group, 1988).

In southern Canada and much of the United States, many of the original wetlands have been drained. The wetland loss near urban and agricultural areas of Canada ranges from 10 per cent to over 90 per cent. To conserve the remaining wetlands in southern Canada and to prevent destruction of the largely unaffected northern wetlands, the federal and provincial governments have developed legislation and policies for wetland evaluation, conservation, and management (Lynch-Stewart et al., 1993). Wetlands deemed socially, biologically, or hydrologically significant are now protected in some areas of Canada.

The most prominant feature of wetlands, which distinguishes them from other terrestrial ecosystems, is the presence of water at or near the ground surface. The availability of water so close to the surface has a profound effect on many biological, biogeochemical, hydrological, and climatological processes. Wetlands are defined as "land that has the water table at, near, or above the land surface or which is saturated for a long enough period to promote wetland or aquatic processes as indicated by hydric soils, hydrophytic vegetation, and various kinds of biological activity that are adapted to a wet environment" (Tarnocai, 1980).

In water-logged soils, downward diffusion of oxygen is very limited, and subsurface anaerobic conditions usually develop. This leads to incomplete decomposition of plant material and slow accumulation of partially decomposed plants, which develops into a peat layer on the land surface. Peatlands are wetlands that have peat thicker than 0.4 m (National Wetlands Working Group, 1988).

It is misleading to think of wetlands as a single type of ecosystem. With the exception of the near-continuous presence of water, structure and function are as varied as those of non-wetland ecosystems, such as grasslands and forests. Variations in the structure of wetlands produce differences in the absorption, reflectance, and emittance of radiant energy, in transfer of heat and vapour, and in exchange of certain greenhouse gases, particulary carbon dioxide (CO_2) and methane (CH_4). Three factors – vegetation, soil, and water – combine to influence the surface climate of wetlands. Vegetation controls the radiative properties of the surface and exchange of water through the soil–plant–atmosphere continuum. Soils, organic or mineral, control storage and exchange of water, which affect evaporation and the ground thermal regime.

How the surface climate of several types of wetlands in Canada is controlled by differences in their structure and function is the focus of this chapter.

WETLAND DISTRIBUTION AND CHARACTERISTICS

Wetlands cover 1,279,900 km^2, or 14 per cent, of Canada's landscape (National Wetlands Working Group, 1988). Approximately one-half of the world's northern wetlands (> 40°N) are situated in Canada (Gore, 1983). Close to 90 per cent of Canada's wetlands are also peatlands.

Wetlands are found throughout Canada, but they are most prominent in the boreal and low subarctic regions, where the climate and physiographic conditions favour their development (Figure 7.1). In these two regions, wetlands, which are almost exclusively peatlands, cover over half of the land. The Hudson Bay Lowland (Figures 7.2 and 7.3) – the second-largest continuous peatland in the world (320,000 km^2) – alone comprises over 25 per cent of all of Canada's wetlands. Other large, continuous peatlands occur north of Lakes Winnipeg and Manitoba and in northern Alberta.

The wetlands of Canada are divided into one of five classes, based on vegetation, water quality, and hydrological criteria (National Wetlands Working Group, 1988) – shallow open water, marsh, swamp, fen, and bog (see Table 7.1). Minerotrophic wetlands receive water from precipitation and mineral-rich groundwater, whereas ombrotrophic (mineral-poor) wetlands receive water only from precipitation. The terms "eutrophic," "mesotrophic," and "oligotrophic" describe the productivity of

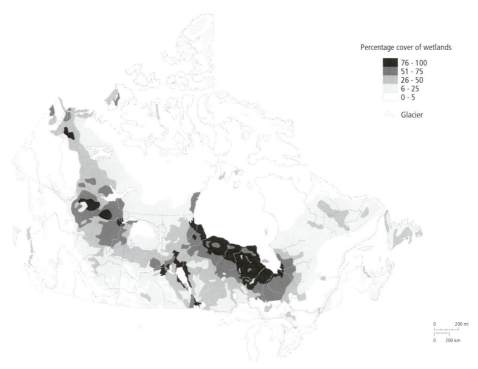

Figure 7.1
Percentage cover of wetlands in Canada.
Source: National Wetland Working Group (1988).

various wetlands. The most productive, nutrient-rich, high-pH (alkaline) systems are called eutrophic, while the oligotrophic systems are the least productive.

Shallow-water wetlands feature large, stable expanses of open, standing, or flowing water, which is semi-permanent to permanent and has midsummer water depths of less than 2 m. The surface climate of these wetlands is similar to that of lakes, discussed above in chapter 6.

Marshes are nutrient-rich, non-acidic wet areas that are periodically inundated with standing or slowly moving water. They are characterized by pools or channels, interspersed with clumps of emergent vegetation – sedges, grasses, rushes, and reeds. Grassy meadows and bands of shrubs or trees border these areas, while submerged and floating aquatic plants occur in open-water areas (National Wetlands Working Group, 1988).

Swamps are wooded wetlands where the water table is at or slightly below the soil surface throughout much of the year. The waters range from neutral to moderately acidic and nutrient-rich. Vegetation is characterized by a dense cover of deciduous or coniferous trees, tall shrubs, herbs, and some mosses. Peat accumulation is found in many swamps in Canada. Trees that characterize swamps include black ash, eastern white cedar, and red maple (National Wetlands Working Group, 1988).

Fens and bogs are the most abundant type of wetland in Canada. They are also exclusively peatlands. Fens are generally more nutrient-rich and have near-neutral

Figure 7.2
Mineral-poor, open fens with raised-bog, treed islands in the Hudson Bay Lowland. The open fen is surrounded by treed fen (tamarack).

Figure 7.3
An ombrotrophic, raised, low-shrub bog, interior of the Hudson Bay Lowland, 100 km from James Bay. The dominant surface vegetation comprises lichen and sphagnum moss.

acidity, since they are supplied with groundwater, even though they have developed in areas of restricted drainage. Fens are dominated by sedges, with reeds and grasses occurring in associated pools. Mosses and wildflowers are also important. Common tree species in fens include eastern white cedar and tamarack.

Table 7.1
Attributes of the major wetland classes, as defined by the Canadian Wetland Classification System

Wetland class	Dominant type of vegetation	Peat depth (m)	NPP	Hydrology	Nutrient status	Acidity
Shallow, open water	Aquatic	0	Low to high	$GW + P$	Oligotrophic to eutrophic	Alkaline to acidic
Marsh	Graminoids: Sedges Reeds Grasses Rushes Aquatic Trees Shrubs	0–0.4 (mineral to peat)	High	$GW + P$	Mesotrophic to very olig- otrophic	Alkaline to neutral
Swamp	Trees: Conifers Deciduous Shrubs: tall	0–2.0 (mineral to peat)	High to moderate	$GW + P$	Mesotrophic	Neutral to acidic
Fen	Graminoids: Sedges Shrubs: Low Mosses: Brown Trees: Conifers	> 0.40 (peatland)	Low to high	$GW + P$	Oligotrophic to mesotrophic	Alkaline to acidic
Bog	Mosses: *Sphagnum* Shrubs: *Ericaceous* Graminoids: Sedges Trees: Conifers	> 0.40 (peatland)	Low to moderate	P	Oligotrophic	Acidic

Note: Wetlands can receive water input from precipitation (P) and sometimes groundwater (GW). This factor affects nutrient status, which ranges from oligotrophic (lack of nutrients) to eutrophic (no nutrient limitations for plant growth). Productivity is a relative measure: "high" indicates high net annual plant production (NPP), and "low," low NPP.

The surfaces of bogs are virtually unaffected by nutrient-rich groundwater and drain-age from surrounding mineral soils because of the accumulation of peat. As a result, their waters are nutrient-poor and very acidic. Bogs are dominated by lower plant forms such as bryophytes, especially *Sphagnum* moss, and by shrubs of the Ericaceae family (National Wetlands Working Group, 1988). At later stages of succession, trees such as black spruce are also common (Figure 7.4).

Three wetland types – swamps, fens, and bogs – have peat substrates. Peat has several unique characteristics because of the way it is formed. Unlike the pedogene-sis of mineral soils, the structure of peat is largely controlled by the composition and

Figure 7.4
A treed bog in the Walley Creek Experimental
Drainage Area near Cochrane, Ontario. The trees
are black spruce, while the understorey com-
prises sphagnum moss and shrubs such as
Chamaedaphne calyculata and *Myrica gale.*

structure of the plants that inhabit the wetland surface. Since the peat is formed
from the remains of dead plants, it is extremely difficult to determine the point
where the vegetation layer stops and peat soil starts. Hence the vertical structure of
peatlands is characterized by functional, rather than physical, units. In peatlands,
density and hydraulic conductivity range between 2 and 10 in the upper 0.2 m of a
peat profile, and this has a significant effect on the surface exchange of energy and
water. The top 0.1 m of peat comprises recently dead plant material (< 25 years),
while the base of the profile may have undergone very slow decomposition for
several thousand years. This profile results in a very hydrologically active upper
0.25 m, called the acrotelm, and a deeper, inactive layer, called the catotelm
(Ingram, 1978). Another consequence of the accumulation of peat is that the surface
elevation of the peatland is increasing, albeit very slowly (≈ 2 to 20 mm y^{-1}), until
the annual rate of decomposition through the entire peat profile equals the amount
of carbon sequestered annually by the living biomass. Clymo (1991) shows that this
limit to peatland growth is several tens of thousands of years – far more than the
time required for changes in climate to influence the water balance of peatlands.

RADIATION BALANCE

The radiation balance of the wetland surface depends on the magnitudes of its components, which vary in time according to the same controls that apply to any other

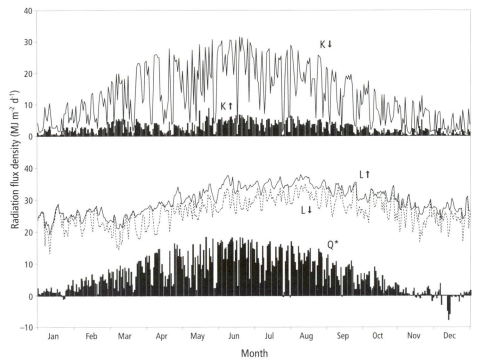

Figure 7.5
Daily total energy flux density for a treed swamp in southern Ontario, net radiation and its components: $K\downarrow$, $K\uparrow$, $L\downarrow$, (Q^*), and $L\uparrow$.
Source: Munro.

type of surface. Large variations in total radiation occur from one day to the next, as well as seasonally (Figure 7.5). This pattern is not unique to wetlands. During winter, wetlands tend to be covered by snow and ice and thus differ little from drier areas in the same location. Summer often brings a thick canopy of vegetation, which, though better supplied with water than drier areas, is much like other vegetation covers. It is only during the transition seasons of spring and autumn that the darkness of the peat substrate and wetland surface is exposed. The surface is often covered with a veneer of water, which further darkens the surface.

Examination of field data reveals that, though contrasts between wetlands and other surfaces may be small, temporal changes in their radiation balance do make wetland surfaces distinctive.

Solar Radiation

Global solar radiation – the primary energy supplier to the environment – is governed by season, location, and weather (chapter 2). Season determines the maximum possible daily radiation, which is greatest at the summer solstice, least at the winter solstice. Location establishes the difference between seasonal extremes, and this difference increases northward, where summer daylight periods and winter nights become longer. Such maxima are achieved only during cloudless weather. When clouds obscure the sun, global radiation drops to as little as 15 per cent of the cloudless maximum.

The difference between the streams of global and reflected solar radiation determine the net solar radiation absorbed at a surface (Figure 7.5). In a southern swamp, the difference between $K\downarrow$ and $K\uparrow$ is small when clouds reduce $K\downarrow$, and large under cloudless conditions. However, there is a considerable range in solar reflectance, from a high of 0.4 on days with snowcover to below 0.1 on snow-free winter days. Nevertheless, the swamp acts as a very effective solar energy collector throughout the year. It would be even more effective throughout summer if the leaf canopy were absent, but removal of the trees would result in much higher reflectances during winter, a situation found in northern wetlands with shorter vegetation.

Water status controls the solar radiation regime, because flooding darkens the surface and reduces its reflectance. This has been demonstrated for subarctic bogs by Petzold and Rencz (1975), who report a reflectance of 0.18 for the raised edge of a string bog and 0.13 for the depressed interior, where 42 per cent of the surface was covered with water. They found a similar, low reflectance of 0.11 for a sedge-moss bog with standing water. Such low reflectances imply that wetlands are relatively high-energy environments. Since there usually is an excess of water at or near the surface of a wetland, it might be expected that it would possess a much lower reflectance than a non-wetland surface. However, it is only in the period between full snowcover and full leaf cover that this is true. Lafleur, Rouse, and Hardill (1987) show that wetlands have markedly different reflectances prior to leaf emergence but that these differences dimish once a full leaf canopy establishes itself (Table 7.2), when reflectances reach values expected for a continuous forest cover. A similar transition is seen in reflectances extracted from the data collected for a temperate swamp (Table 7.2).

Longwave Radiation

Most natural surfaces have an emissivity close to unity, especially when either snow or water covers them. Therefore the longwave radiation balance (L^*) is controlled principally by temperature and atmospheric emissivity. Atmospheric emissivity is about 0.75 under cloudless conditions and close to 1.0 when the sky is overcast. The annual trends in $L\downarrow$ and $L\uparrow$ respond to seasonal temperature, with daily variations superimposed. $L\downarrow$ approaches $L\uparrow$ when the sky is cloudy (Figure 7.5).

Outgoing longwave radiation persistently exceeds that delivered from the sky, resulting in a net longwave radiation loss, which varies little among wetlands (Table 7.2). Differences are more likely the result of differences in cloud cover between measurement periods than of contrasting surface conditions. This is also the case for non-wetland surfaces.

In contrast to the solar radiation regime, the seasonal maxima and minima of the longwave radiation components in the southern Ontario treed swamp are characterized by a much smaller range and are lagged (Figure 7.5). Longwave radiation increases from low winter values in late March to a maximum around mid-August. The peak in longwave radiation occurs well after the late-June peak in global radiation. From

Table 7.2
Components of the surface radiation balance for various wetlands

Author(s)	Location	Period	Wetland type	Q^*	α	L^*	$Q^*/K\downarrow$
Lafleur, Rouse, and Hardill (1987)	Southern Hudson Bay Lowland (51°10′N)	Snow-free	Sedge fen:				
			Pre-growing	11.8	0.11	−3.93	0.66
			Growing	4.0	0.14	−4.08	0.61
			Post-growing	10.6	0.19	−3.94	0.56
Rouse, Mills, and Stewart (1977)	Hudson Bay Coast (57°45′N)	Summer	Sedge fen		0.11		0.70
Rouse and Bello (1983)	Northern Hudson Bay Lowland (58°45′N)	Summer	Peat fen	11.1	0.15		0.59
				4.0	0.12		0.62
Petzold and Rencz (1975)	Northern Quebec (54°43′N)	Midsummer	String bog: ridge		0.18		
			String bog: valley		0.13		
			Moss bog		0.11		
			Sedge fen		0.18		
Berglund and Mace (1976)	Minnesota (47°31′N)	Snow-free	Black spruce		0.06		
			Bog		0.07		
			Sphagnum		0.12		
			Sedge bog		0.18		

Note: Units of Q^* and L^* are MJ m^{-2} d^{-1}; α and $Q^*/K\downarrow$ are dimensionless.

August onward, longwave radiation decreases until November. There is marked day-to-day variation, but there is little trend from November to March.

Net Radiation

Net radiation shows a marked seasonal range and considerable day-to-day variation. However, the magnitude of the flux is well below that for global radiation, despite the small reflectance of wetland surfaces. The seasonal range is dominated by the global radiation input. It is the continuous loss of energy through L^* that accounts for most of the reduction. Because net longwave radiation changes little with season, its relative importance to wetland radiation balance is greatest during winter, when Q^* can take on negative values. This is more likely to be the case the further north the wetland is.

Given the small variability in L^* found in Table 7.2, and the dependence of solar radiation on latitude and weather, it is solar reflectance that determines the efficiency of the wetland in capturing radiation. It is convenient to express this variable in terms of radiative efficiency, $Q^*/K\downarrow$ (Lafleur, Rouse, and Hardill, 1987). Radiative efficiency is closely related to surface albedo, such that the greater the albedo, the lower the efficiency (Table 7.2).

The dependence of radiative efficiency on albedo also illustrates the dynamic nature of the surface's capacity to capture solar energy. Prior to leaf emergence, a vegetated wetland is likely to exhibit the lowest reflectances of the year and the greatest radiative efficiency in excess of 0.6 (Table 7.2). Leaf growth increases the albedo by a factor of

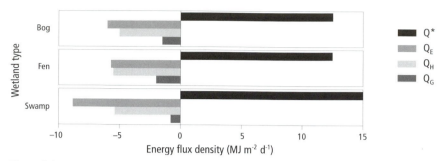

Figure 7.6
Summer surface energy balance for three types of wetland: raised shrub bog, interior Hudson Bay Lowland; sedge fen, Hudson Bay Lowland; and forested swamp, southern Ontario.
Sources: (Bog) den Hartog et al. (1994); (fen) Rouse and Bello (1985); (swamp) Munro (1979).

nearly two in some instances and consequently reduces the proportion of global radiation expressed as net radiation. Thus vegetated wetlands, in which leaf emergence is part of the seasonal pattern, are best able to absorb energy when plants require it most, at the start of the summer growth period.

The end result of leaf development is that wetlands have a radiative response more similar to that of non-wetland, continuous vegetation covers, such as grass or forest, than to that of open water.

ENERGY BALANCE

The components making up the surface energy balance of wetlands are very similar to those for land surfaces (chapter 2) – i.e., $Q^* = Q_H + Q_E + Q_G$. Heat transfers resulting from precipitation and photosynthesis are not documented for wetlands, but, according to limited measurements they should not be significant components of the energy balance. Freezing and thawing involve significant amounts of energy at certain times of the year, but it is best to consider them as special categories of heat storage in the ground.

The presence of so much water suggests an environment in which latent heat transfer dominates the energy balance. Heat storage might also play a prominent role because of the large heat capacity of water. That is the case for lakes (chapter 6), but wetlands respond to Q^* like other land surfaces (Figure 7.6). Bog, fen, and swamp each show a balance, with Q_E using the greatest proportion of Q^* during summer. However, Q_H is not much smaller and becomes the larger of the two turbulent terms for treed wetlands, when the trees lose their leaves in September. In all cases, Q_G is relatively small.

Despite the soil's high water content, which suggests greater capacity for surface heat storage, the atmosphere makes the strongest demand on the energy supply at the surface. Bowen ratios slightly less than 1.0 are common, but β does exceed unity during parts of the day or over periods of days (Lafleur, Rouse, and Hardill, 1987; Price, 1991). Field investigations show that though Q_G can be substantial in some cases, it rarely amounts to more than 15 per cent of Q^*.

Table 7.3
Thermal properties of peat soils

Soil type	Porosity	Volumetric moisture content (%)	Density (kg m⁻³)	Heat capacity $(J\,m^{-3}\,K^{-1} \times 10^6)$	Thermal conductivity $(W\,m^{-1}\,K^{-1})$
Peat	0.8	Dry	300	1.46	0.06
		40	700	3.14	0.29
		80	1,100	4.81	0.50
		Dry	400	0.85	0.17
		Dry	500	0.84	0.06
		Saturated	1,000	4.15	0.54
Frozen peat	–	Saturated	900	1.50	1.88
Sand	0.4	Saturated	2,000	2.96	2.20
Ice	–	–	917	1.93	2.24
Still air	–	–	1.2	0.0012	0.025

Sources: Data from Farouki (1981); Lunardini (1981); Oke (1987).

Ground Heat Storage

Heat storage in a forested swamp in southern Ontario amounted to approximately 7 per cent of the dispensation of net radiation (Munro, 1979). This figure is small in comparison to ground heat fluxes in open bogs and fens. Larger fluxes of ground heat might be expected in most wetlands because the soil has high water content, regardless of whether it is in a liquid or frozen state. But the thermal conductivity of saturated peat is only half that of wet mineral soil, and the heat capacity is twice as great (Table 7.3). Therefore peat takes considerable time to warm up and cool down, a feature that makes wetlands good thermal reservoirs, but not quite as good as lakes.

The thermal conductivity of peat drops by 50 per cent when it becomes drier, thereby inhibiting transfer of heat (Table 7.3). Furthermore, in areas where the peat freezes in winter, the latent heat of fusion is a major component of Q_G. A large heat deficit due to the soil's large ice content (> 80 per cent) must be satisfied before the soil temperature can rise above 0°C. Consequently, wetlands thaw and freeze later, and are cooler in summer and warmer in winter, than mineral soils in the same climate region. Where winter flooding protects the soil itself from freezing, the same effects occur as a consequence of ice formation and break-up. In the region of continuous permafrost, the active layer is shallower in wetlands. Relic permafrost is often found preserved in peatlands far south of the region of discontinuous permafrost (Brown, 1973).

Soil heat storage may become relatively small during summer, when vegetation shields the soil from solar input during the day and reduces $L\uparrow$ at night. Summer is most likely to be the period when the wetland surface is driest. Even if the water table is not far below the wetland surface, the unsaturated peat and surface vegetation layer (for example, mosses and in some cases lichens) can be sufficiently dry that they act as a very effective insulator and thus reduce heat flow from the surface. Experience shows that bare peat exposed to direct sunlight is painfully hot to touch, despite the presence of the water table only a few millimetres below. In forested wetlands, over half the heat storage can occur within the biomass and the canopy air space (Munro, 1979).

It is instructive to look at the components of daytime heat storage for a forested swamp in greater detail. Heat is stored in the forest canopy air space as well as in the vegetation and soil. Storage in the air space results in changes to both temperature and humidity, and the amounts of energy used to effect each change are approximately the same proportions as Q_H to Q_E. Since the heat does not readily escape to the atmosphere above the trees, it is released from storage by surface cooling during the night. Non-forested wetlands have a much lower standing biomass, and consequently Q_G is almost exclusively heat storage in the soil alone. However, they tend to store more energy as Q_G than do forested wetlands, because of the direct absorption of $K\downarrow$.

Sensible and Latent Heat

The greater the surface roughness length (z_o), the stronger the link between the surface and the air through turbulent exchanges. Wetlands vary greatly in their ability to mix the air above their surface, in response to their surface roughness. Generally, the taller the vegetation, the greater the roughness length. However, heterogeneity in the vegetation structure can increase the roughness length to a value greater than expected for uniformly tall vegetation. Munro (1987) found a z_o of approximately 2 m for a canopy with an average vegetation height of 2 m. This figure is well in excess of the 0.2 m expected for a homogeneous canopy of similar height. The extra roughness was attributed to the portion of the tree cover that reaches to heights that are four to five times the average, and it was the taller individuals that established the roughness regime. Such heterogeneity, and the consequent effects on aerodynamic mixing, are expected for shorter vegetation as well, because uniform plant cover is rare in the wetland environment.

Regardless of vegetation height, turbulent transfer is the dominant mode of heat exchange throughout the daylight hours, far exceeding the soil's heat storage (Figure 7.7). At night, the energy-balance components are comparable in size, but fluxes are also very small. Hence it is the daylight hours that determine how the total transfer of sensible and latent heat varies from one day to another. The example of the blanket bog shows an unexpected feature of the wetland environment; Q_H exceeds Q_E throughout most of the day. However, the relative importance of the two turbulent transfer terms depends very much on local weather conditions. On two other days, when fog occurred, Price (1991) found $Q_E > Q_H$ for much of the day. Measurements taken in a forested swamp (Munro, 1979) and in fens and marshes in the Hudson Bay Lowland (Lafleur and Rouse, 1988) show just how variable the daily turbulent exchanges are. Bowen ratios range from 0.4 to greater than 1.0. Thus, like their radiative environment, wetlands have a pattern of turbulent transfer that is closer to that of other vegetated land than to that of open water.

The generally conservative nature of energy partitioning in wetlands is shown by comparing the hourly Bowen ratio trends for blanket bog, forested swamp, and marsh (Figure 7.8). In all but one case – the Hudson Bay Lowland marsh – Bowen ratio

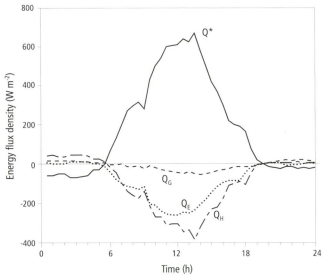

Figure 7.7
Energy balance on a blanket bog in Newfoundland during cloudless conditions.
Source: Price (1991).

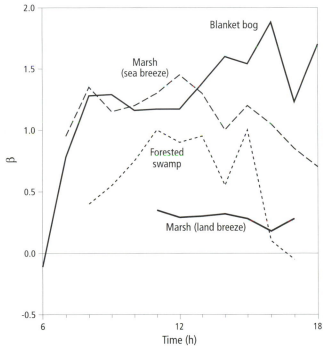

Figure 7.8
Hourly Bowen ratio (β) values (dimensionless) in three wetlands for selected days.
Sources: Lafleur and Rouse (1988); Munro (1979); Price (1991).

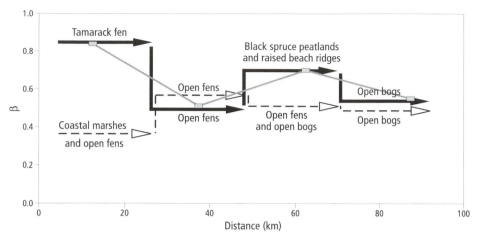

Figure 7.9

Partitioning of turbulent transfer between the fluxes of sensible and latent heat, as shown by average Bowen ratio (β) for 25-km segments of two 100-km transects (solid and dashed arrows) across a portion of the southern Hudson Bay Lowland. Poor fens are dominated by sedges and open bogs by *Sphagnum* moss and lichens; beach ridges are forested. Coastal marshes are salt and freshwater types. Data on flux density from which β was calculated were measured by airborne eddy correlation at approximately 50 m above the wetland surface.

Source: Desjardins et al. (1994).

values lie between 0.5 and 1.5 throughout the middle of the day. The β value of the Hudson Bay Lowland marsh is depressed because local advection of dry air creates a greater capacity for evaporation (Lafleur and Rouse, 1988). In the latter case, it appears that air mass conditions have a greater influence on energy partitioning than for the other wetland surfaces. To the extent that surface conditions help explain the Bowen ratio, they may account for conditions early in the day, when β tends to be low. Such values may arise from the need to evaporate dew before any significant warming of the air can begin. Also, β may increase slightly over the day if evaporation is restricted because of a lack of water. Caution is advised in the interpretation of early-morning and late-afternoon Bowen ratios, however, because the turbulent transfer terms tend to be small at such times and prone to error.

For most wetlands the largest demand on the $Q*$ energy supply is made by the latent heat flux. Hence Bowen ratios are less than unity, but greater than the usual values for lakes. Values greater than 0.20 to 0.25 are observed most of the time over a wide variety of wetland types. For example, β varied between 0.4 and 0.85 along a transect across the southern portion of the Hudson Bay Lowland (Figure 7.9). The energy partitioning across this mosaic of wetlands is much more complex than that of a single, isolated wetland. Direct measurements of sensible and latent heat flux from a low-flying aircraft (Desjardins et al., 1992) confirm the dominance of Q_E over Q_H. Also, as vegetation changed, so did the ratio; β increased as the elements in the mosaic went from coastal marshes, through various types of open bogs and fens, to treed peatlands and fens (Figure 7.9). Such a progression suggests the need to study plant physiology, canopy architecture, and water-supply as controls on evaporation rates from wetlands.

Table 7.4
Rate of evaporation from various wetlands

Author	Location	Period	Wetland type	E (mm d^{-1})
Koerselman and Beltman (1988)	Netherlands	Late April– early October	Sedge fens	3.6[a]
Roulet and Woo (1986)	Low arctic Canada	Summer	Fen	4.5[a] (2.2–7.3)[b]
Lafleur (1990)	Hudson Bay Lowland	Summer	Wet fen Dry fen	3.1[a] 2.6[a]
Dolan et al. (1984)	Florida	All year	Marsh	1.2–5.6[a]
Munro (1979)	Southern Ontario	Late August– early September	Treed swamp	2.1–4.6[b]
Price (1991)	Newfoundland	May–July	Coastal blanket bog	0.2–3.4[b]
Virta (1996)	Finland	Summer	Treeless fen Open fen Raised bog	3.3–3.5[a] 1.8–2.6[a] 1.5–2.1[a]

[a] Mean daily values.

[b] Range in daily rates.

CONTROLS ON WETLAND EVAPORATION

The evapotranspiration rate – the sum of evaporation and transpiration – is highly variable among wetlands, depending on wetland type and location (Table 7.4). Fens appear to have the maximum daily evaporative loss of water, followed by treed swamps, and then bogs.

Differences in evaporation rates can be explained by the factors that control the exchange of vapour from the soil, through the plants and soil surface, to the atmosphere. The three controls on evaporation from wetlands are the amount of energy available at the surface ($Q^* - Q_G$), the ability of the atmosphere to hold and transport vapour (vpd, K_V, r_a, and z_o), and the ability of surface vegetation and soils to conduct water through to the atmosphere (r_c). The importance of these factors varies among wetland types.

The atmospheric controls are not a direct consequence of the wetland itself, but they help determine the partitioning of energy and therefore ultimately control the maximum possible rate of evapotranspiration – potential evapotranspiration. The influence of the atmosphere is clearly shown by differences in the proportion of energy going to evaporation from coastal wetlands as the direction of sea breezes changes (Rouse, Hardill, and Lafleur, 1987). When cool, moist air from Hudson and James bays is advected over coastal fens and marshes, the ratio Q_E/Q^* is lower than when warmer, drier air comes from the south (Figure 7.10). This change is due to differences in the heat content and vapour pressure deficit (vpd) of the air carried by onshore and offshore winds. Evaporation rates are suppressed from coastal blanket bogs during fog events because the atmospheric humidity approaches saturation

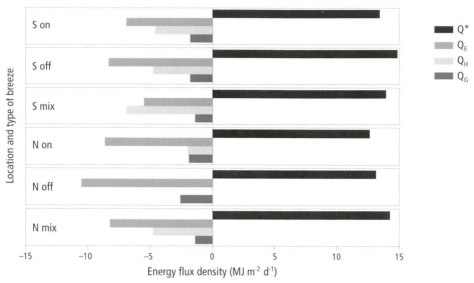

Figure 7.10
Average surface energy balance for sedge fens in the southern (S) and northern (N) regions of the Hudson Bay Lowland, under the influence of onshore (on), offshore (off), and mixed (mix) sea breezes.
Source: Rouse, Hardill, and Lafleur (1987).

(Price, 1991). The influence of sea breeze advection is not necessarily isolated to a narrow coastal region. In the Hudson Bay Lowland, sea breezes can penetrate as far as 120 km inland (McKendry and Roulet, 1994).

The combination model of evaporation (equation 2.18) provides a way to conceptualize how the atmosphere and the surface of a wetland interact to control evaporation. The model's second term balances the demand for vapour by the atmosphere against a series of resistances – to transport of water from the soil, through the plants, to the atmosphere above the wetland. The atmosphere provides the aerodynamic resistance that is intimately connected with the architecture of the vegetation canopy. Thus the tall vegetation canopies of wooded swamps create a rough surface, which reduces aerodynamic resistance through strong turbulent mixing, to produce an r_a of between 5 to 10 s m^{-1} (Munro, 1987). The short vegetation cover of bogs and fens is smoother, creating less turbulence and, consequently, a larger r_a, of between 20 and 35 s m^{-1} (Lafleur and Rouse, 1988; Price, 1991).

The surface resistance characterizes the control that plant physiology and soil moisture place on evaporation, although the effect of the latter is normally small in wetlands. It treats the entire wetland surface as one large leaf. Physiological control of transpiration is expressed through stomatal resistance, which characterizes the control of the exchange of vapour from individual leaves to the atmosphere. There is a very close connection between stomatal (r_s) and canopy resistance (r_c), as demonstrated by the similarity of their diurnal patterns (Figure 7.11). Both are lowest in the morning, when moisture reserves are at their maximum. They increase in the afternoon, when atmospheric demand for moisture tends to be much larger. Stomatal resistance values are larger than those for canopy resistance. Stomata provide many parallel paths for

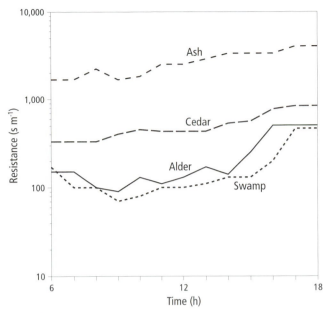

Figure 7.11
Stomatal resistances (r_s) for three tree species and bulk canopy resistance
(r_c) for swamp in Southern Ontario.
Source: Munro (1987).

transfer of water vapour. The leaves that constitute a canopy transport more water in combination than would be suggested by the water vapour transport rates of individual leaves. Species, relative numbers, and leaf area coverage are therefore crucial in determining the surface resistance of a given wetland. Because these characteristics vary so much across the wetland mosaic, it is difficult to generalize according to wetland type.

The difficulty in generalizing can be seen by comparing surface resistance patterns for a bog, a fen, a marsh, and a swamp (Figure 7.12). Wooded swamps exhibit a pattern much like that of forests in general. The wooded marsh has much lower resistance, perhaps because it contains mostly alder species, which have among the lowest stomatal resistances of any wood species. The bog, fen, and marsh are similar. Weather conditions also play a role. Price (1991) demonstrated r_c to be close to 0 for a fog-covered Newfoundland bog. Many more field studies will be required before we will have a climatology of surface resistance for wetlands.

Unlike trees and graminoids (such as sedges and grasses), the moss and lichen cover of bogs lacks an internal vascular system and is thus unable to transpire water in the conventional way. Water is conducted over the plant surface by capillarity and within the *Sphagnum* mosses through internal hyaline cells (Hayward and Clymo, 1982). This form of conductance can be effective only if there is a source of water in immediate contact with the plant. The structure of the surface layers of peat, however, is not conducive to this type of transport unless the water table is very close to the sur-

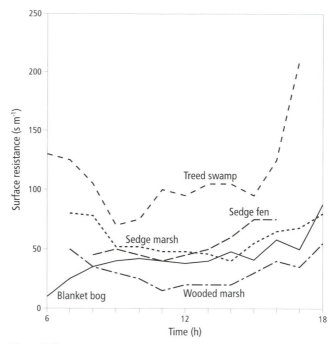

Figure 7.12
Diurnal pattern of surface (bulk canopy) resistance (r_c) for a wooded
marsh, a sedge fen, a sedge marsh, a treed swamp, and a blanket bog.
Sources: Lafleur and Rouse (1988); Munro (1979); Price (1991).

face. Relatively large pores that are ineffective at exerting significant capillary ten-
sions occur in the top 0.15 m of peat in most bogs (i.e., the acrotelm). As a result,
evaporation from bogs decreases significantly once the water table is 0.10 to 0.15 m
below the peat surface (Verry, 1988).

The supply of water to the surface of a peatland is further complicated by its being
a compressible, and sometimes buoyant, soil. The surface of some bogs and fens
moves up and down, leading to the label "quaking" bogs or fens. In some wetlands,
as the water table drops, so does the surface of the wetland. This maintains a high
water content in the surface layers of the peatland, even during periods of little rain
(Figure 7.13). Fens with an adjusting, or a quaking, surface have relatively constant
r_c. The evaporation rate is directly proportional to the energy available at the surface
and atmospheric demand for vapour, while water loss from non-adjusting fen surfaces is controlled by a variable r_c (Lafleur and Roulet, 1992).

WETLANDS AND CHANGING CLIMATES

Wetlands, particularly peatlands, participate in the exchange of radiatively active greenhouse gases. Peatlands are a large reservoir of the world's soil carbon (Figure 7.14)
and consume approximately 28 g C m^{-2} y^{-1} of atmospheric CO_2 (Gorham, 1991). Wetlands also contribute approximately 20 per cent of the world's annual input of methane

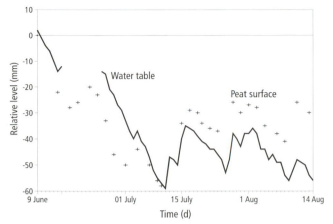

Figure 7.13
Elevation of the surface and water table in a subarctic fen in northern
Quebec. The surface of the peatland drops over 60 mm, corresponding to
loss of water by drainage and evaporation.
Source: Roulet (1991).

(CH$_4$) to the atmosphere (Fung et al., 1991). Global tropospheric concentrations of CO$_2$
and CH$_4$ are currently increasing at a rate of 0.3 and 0.8% y^{-1}. Since the beginning of
the sixteenth century, CO$_2$ and CH$_4$ have increased 20 per cent and 100 per cent, respectively, to their present concentrations of approximately 360 and 1.76 ppm(v), respectively (Intergovernmental Panel on Climate Change, 1990). Even though the
concentration of CH$_4$ is much lower than that of CO$_2$, as a greenhouse gas it is twenty-
one times more radiatively efficient mole per mole (fifty-seven times more efficient by
mass) than CO$_2$.

The role of peatlands in the global carbon cycle and production of CH$_4$ stems
from their nearly constant saturated conditions. Peatlands are not the most productive
ecosystems in the world. The net primary productivity (NPP) for northern peatland
ecosystems is approximately 360 g C m^{-2} y^{-1} (Bartsch and Moore, 1985), compared to
500–900 g C m^{-2} y^{-1} for temperate-to-tropical forests (Schlesinger, 1991). However,
because of the anaerobic conditions (lack of oxygen) of peatlands, decomposition is
inhibited and between 5 and 10 per cent of the carbon assimilated by plants each year
is buried in the form of peat. The rate of carbon removal varies greatly by type of wetland, age of wetland, and climate region, but the presence of peat is a clear indication
that the wetland did, and still may, act as a sink for atmospheric CO$_2$.

The rate of CO$_2$ sink and CH$_4$ source in wetlands depends on the level of saturation
of the wetland substrate. General circulation models (GCMs) run for 2 × CO$_2$ scenarios predict an increase in temperature and soil moisture deficit for the regions that
contain most peat-forming wetlands (Intergovernmental Panel on Climate Change,
1990). If the northern climate were to become warmer and drier, it is not unreasonable
to expect that the sink of carbon in northern wetlands could be reduced. If extreme
drying occurs, northern wetlands could become a net source of CO$_2$ to the atmosphere. At least 2 per cent of the increase in atmospheric CO$_2$ over the last one hundred years is attributable to oxidation of peat from drained wetlands (Armentano and

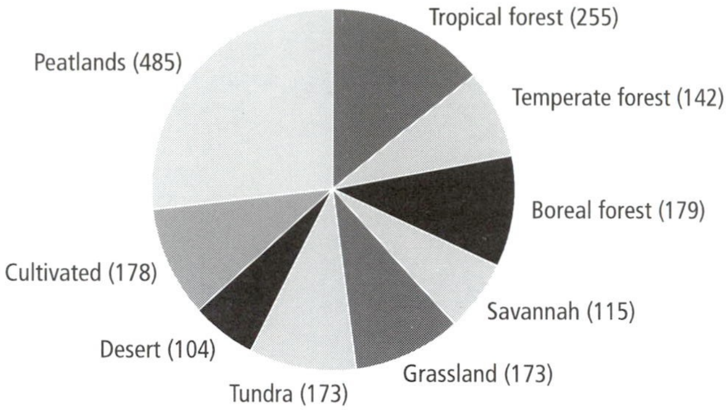

Figure 7.14
Storage of carbon in the world's soils, in gigatonnes ($1 \text{ Gt} = 10^{15}$ g).
Sources: Data for peatlands from Gorham (1991) and for all other ecosystems from Schlesinger (1991).

Menges, 1986). CH_4 fluxes from wetlands are correlated with the mean water-table depth and peat temperature: a warm, saturated wetland emits more CH_4 than a drier, or colder, wetland (Roulet, 1991). In a warmer, drier climate, moisture and temperature are forced in opposite directions. Using a simple hydrothermal model for a northern fen, Roulet et al. (1992) show that the flux of CH_4 is more sensitive to a decrease in the moisture content than to an increase in soil temperature.

Much of the research on the surface climate, ecology, and trace-gas exchange of wetland ecosystems has been done on small, relatively isolated wetlands near the populated areas of Canada, the United States, the Nordic countries, and Russia. However, much of the area of northern wetlands consists of a few, very large wetland complexes, such as the Hudson Bay Lowland and the Western Siberian Lowland (approximately 35 per cent of wetlands are north of 45°N). Unfortunately, much of our current knowledge on the climate of wetlands is not directly transferable from small, isolated wetlands to the large northern wetland complexes (Gorham, 1995; Oquist and Svensson, 1995). Early global methane budgets (for example, Matthews and Fung, 1987) estimated that wetlands the size of the Hudson Bay Lowland emitted between 4 and 8 Tg CH_4 m^{-2} yr^{-1}, but recent direct measurements of CH_4 flux from the Hudson Bay Lowland (Roulet et al., 1994) reveal that the flux is ten to fifteen times lower. Relative to knowledge of the surface climates of other, less-extensive landscape units of Canada, such as lakes and agricultural areas, the surface climate of wetlands is very poorly understood.

REFERENCES

Armentano, T.V., and Menges, E.S. 1986. "Patterns of Change in the Carbon Balance of Organic Soils of the Temperate Zone." *Journal of Ecology* 74: 755–74.
Bartsch, I., and Moore, T.R. 1985. "A Preliminary Investigation of Primary Production and Decomposition in Four Peatlands near Schefferville, Quebec." *Canadian Journal of Botany* 63: 1241–8.

Berglund, E.R., and Mace, A.C. 1976. "Diurnal Albedo Variation of Black Spruce and Sphagnum-Sedge Bogs." *Canadian Journal of Forest Science* 6: 247–52.

Brown, R.J.E. 1973. "Influence of Climate and Terrain Factors on Ground Temperature at Three Locations in the Permafrost Region of Canada." *Proceedings of Second International Permafrost Conference,* Washington, DC, 27–34.

Clymo, R.S. 1991. "Peat Growth." In L.C.K. Shame, and E.J. Cushing, eds., *Quaternary Landscapes*, Minneapolis: University of Minnesota Press, 76–112.

den Hartog, G., Neumann, H.H., King, K.M., and Chipanshi, A.C. 1994. "Energy Budget Measurements Using Eddy Correlation and Bowen Ratio Techniques at the Kinosheo Tower Site during the Northern Wetlands Study." *Journal of Geophysical Research* 99(D1), 1539–50.

Desjardins, R.L., MacPherson, J.I., Schuepp, P.H., and Hayhoe, H. 1994. "Airborne Flux Measurements of CO_2 and H_2O over the Hudson Bay Lowland." *Journal of Geophysical Research* 99(D1): 1551–62.

Dolan, T.J., Hermann, A.J., Bayley, S.E., and Zoltek, S.E. 1984. "Evapotranspiration of a Florida, u.s.a., Freshwater Wetland." *Journal of Hydrology* 74: 355–71.

Farouki, O.T. 1981. *Thermal Properties of Soils.* us Army Corp of Engineers, Cold Region Research and Engineering Laboratory Monograph 81–1.

Fung, I., John, J., Lerner, J., Matthews, E., Prather, M., Steele, L.P., and Fraser, P.J. 1991. "Three-Dimensional Model Synthesis of the Global Methane Cycle." *Journal of Geophysical Research* 96(D7): 13033–65.

Gore, A.J.P., ed. 1983. *Ecosystems of the World. 4A Mires: Swamp, Bog, Fen and Moor.* Amsterdam: Elsevier.

Gorham, E. 1991. "The Role of Northern Peatlands in the Carbon Cycle and Their Probable Response to Climate Warming." *Ecological Applications* 1: 182–95.

Hayward, P.M., and Clymo, R.S. 1982. "Profiles of Water Content and Pore Size in Sphagnum and Peat, and Their Relation to Peat Bog Ecology." *Proceedings of the Royal Society of London* 215: 299–325.

Ingram, H.A.P. 1978. "Soil Layers in Mires: Function and Terminology." *Journal of Soil Science* 29: 224–7.

Intergovernmental Panel on Climate Change. 1990. *Climate Change: The IPCC Scientific Assessment.* Ed. J.T.T. Houghton, G.J. Jenkins, and J.J. Ephrams. Cambridge: Cambridge University Press.

Koerselman, W., and Beltman, B. 1988. "Evapotranspiration from Fens in Relation to Penman's Potential Free Water Evaporation (E_o) and Pan Evaporation." *Aquatic Botany* 31: 307–20.

Lafleur, P.M. 1990. "Evapotranspiration from Sedge-Dominated Wetland Surfaces." *Aquatic Botany* 37: 341–53.

Lafleur, P.M., and Roulet, N.T. 1992. "A Comparison of Evaporation Rates for Mineral Poor and Mineral Rich Fens of the Hudson Bay Lowland." *Aquatic Botany* 44: 59–69.

Lafleur, P.M., and Rouse, W.R. 1988. "The Influence of Surface Cover and Climate on Energy Partitioning and Evaporation in a Subarctic Wetland." *Boundary-Layer Meteorology* 44: 327–47.

Lafleur, P.M., Rouse, W.R., and Hardill, S.G. 1987. "Components of the Surface Radiation Balance of Subarctic Wetland Terrain Units during the Snow-free Season." *Arctic and Alpine Research* 19: 53–63.

Lunardini, V.J. 1981. *Heat Tranfers in Cold Climates.* New York: Van Nostrand Reinhold.

Lynch-Stewart, P., Rubec, C.D.A., Cox, K.W., and Patterson, J.H. 1993. *A Coming of Age: Policy for Wetland Conservation in Canada.* North American Wetlands Conservation Council (Canada), Report No. 93–1. Ottawa.

McKendry, I.G., and Roulet, N.T. 1994. "Sea Breezes and Advective Effects in Southwest James Bay." *Journal of Geophysical Research* 99(D1): 1623–34.

Matthews, E., and Fung, I. 1987. "Methane Emissions from Natural Wetlands: Global Distribution, Area, and Environmental Characteristics of Sources." *Global Biogeochemical Cycles* 1: 61–86.

Mitsch, W.J., and Gosselink, J.G. 1993. *Wetlands.* New York: Van Nostrand, Reinhold.

Munro, D.S. 1979. "Daytime Energy Exchange and Evaporation from a Wooded Swamp." *Water Resources Research* 15: 1259–65.

– 1987. "Surface Conductance to Evaporation from a Wooded Swamp." *Agricultural and Forest Meteorology* 41: 249–58.

National Wetlands Working Group. 1988. *Wetlands of Canada.* Ecological Land Classification Series, No. 24. Sustainable Development Branch, Environment Canada, Ottawa and Polyscience Publications Inc., Montreal.

Oke, T.R. 1987. *Boundary Layer Climates.* 2nd edn. London: Routledge.

Oquist, M.G., and Svensson, B.H. 1995. "Non-tidal Wetlands." In R.R. Watson, M.C. Zinyowera, R.H. Moss, and D.J. Dokken, eds., *Climate Change 1995: Impacts, Adaptations and Mitigation of Climate Change – Scientific Technical Analyses.* Cambridge: Cambridge University Press, 215–39.

Petzold, D.E., and Rencz, A.N. 1975. "The Albedo of Selected Subarctic Surfaces." *Arctic and Alpine Research* 7: 393–8.

Price, J.S. 1991. "Evaporation from a Blanket Bog in a Foggy Coastal Environment." *Boundary-Layer Meteorology* 57: 391–406.

Roulet, N.T. 1991. "Surface Level and Water Table Fluctuations in a Subarctic Fen." *Arctic and Alpine Research* 23: 303–10.

Roulet, N.T., Jones, A., Kelly, C.A., Klinger, L.F., Moore, T.R., Protz, R., Ritter, J.A., and Rouse, W.R. 1994. "Role of the Hudson Bay Lowland as a Source of Atmospheric Methane." *Journal of Geophysical Research* 99(D1): 1439–54.

Roulet, N.T., Moore, T., Bubier, J., and Lafleur, P. 1992. "Northern Fens: Methane Flux and Climate Change." *Tellus* 44B: 100–5.

Roulet, N.T., and Woo, M-K. 1986. "Wetland and Lake Evaporation in the Low Arctic." *Arctic and Alpine Research* 18: 195–200.

Rouse, W.R., and Bello, R.L. 1983. "The Radiation Balance of Typical Terrain Units in the Low Arctic." *Annals of the American Association of Geographers* 73: 538–49.

– 1985. "Impact of Hudson Bay on the Energy Balance in the Hudson Bay Lowlands and the Potential for Climate Modification." *Atmosphere-Ocean* 23: 375–92.

Rouse, W.R., Hardill, S.G., and Lafleur, P. 1987. "The Energy Balance in the Coastal Environment of James Bay and Hudson Bay during the Growing Season." *Journal of Climatology* 7: 165–79.

Rouse, W.R., Mills, P.F., and Stewart, R.B. 1977. "Evaporation in High Latitudes." *Water Resources Research* 13: 909–14.

Schlesinger, W.H. 1991. *Biogeochemistry: An Analysis of Global Change.* San Diego: Academic Press.

Tarnocai, C. 1980. "Canadian Wetland Registry." In C.D.A. Rubec and F.C Pollett, eds., *Proceedings, Workshop on Canadian Wetlands, Lands Directorate, Environment Canada,* Ecological Land Classification Series No. 12, Ottawa, 9–39.

Verry, E.S. 1988. "The Hydrology of Wetlands and Man's Influence on It." *Proceedings International Symposium on the Hydrology of Wetlands in Temperate and Cold Regions*, Vol. 2, Helsinki: Suomen Akatemian Julkaisuja, 41–61.

Virta, J. 1966. "Measurement of Evapotranspiration and Computation of Water Budget in Treeless Peatlands in the Natural State." *Physico-Mathematicae Fennica* 32: 1–69.

CHAPTER EIGHT

The Arctic Islands

MING-KO WOO AND ATSUMU OHMURA

INTRODUCTION

The Canadian Arctic Archipelago extends from Resolution Island (61°N) to the northern tip of Ellesmere Island (83°N), about 800 km from the North Pole. Other than the southern part of Baffin Island and several small islands, the archipelago lies within the Arctic Circle (Figure 8.1). Many deep sounds and channels separate the islands, but the sea's surface is covered by ice for most of the year, with multi-year ice cover present over the Arctic Ocean. The general topography of the western islands tends to range from undulating to flat-lying, with elevation averaging below 300 m, so the area is classified as the Arctic Lowland (Figure 8.1). In the east is the Arctic Upland, where rugged mountains, some rising over 2,000 m, are dissected by deep valleys and fiords and sometimes enclose low, intermontane basins. These uplands separate the areas to the west from the direct influence of Baffin Bay. The entire region lies north of the treeline, and in summer the surface cover consists of tundra vegetation, barren ground, semi-permanent snowfields, and glaciers (Edlund and Alt, 1989).

The Arctic Islands region, including the sea surfaces, is about 2.0×10^6 km^2, with land occupying 1.3×10^6 km^2 (about 13 per cent of the total terrestrial surface of Canada). Baffin Island (0.51×10^6 km^2) is the largest island, followed by Victoria (0.22×10^6 km^2) and Ellesmere (0.20×10^6 km^2). Despite the large area, there is only a sparse network of climatological stations. For example, only four stations service the Queen Elizabeth Islands, which have a combined land area of 0.41×10^6 km^2. (In comparison, the area of France is 0.55×10^6, and that of the United Kingdom, 0.24×10^6 km^2.) In the past decades, field research conducted by universities and government agencies has augmented climatological information for this region (Figure 8.2), although in disparate areas.

Several review articles have been written on the climate of the Arctic. Barry and Hare (1974) emphasized the energy balance and circulation patterns for different arctic regions. Hare and Hay (1974) examined various climatic phenomena of Canada and Alaska, including the Arctic Islands. Maxwell (1980; 1981) provided a systematic study of climatic zones, summarizing previous work on the regional climate of the Arctic Islands. These studies rely heavily on weather station data, supplemented by short-term information gathered during scientific experiments. The spatial database is

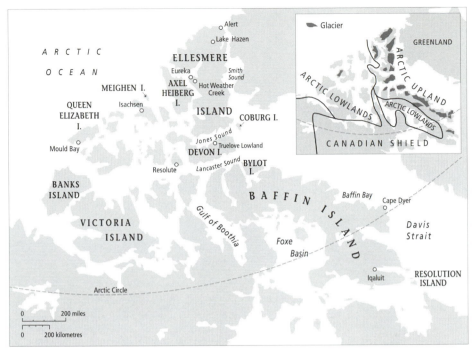

Figure. 8.1
Location of Canada's Arctic Islands. Inset shows major physiographic regions and distribution of glaciers.

Figure 8.2
Interior of Banks Island (73°14′N, 119°32′W), with fair-weather cumulus over the tundra plateau. Field camps, like the one shown in the photograph, provide weather observations during summer to augment sparse data for the Arctic Islands.

Figure 8.3
Midnight on 25 May 1979 at Eidsbotn, Devon Island (76°10′N, 91°30′W), with mountains casting long
shadows across the snow-covered fiord.

limited, since all weather stations are located along the coastal zones (Maxwell, 1981).
The present chapter is not an attempt to present the regional climatology of the Arctic
Islands. Rather, it emphasizes the surface climatological processes and is based on field
observations and draws on weather-station records to provide illustrative examples.

THE ENVIRONMENT

Radiation Regime

The inclination of the earth's axis to the plane of the ecliptic (the plane on which the
earth and the sun revolve) gives rise to twenty-four–hour winter nights and equally
long summer daylight periods for most of the Arctic Islands. Thus the radiation regime is one of extreme seasonal contrasts. The Queen Elizabeth Islands, for example,
receive no solar radiation for over two winter months but experience continuous daylight for several months in summer (Figure 8.3). Further south, several hours of solar
radiation input are gained during winter, and there are fewer days of twenty-four–hour
solar radiation input, as is shown by comparison of the radiation regimes of Alert
(82°30′N) and Iqaluit (63°45′N) in Figure 8.4.

The intensity of radiation is weakened by the low angle at which the sun's rays strike
the polar surfaces. There is a poleward decrease in annual total solar radiation, from
3.5 GJ m^{-2} y^{-1} at Iqaluit to 2.9 GJ m^{-2} y^{-1} at Alert. Locally, solar radiation is also affected by cloudiness and other factors such as multiple reflections from sea ice. Woo

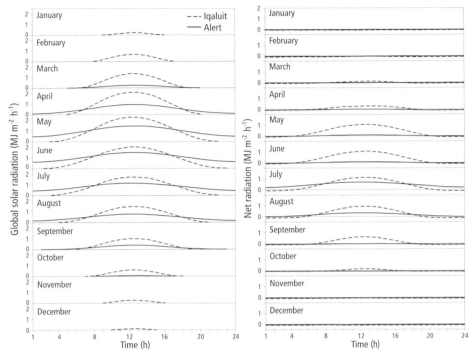

Figure 8.4
Mean hourly global solar and net radiation at Alert and Iqaluit, for all months.

and Young (1996) used a cloud layer model to produce regional maps of the distribution of solar radiation. The annual solar radiation input to the Arctic Islands always amounts to more than half that received in southern Canada – for example, 4.9 GJ m^{-2} y^{-1} at Toronto (Barry and Hare, 1974). The annual net radiation for the Arctic Islands is much smaller, however; it is still 0.3 GJ m^{-2} y^{-1} for Alert and 0.9 GJ m^{-2} y^{-1} for Iqaluit, compared with 1.9 GJ m^{-2} y^{-1} for Toronto. Expressed as a ratio of solar radiation ($Q*/K{\downarrow}$), the values are 0.10, 0.26, and 0.39, at Alert, Iqaluit, and Toronto, respectively. Low net radiation (Figure 8.4) is largely attributed to the high snow albedo that prevails throughout the snow-covered period, as discussed above in chapter 4.

Temperature

Prolonged negative net radiation during winter, and the low level of energy gain in summer, lead to persistent cold conditions. Above-freezing temperatures occur for only two months at Alert and four months at Iqaluit. February is the coldest month for the Arctic Ocean and on most of the Arctic Islands, except for southern Baffin Island. The lowest mean January sea-level temperature in the northern hemisphere, of $-35\,°C$, has been recorded for northern Ellesmere Island. Temperature in winter increases southward to $-25\,°C$ in southern Baffin Island and southern Baffin Bay. Winter temperatures are strongly influenced by proximity to open water and by surface topography. Anomalously high temperatures on the shores of Smith Sound, Jones Sound, and Lancaster

Table 8.1
Diurnal amplitudes of air temperature (Celsius degrees) for selected stations of the Arctic Islands

	J	*F*	*M*	*A*	*M*	*J*	*J*	*A*	*S*	*O*	*N*	*D*
Alert	0.6	0.6	1.1	2.0	2.2	1.1	1.1	1.5	1.1	0.6	1.1	1.1
Lake Hazen	0.6	1.0	2.4	4.1	3.5	2.8	2.6	m	1.1	0.9	0.8	0.8
Fort Conger	0.7	1.1	1.7	5.0	3.9	2.3	1.7	1.5	1.6	0.8	0.8	0.6
Eureka	0.8	0.0	2.2	4.3	3.9	2.4	2.1	1.7	2.2	0.8	0.6	1.1
Axel Heiberg I.	m	m	m	5.2	3.5	1.6	2.1	2.6	m	m	m	m
Isachsen	0.6	0.6	1.7	3.3	2.6	2.3	2.3	1.9	0.8	0.8	0.6	0.6
Mould Bay	0.8	0.8	1.7	4.7	3.6	2.2	3.1	1.9	1.1	1.1	0.6	0.6
Resolute	0.3	0.2	1.9	3.4	3.3	2.1	2.4	2.2	0.9	0.3	0.4	0.3

m: Missing data.

Sound at the northern end of Baffin Bay are due to the heating effect of North Water – one of the largest recurring polynyas (area of open water in sea ice) in the Arctic. The February mean temperature at Coburg Island at the western edge of North Water is about the same as that at Baffin Bay, more than 1,000 km south. Extreme low temperatures are observed in valley bottoms, which receive cold air drainage. This is why the lowest minimum temperatures at standard meteorological stations were recorded at Isachsen (−53.9°C) and at Eureka and Pond Inlet (−53.3°C), not at the northernmost Canadian station of Alert (−49.4°C). The absolute lowest temperature, of −55.8°C, was observed in January during Operation Hazen at Lake Hazen (Jackson, 1958). This station, 60 km inland from the coast, was at 163 m above mean sea level on the north shore of the lake, which lies on the bottom of a valley nestled between two mountain ranges.

July is usually the warmest month in the Arctic. Air temperatures over the Arctic Ocean rise above 0°C and are about 2°C along the ocean's coast. Temperature increases to the south, reaching 9°C along the southern shores of Victoria and Baffin islands. Besides the latitudinal trend, summer temperatures are influenced by distance from coasts. The interior of the Arctic Islands show much higher temperatures than registered at the official stations along the coast.

Most arctic stations show maximum diurnal temperature amplitudes in spring, usually April, with autumn showing a secondary maximum (Table 8.1). This phenomenon is sometimes referred to as the "Fram-type temperature change" and is widespread, being observed at the standard meteorological stations on the coasts, as well as at the interior sites of the islands, and on the sea ice and glaciers. Indeed, one of the largest diurnal temperature amplitudes was reported for Eismitte on Greenland in spring, when the albedo is extremely large.

The temperature contrast between coastal and interior locations was first treated systematically by Jackson (1958) for the Lake Hazen area of Ellesmere Island. Other interior stations also show significant differences from the coastal locations in terms of maximum, mean, and minimum temperatures. During winter, daily minimum temperature is often up to 10 Celsius degrees lower at interior sites, compared with the coastal stations. Even for the monthly mean temperature, the interior sites are colder by 2 to 8 Celsius degrees. The main cause of warmer conditions along the coast is the heat released by the sea water, primarily through convection from leads and cracks in the sea ice. Heat conduction through the sea ice plays only a secondary role. The magnitude of the heat flux from the sea water is not well known. For the central Polar Ocean, Vowinckel and Orvig (1966) estimated the ocean heat flux for the winter months of December, January, and February at 4 W m^{-2}. Since ice concentration in the inter-island channels is higher than in the central Polar Ocean, this value can be regarded as the upper limit.

The interior warms up faster than the coast in May or June, as soon as snowmelt begins. Afterward, the interior becomes significantly warmer. A comparison of the air temperature at Eureka, a coastal weather station (80°00′N, 85°56′W), with Hot Weather Creek 30 km inland shows that the coastal station is cooler in summer (Figure 8.5). The temperature contrast is the result of differences between the heat balance of the land and that of the sea. While melting sea ice withdraws sensible heat from the atmosphere above, the land emits relatively large amounts of sensible heat into the lower atmosphere. Where the effect of sea ice is enhanced by the influence of cold currents, the summer temperatures at coastal stations are reduced further. For example, Maxwell (1981) noted that Resolution Island has almost the lowest July mean temperature in the Arctic Islands, even though this station is at the southernmost point in the Arctic Archipelago. This is due to maintenance of the near-shore ice pack during most of the summer season.

The summer heat flux into the ground is insufficient to thaw the frozen soil except in the surface zone. Thus only the surface layer experiences annual freeze and thaw, and it is known as an active layer. Below it lies the permafrost, which is defined as the ground that remains at or below 0°C for at least two consecutive years. Permafrost in the Arctic Islands is thicker than 100 m and occurs continuously under all land. Only below large water bodies such as deep lakes can non-permafrost zones, known as taliks, be found.

Precipitation

Though precipitation in the Arctic Islands is generally low, high precipitation is recorded in the eastern Arctic. This latter occurs because of the frequent passage of cyclones along the Davis Strait and Baffin Bay (Maxwell, 1981), which, on encountering the mountainous areas, produce orographic precipitation. Maximum values are noted in the southeast, with a tendency for precipitation to increase with altitude. For example, at Resolution Island, the annual precipitation averages 313 mm at an elevation of 40 m and is 404 mm at 370 m. Cape Dyer (66°35′N, 61°37′W), at 390 m, receives 663 mm y^{-1}, the region's highest recorded average. The glaciers on the eastern flank of the mountains of Devon and Ellesmere Islands may have precipitation exceeding 400 mm, as shown by data from snow cores (Koerner, 1979). The rest of the region has under 250 mm y^{-1}, making it a polar desert or polar semi-desert (Bovis and Barry,

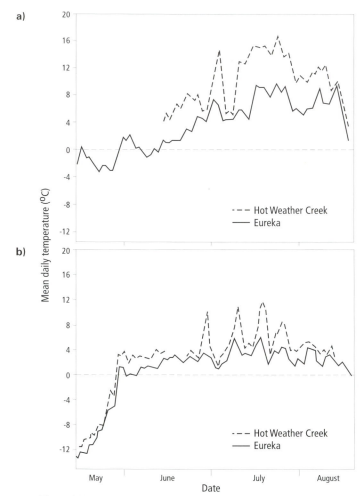

Figure 8.5
Mean daily air temperatures at a coastal (Eureka) and an inland (Hot Weather Creek) station during two summers: (a) 1988 (hot, day); (b) 1989 (cool, wet).
Source: Edlund, Woo, and Young (1990).

1974) – here defined as a non-glacierized area with mean annual precipitation of less than 250 mm and mean temperature in the warmest month below 10°C (Péwé, 1974).

Arctic weather stations usually underestimate precipitation, causing problems in water balance computations, as recorded precipitation may be even less than runoff. Errors in precipitation measurement are attributed to frequent trace events that escape being registered numerically by the gauges (Woo and Steer, 1979) and especially to the under-catch of snowfall because of windy conditions in the open tundra or barren ground. Since snowfall accounts for over half of the annual precipitation, its underestimation causes inaccuracies in weather-station data. Woo, et al. (1983) found that total snow ac-cumulation in drainage basins, as determined by end-of-winter snow surveys, can be twice as much as the total winter snowfall recorded by weather stations located near

Table 8.2
Variations in precipitation at Hot Weather Creek, Ellesmere Island

| | Plateau | Slope aspect | | | |
		North	East	South	West
SWE (mm)					
Winter 1988–89	89	181	109	51	26
Winter 1989–90	118	247	72	63	62
Rain (mm)					
1 July–1 Aug. 1989	91	168	94	130	66
19 May–12 Aug. 1990	11	14	11	12	1

Source: Woo, Young, and Edlund (1990).

these basins. For example, Resolute recorded 137 mm y^{-1} of precipitation as the average for the period 1975–82. A more accurate estimate, based on snow-survey data from the nearby McMaster basin, gave mean annual precipitation of 208 mm.

Microclimatic Conditions

The microclimate of the Arctic Islands exhibits phenomena covering a range of scales. Slope aspect causes local variations in precipitation and radiation receipt. At Hot Weather Creek, four slopes with an elevation range of 8 to 16 m and a flat surface, all located within an area of 1 km^2, yielded substantially different amounts of winter snow accumulation and summer rainfall (Table 8.2). For the winter of 1988–89, the north-facing slope had the most snow, because of winter drift, followed by the east-facing slope, and both had more snow than the flat site. In summer, the north- and south-facing slopes received more rainfall, possibly associated with the southerly winds that accompanied the rain events. Net radiation after snowmelt was highest for the south-facing slope and least for the north-facing, with the east and west slopes receiving intermediate amounts (Woo, Young, and Edlund, 1990). These aspect controls create a situation where the water balance of the slopes differs substantially, with the north slope having more precipitation and less evaporation and the south slope, more precipitation and evaporation.

Mountainous terrain gives rise to large microclimatic variability. Elevation differences produce contrasts in temperature, cloudiness, and precipitation between uplands and lowlands. Ohmura (1982) found substantial increases in orographic precipitation on Axel Heiberg Island. Courtin and Labine (1977) described chinook effects that hasten snowmelt on northern Devon Island. Normally, higher elevations have colder and harsher climates, which can delay snowmelt on plateau sites even of moderate elevation (Courtin and Labine, 1977). Temperature inversions are also common in winter because of large heat losses from the ground and because of cold-air drainage (for example, in the intermontane area around Eureka).

Figure 8.6 illustrates vertical temperature gradients for adjacent tundra and glacier surfaces on Axel Heiberg Island. In late winter (May), inversion conditions prevailed

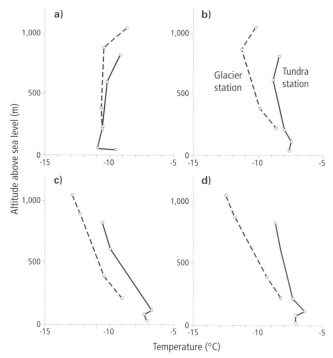

Figure 8.6
Vertical distribution of air temperature for glacier and tundra stations on
Axel Heiberg Island: (a) May; (b) June; (c) July; (d) August.

over both tundra and glacier, except very near the tundra surface. The mean July lapse
rates were similar over tundra and glacier surfaces, except that the strong surface heating layer above the tundra was capped by a low-level inversion. In August, the pattern
was similar to July's, except that lapse rates above the glacier became substantially
greater than over tundra. When lapse rates are relatively large over tundra, they are
caused by solar heating of the ground, while large mid- and late-summer lapse rates
over the glacier are attributed to downslope (katabatic) drainage, of surface air which
creates pseudo-adiabatic warming in the lowest layers.

Intermontane areas have their own local climates. Based on temperature data from
weather stations and from short-duration field measurements, Edlund and Alt (1989)
produced maps that clearly indicate warmer intermontane regions in the mountainous
zones of Ellesmere and Axel Heiberg islands. These areas generally have less cloud
cover and are estimated to have less precipitation than their surroundings. Eureka, for
instance, averages 3/10 cloud cover in winter and 6/10 in summer, both of which are
lower than other coastal stations.

Advection of moist sea air, particularly in late summer, spreads clouds and coastal
fog to the islands (Hare and Thomas, 1979; Maxwell, 1981) (Figure 8.7). Temperature
and humidity contrasts between coastal and interior sites raise the question of the fea-
sibility of using information from the coastal weather-station network to generalize on
regional trends of the surface climate.

Figure 8.7
Aerial view of the southern coastal zone of Devon Island (74°35′N,
86°05′W) in early August, showing the open water of Lancaster Sound
and outliers of the Devon Island Ice Cap. Low stratus clouds and fog seen
in the background are common during this time of the year.

Variations in local climate are broadly reflected in vegetation cover. Edlund (1992) suggested the following relationships from south to north. A low, erect-shrub zone, characterized by thickets of deciduous woody species, occupies the zone north of the treeline and is bounded by the 10 and 7°C mean July isotherm along its southern and northern boundaries, respectively. In like fashion, a zone of dwarfed and prostrate shrubs corresponds roughly to the 7 and 6°C mean July isotherms, respectively. The 6 and 4°C July isotherms contain the prostrate-shrub zone, where woody plants are much reduced and vascular species less diverse. A zone of herbs and prostrate shrubs, corresponding approximately with July temperatures of 4 to 3°C, is dominated by herbs, with arctic willow being the main woody plant. Finally, the areas with mean July temperatures of 3 to 1°C form the herb zone, where the only vascular plants are herbaceous. The vegetation pattern thus presented is closely linked to the summer temperature, indicating that arctic vegetation is controlled strongly by heat supply during the growing season.

THE SNOW-COVERED PERIOD

Snowcover persists from early October until mid-June in the southern Arctic Islands and stays for over ten months per year in the northern sector. Perennial snowcover occurs at high elevations, to form semi-permanent snowbanks or accumulations on ice caps. Lake

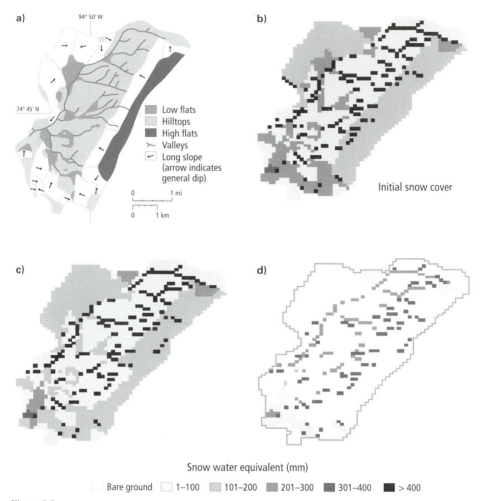

Figure 8.8
Melting of snow in McMaster basin: (a) map; (b) premelt snow distribution (30 June 1978); and the modelled spatial variation of snow during the melting period: (c) 4 July; (d) 9 July.
Source: Woo (1983).

ice tends to last even longer than snow at similar localities. Fresh lake ice often develops by mid-September or early October and remains for some time after the snow has disappeared. Some lakes retain part of their ice cover throughout the year, especially if summers are cooler than usual. Inter-island channels are ice-bound in winter, except where polynyas occur, as in Smith Sound and Foxe Basin. During summer, sea ice in all channels melts, to the extent that snowmobile traverse between islands becomes extremely hazardous. As ice volume reaches its minimum in September, most of Baffin Bay, Jones Sound, Lancaster Sound, Gulf of Boothia, and Foxe Basin becomes ice-free. The channels in the northwest of the archipelago, however, usually remain more than 50 per cent ice-covered until the onset of freezing. The lengthy presence of snow and ice causes much of the global solar radiation to be reflected, because of high albedo, and reduces the time during which soil and water bodies contribute to evaporation losses.

Snow Distribution

Spatial variations in snowfall, and a prolonged period of snow drifting, produce highly uneven snowcover. During the long winter, strong winds transport the snow from exposed locations and deposit it in more sheltered positions, such as ravines and leeward slopes. The interaction of terrain, wind speed, and direction influences the pattern of snow distribution, as can be seen in end-of-winter snow conditions for different types of terrain in the McMaster basin near Resolute (Figure 8.8). Snow surveys carried out for seven years in this basin (Woo, Marsh, and Steer, 1983) show that hilltops consistently have less snow than do exposed plateaus, and these in turn have less than does low-lying flat ground. Gullies and valleys have thick accumulations, while amounts of snow on slopes are highly variable, depending on local concavities and exposure to prevailing wind.

Energy Budget

Most of the winter is a period with little solar radiation input, and net radiation during very cold months indicates a strongly negative longwave balance. Ground heat flux is directed upward, towards the surface, until after the return of the sun, when sufficient solar energy is received at the snow surface to reverse the heat flow. Net radiation remains negative until well after the March equinox, because most of the solar radiation is reflected by the snow and ice. Figure 8.9 shows that during the pre-melt period of 1978, the albedo at Resolute averaged 0.7 until early June. For that year, average daily net radiation was negative or close to zero even after mean daily solar radiation began to exceed 25 MJ m^{-2} d^{-1}. Air temperature stayed below freezing until the summer solstice, though the average temperature of the snowcover increased slightly, indicating some heat flux into the snow.

Net radiation is a major contributor to snowmelt. In the initial phase, most of the energy is used to warm the snow, which may have temperatures below $-15°$C. Once melt begins, snow grains enlarge, water content in the snow increases, and surface roughness also increases. These processes cause the albedo to decrease (Woo and Dubreuil, 1985).

Latent heat flux controls condensation and sublimation. Its importance varies with location. While its magnitude tends to be small at Resolute, this flux is significant in drier areas such as the intermontane zones of Ellesmere and Axel Heiberg islands (Figure 8.10). In the latter areas, sublimation may be a major form of snow loss before the main melt period, in a fashion similar to that on the Prairies discussed in chapter 4. Sensible heat flux increases as the air becomes warmer. As the melt season advances, some areas become snow-free, and there is often strong advection of sensible heat from the bare ground to the patchy snow. Enhancement of sensible heat, accompanied by a decline in snow albedo, accelerates ablation of residual snowcover towards the end of the melt period.

In most years and for large parts of the Arctic Islands, large rainfall events are infrequent during melt, and rain-on-snow melt is insignificant. The snowmelt energy balance varies according to slope direction, inclination, and elevation. An example of systematic melt progression with elevation is the manner in which snow temperature rose at several sites in the McMaster basin during 1981 (Figure 8.11). The temperature near the base of the snow at the 10-m elevation reached $0°$C (or ripened) on 9 June,

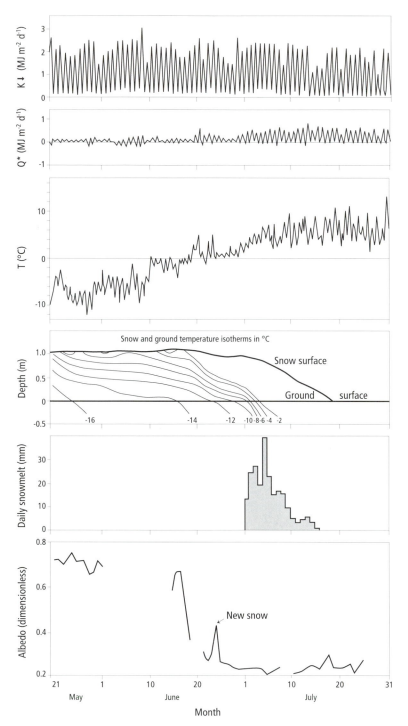

Figure 8.9
Hourly global solar radiation, net radiation, air temperature, snow temperatures, melt, and daily albedo, at a site near Resolute, NWT, 1978.

Figure 8.10
Large "ploughshares" (rough snow surface caused by sublimation) in the snow on a valley slope in the "polar desert" area of Fosheim Peninsula, Ellesmere Island (79°58′N, 84°28′W). These features are produced by sublimation of the snow.

but at 190 m it was delayed until 18 June. The pattern was disrupted in 1980, when several rain events occurred simultaneously at all elevations, obliterating the spatial differences caused by other components of the energy balance. The melt energy contributed by rainfall depends on the amount of rain and the rainwater temperature. Very often, spring rain is fairly cool, and the amount may not be large as a source of energy for surface melt. However, the infiltrated rainwater can refreeze in the snow releasing latent heat, which quickly warms and ripens snowcover.

   Snowmelt at most lowland sites, once begun, proceeds rapidly. The principal reason is that the melt season coincides with the period around the summer solstice when ample radiation energy is available. The usually shallow snowcover in many Arctic Island sites disappears within two to three weeks in most years.

Spatial Melt Patterns

The initial uneven snow distribution and differential melt rates at various locations cause bare patches to emerge as soon as melt season arrives (Figure 8.12). Topography gives rise to contrasts in the melt energy supply, with aspect and elevation both playing important roles. For example, in 1990, on a west-facing slope (gradient of 0.35) at Hot Weather Creek, snowmelt was more intense than on all other slopes (Woo, Edlund, and Young, 1991); the south-facing slope came second (gradient of 0.12). Slopes with favourable orientations usually lose snowcover earlier than others in the vicinity. Snow in shaded locations or in deep gullies tends to linger long after the nearby areas become

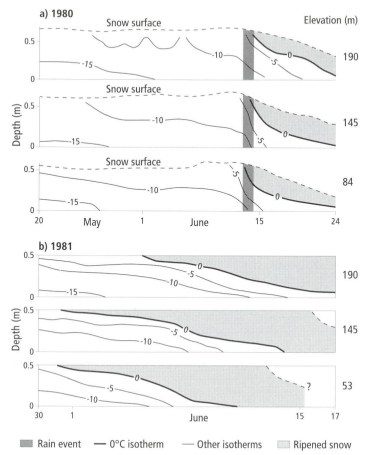

Rain event   — 0°C isotherm   — Other isotherms   Ripened snow

Figure 8.11
Changes in snow temperature in McMaster basin at three sites of various
elevations during the melt periods of (a) 1980 and (b) 1981.
Note: In the 1981 cross-sections only the lowest 0.5 m of the snowcover is shown.

bare. Snowcover also persists on uplands, where, besides there usually being higher snowfall, snowmelt is retarded by more persistent cloud cover and low temperatures. In mountainous areas, the snowline often recedes upslope during melt, but in cooler years the snow at high altitudes may remain throughout the summer.

The changing snowcover pattern for a drainage basin during the melt season can be modelled if one takes into consideration pre-melt snow distribution and the differential melt rate for various facets of the terrain (Woo, 1983). Figure 8.8 shows several stages of snowcover depletion in the McMaster basin. Low-lying, flat areas, with an initial thin snowcover and a relatively high melt rate, became bare early. Gullies and valleys with abundant snow remained snow-filled after most of the other basin surfaces were exposed.

On a broader scale, large tracts of the Arctic Islands have microclimates favouring earlier melt than their surroundings. The intermontane areas of Fosheim and Raanes peninsulas provide examples where snowmelt occurs five to ten days ahead of other locations at similar latitudes (Edlund and Alt, 1989). On a regional scale, there is a

Figure 8.12
Large spatial contrasts in heat and water balances between snow-covered and snow-free sites result from uneven snowcover during the melt period. At the centre of the photograph is a lake. The lake's ice is beginning to melt, but its edges are flooded by snowmelt runoff from slopes.
Photographed 11 July 1978, near Resolute.

latitudinal delay of melt. The mean date of snowcover loss is 15 June for western coastal Victoria Island. Loss of snowcover is delayed to 1 July for islands south of the Queen Elizabeth group, except for mountainous areas of Baffin Island (Maxwell, 1981). The Queen Elizabeth Islands generally lose their snowcover after 1 July, except for the intermontane areas mentioned above.

   Spatial variations in melt patterns affect the evaporation regime. Where the ground is snow-covered, sublimation may occur, but the amount of moisture loss is orders of magnitude lower than evaporation from saturated soils. During the melt season, soils freshly exposed from under the snow remain largely frozen. For most soils, meltwater infiltration is limited (Woo and Marsh, 1990), leaving ponded water. Abundant solar radiation at this time ensures high evaporation, which, together with rapid runoff losses, dries the soil quickly. It is common to have simultaneous occurrences of snow-melt and sublimation, high and low rates of evaporation from bare ground, and various magnitudes of infiltration and runoff on the slopes. The energy and water balances for such a mosaic are usually complex during the arctic melt season.

THE SNOW-FREE PERIOD

From the beginning to the end of the short melt period, the albedo of the land surface decreases from about 0.8 to 0.2. The Arctic Islands experience the most active energy exchange immediately after completion of snowmelt. This period is characterized by ample surface moisture, low albedo, and high solar elevation. For example, the mean net radiation during the first week following melt is 150 W m^{-2} for Alert, 160 W m^{-2} for Eureka, 145 W m^{-2} for Base Camp at Axel Heiberg Island (Ohmura, 1981), 165 W m^{-2} for Isachsen, 130 W m^{-2} for Mould Bay, 160 W m^{-2} for Truelove Lowland

(Courtin and Labine, 1977), and 130 W m^{-2} for Resolute. These values are larger than the largest monthly mean net radiation values for southern Canada. This is the period when soil heat flux is at its largest for the year, reaching 20 to 25 W m^{-2} (Ohmura, 1984). Net radiation decreases thereafter and attains negative values, usually during the last week of August at Alert and in the second week of September at Resolute. The outstanding feature of the energy balance during the snow-free period is the large Bowen ratio. The seasonal mean Bowen ratio for the moderately vegetated tundra surface on Axel Heiberg Island was measured at 0.75. The relatively wet sedge-moss meadow at Truelove Lowland yielded a Bowen ratio of 0.74, while a dry, raised beach reached a high value of 3.9 (Addison, 1977). Large Bowen ratios result partly from low air temperature (Ohmura, 1984) and partly from frequently dry surface conditions. On the Arctic Lowland, evapotranspiration often exceeds precipitation during the snow-free season. The summer water deficit is replenished by moisture from snowmelt or water that is stored as soil moisture, having drained from higher slopes. On Axel Heiberg Island, summer precipitation exceeds evaporation for altitudes higher than 500 m above sea level (a.s.l.). The energy balance for two sites on Axel Heiberg Island during the snow-free period is summarized in Figure 8.13.

Frost-Free Period

One significant feature of the summer thermal condition in lowland areas is the long frost-free period. Corbet (1967) reported fifty-five days for Lake Hazen. Hare and Thomas (1979) calculated twenty-nine days for Eureka. In the coastal region of Axel Heiberg Island, three years of observation gave an annual mean of fifty-six frost-free days. The frost-free period became shorter above 150 m and was reduced to only fourteen days at 800 m a.s.l. There are two reasons why the lowlands have a milder frost climate. First, the Arctic Islands are interspersed with extensive sea surfaces. After the commencement of snow and ice melt in June, the uppermost layer of the sea stores energy in the form of latent heat of fusion and enthalpy. Because of the large amount of latent heat required for ice melt and because of the large heat capacity of the mixing layer, the air above the sea surface is kept at or just above 0°C during the rest of the summer. Second, with twenty-four–hour midsummer sun, net radiation on the tundra remains positive for a long period. For example, on Axel Heiberg Island, continuously positive net radiation was recorded for ten and twenty-one days in 1969 and 1970, respectively. With positive net radiation over the land, the advected cold, arctic maritime air is heated well above the freezing point. Considering the above factors, it becomes unusual for arctic surface air to reach subfreezing temperatures in summer, unless there is adiabatic cooling associated with uplift in synoptic-scale depressions, or during the forced ascent of air at higher elevations.

Near-Surface Temperature

Since the ground's surface is the heat source in summer, the probability of frost decreases as the surface is approached. As an example, on Axel Heiberg Island, the frost-free period of the 0.2-m layer of air adjacent to the surface was about one week longer than at the screen-level (Ohmura, 1981).

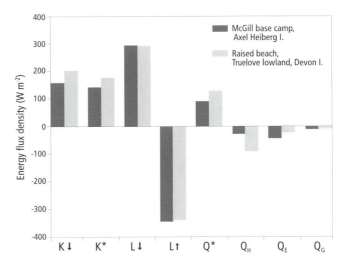

Figure 8.13

Energy and heat balance components at a partially vegetated tundra site (McGill base camp) and at a dry, raised beach, during snow-free period. Flux is positive when directed to the surface.

Ohmura (1981) also investigated air temperature at 0.2 m above the tundra surface at various topographic elevations on Axel Heiberg Island. Air temperature was higher at 0.2 m than at screen-level for most sites, but only by a fraction of one degree. At such locations as an outwash plain where the surface was dry, the 0.2-m temperature was up to 2 Celsius degrees higher than that at screen-level. Right at the tundra surface, temperatures can exceed screen temperatures by 15 Celsius degrees (Ohmura, 1981). It is the longer duration of the frost-free period that most significantly characterizes the mild aspect of the near-surface climate. At high elevations, however, the near-surface temperature can remain lower than at screen-level well into summer because of cooling from the late-lying snowcover.

GLACIER CLIMATE

Glaciers cover an area of 152×10^3 km^2, or 12 per cent of the land surface of the Arctic Islands (Figure 8.1). Most glaciers are found on mountains or high plateaux, but some are located at low altitudes, such as the Barnes Ice Cap on Baffin Island and smaller ice caps on northeastern Baffin Island, Bylot Island, Meighen Island, northern Axel Heiberg Island, and Ellesmere Island. On the northern coast of Ellesmere Island are small ice shelves, collectively called the Ellesmere Ice Shelves, which, though quite small, are the only ones in the northern hemisphere (Figure 8.14).

Main Features

The climate on the glaciers is characterized by lower temperature, lower humidity, and higher wind speed than the adjacent land (Ohmura, 1981). These features are direct consequences of the surface heat balance. There is a strong flux of sensible heat

Figure 8.14
Aerial view of the north coast of Ellesmere Island (83°03′N, 71°50′W) in early August, showing the ice-field sloping down from the mountains to the Ward Hunt Ice Shelf at sea level.

from the atmosphere to the glacier surface. The lower surface temperature of glaciers is maintained by the latent heat of ice melt at lower altitudes, and by evaporation and negative net radiation at the higher altitudes. The higher wind speed on the glaciers is due to the combined effects of downslope, katabatic air flow and the low surface roughness of the ice. Where there are large glaciers, the local climate of their adjacent areas can be modified significantly (Figure 8.15).

Climates in Accumulation and Ablation Areas

Glacier climate varies significantly with altitude. For example, over a 1,300-m altitudinal range, temperature decreases of 6.4 Celsius degrees have been documented (Andrews, 1964; Havens, 1964), as have decreases of 3 Celsius degrees between glacier snout and equilibrium line (Ohmura, 1981).

Cloud conditions can differ significantly between accumulation and ablation areas. Although there are few simultaneous observations of clouds over different sites, available data (for example, Andrews, 1964; Havens, 1964) indicate that the summer

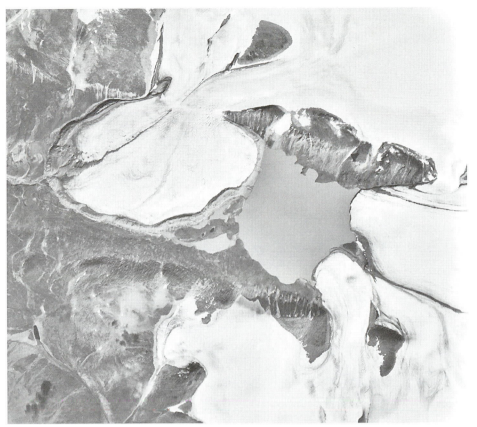

Figure 8.15
Aerial view of glaciers in late August, flowing from the Penny Ice Cap on Baffin Island (66°55′N, 65°50′W). The two glacier lobes have coalesced to dam a valley and have created a lake. Glacier features such as these exert substantial influence on the surrounding climate.

cloud amount over the ablation zone is about 10 per cent greater than that over the accumulation area. The difference is mostly due to the frequent occurrence of stratus and stratocumulus over the lower altitudes of the glaciers. On Axel Heiberg Island, elevational differences in the summer cloud cover can cause regional differences of 6 per cent in sunshine duration and 44 W m^{-2} in solar radiation.

Altitudinal variation in precipitation is likewise large. Annual precipitation rates of the White Glacier for the snout (200 m a.s.l.), the mean equilibrium line altitude (870 m), and the accumulation basin at 1,450 m have been estimated at 170, 300, and 330 mm, respectively (Ohmura, Kasser, and Funk, 1992). Mean annual accumulation for the accumulation area of the Müller Ice Cap–White Glacier System has been calculated at 371 mm. This value yielded an average altitudinal precipitation increase of about +7 per cent per 100 m.

Altitudinal variations of the climate on glaciers imply that the energy balance at the glacier surfaces must differ considerably from the snout to the accumulation area.

Table 8.3
Energy balance of glaciers

Glacier	Site	Altitude (m)	Observation period	Energy budget ($W\ m^{-2}$)					Reference
				Q^*	Q_H	Q_E	Q_M	Q_G	
Accumulation area									
Barnes Ice Cap	70°14′N 73°55′W	1,075	5 July–11 Aug. 1962 and 1963	36	6	2	−29	−15	Sagar (1966)
Devon Ice Cap	73°30′N 83°18′W	1,320	21 May–11 Aug. 1962 and 1963	16	10	−4	−10	−12	Holmgren (1971)
Müller Ice Cap	79°41′N 90°27′W	1,530	5 June–26 Aug. 1960	22	32	3	−46	−11	Havens (1964)
Penny Ice Cap	66°59′N 65°28′W	2,050	13 July–26 July 1953	25	4	12	−40	0	Orvig (1954)
Ablation area									
White Glacier	79°21′N 90°39′W	208	8 July–19 Aug. 1960	109	74	46	−211	−18	Andrews (1964)
Meighen Ice Cap	80°00′N 99°10′W	241		41	−4	−4	−35	−11	Alt (1975)
Sverdrup Glacier	75°40′N 83°15′W	300	9 July–10 Aug. 1963	50	31	14	−84	−11	Keeler (1964)
Barnes Ice Cap	69°43′N 72°13′W	865	25 June–5 Aug. 1950	27	13	0	−40	0	Ward and Orvig (1953)

Table 8.3 indicates that there is no clear trend in the difference between the accumulation and ablation zones, mainly because of the short observation periods. In general, the net radiation and the sensible heat flux appeared to be larger in the ablation area. There are exceptions, such as the Meighen Ice Cap; as a lowland ice cap in a maritime environment, it loses energy through sensible heat – perhaps the cause of its existence at such a low altitude. Table 8.4 shows how individualistic glaciers can be and suggests that maintenance of a glacier must be examined through both the energy and the mass balances.

Climate at the Equilibrium Line

For surface climate, the energy balance at the equilibrium line is of particular importance. The zone around the equilibrium line often experiences the largest year-to-year variation in the annual mass balance, and it is also from such a location that the largest amount of melt-water discharge is produced. Several features of the equilibrium line for selected Arctic Island glaciers are presented in Table 8.4. The equilibrium lines of these glaciers are distributed from sea level to the highest known altitude in the Arctic Islands.

Table 8.4
Climate at the equilibrium line on Arctic Islands glaciers

Glacier, island	Altitude of equilibrium line (m)	Annual precipitation (mm)	Summer temperature (°C)	Summer K↓ (W m⁻²)
Ward Hunt Ice Shelf, Ellesmere I.	0	240	−1.0	207
Meighen Ice Cap, Meighen I.	250	210	−0.8	202
Laika Glacier, Cobourg I.	355	590	0.5	227
Devon Ice Cap (s.e.), Devon I.	700	320	0.8	m
Barnes Ice Cap, Baffin I.	810	590	1.9	202
White Glacier, Axel Heiberg I.	855	310	−1.6	239
Baby Glacier, Axel Heiberg I.	935	350	−1.7	223
Devon Ice Cap (n.w.), Devon I.	1,050	240	−0.3	270
Decade Glacier, Baffin I.	1,175	310	−0.3	196
Gilman Glacier, Ellesmere I.	1,250	170	−1.1	240
Per Ardua Glacier, Ellesmere I.	1,350	190	−1.5	242

m: Missing data.

The summer solar radiation at the equilibrium line is around 200 to 240 W m⁻². Because of the persistently high albedo of the glacier surface near the equilibrium line, solar radiation has limited influence on the elevation of this line. The summer temperature at the equilibrium line also varies within a narrow range: between −1.7°C and 0.8°C. It is precipitation that most strongly influences the altitude of the equilibrium line. The Laika Glacier has a low equilibrium line altitude, despite a relatively high summer temperature – a tendency shared with many glaciers on southeastern Ellesmere Island. Such low equilibrium line positions are largely the result of high winter precipitation, caused by proximity to North Water. However, low precipitation on the glaciers in the United States Range, such as the Gilman Glacier and the Per Ardua Glacier in northern Ellesmere Island, pushes the equilibrium lines to the highest altitudes in the Arctic Islands. Indeed, the United States Range has the driest mountains among those that rise above 2,000 m. Precipitation differences also produce a contrast in equilibrium line altitudes between the northwestern and southern sectors of the Devon Ice Cap. The equilibrium line climates of the Arctic Island glaciers, from a global perspective, fit into a low-temperature, low-precipitation category.

Cooling Effect of Glaciers

During the melt period, glaciers have a cooling effect on the overlying air. Temperature gradients commonly show an inversion profile. The cooling is caused by energy lost to ice melt. The glacier surfaces in the dry-snow zone also cool the atmosphere through negative net radiation and through relatively large losses of latent heat during sublimation. Usually, the cooling effect is strongest at the altitude of the equilibrium line. This is partly because of cold-air drainage from the accumulation zone towards the equilibrium line and partly because of warming of the lower ablation zone, by the large flux of sensible heat from its adjacent tundra, which has less cloud and limited evaporation (Ohmura, 1981).

Advection from the glacier frequently cools nearby land, usually the outwash plain, but the effect is weak. For example, a station 500 m away from the snout of the White Glacier was measured to be only 0.8 Celsius degrees cooler than sites located far from the glacier – glacier-cooled air is rapidly transformed by a relatively large sensible heat flux from the bare ground.

DISCUSSION

This chapter illustrates that the surface climates of Canada's Arctic Islands are highly varied, in response to latitude, topography, and proximity to the sea. The snowcover period – most of the year – is not only important in the present-day climate but provides a glimpse into what the climate of the Pleistocene Glaciation period may have been like, particularly in areas where glaciers remain today. The short arctic growing season is strongly influenced by the legacy of winter snow accumulation, which at the time of melt may be non-existent in some windswept areas and be very deep in depressions or slope concavities. Spatial variability of the snowcover sets the scene for much of the surface's energy balance and water balance in the snow-free period.

The large susceptibility of this high-latitude environment to climate change is the result partly of the long temporal overlap between the snow-covered period and the high-sun season. Any perturbation, such as greenhouse warming, that diminishes the longevity of the snowcover and lake-ice cover automatically increases radiation receipt at the surface by about an order of magnitude. The effect of greenhouse warming on surface climates would be immense, as revealed in the numerical experiments of General Circulation Models, which indicate potential major warming for high-latitude regions under higher atmospheric concentration of carbon dioxide (CO_2) and other greenhouse gases. This issue is explored below, in detail, in chapter 14. Whether such warming could translate into more vigorous plant growth and consequently the spread of subarctic plant species into the Arctic Islands would depend on the nature of any precipitation change, the rate of soil development, and the extent to which the permafrost is thawed (Edlund, 1992).

Climatic change can have a powerful influence on arctic ice caps. Even contemporary climatic variability has caused pronounced year-to-year fluctuations in glacier mass balance. Climate warming accompanied by lower precipitation will move the equilibrium line higher, but climate warming with more snowfall might either preserve the present status or even drop the equilibrium line to lower altitudes. Cooling with increased precipitation would certainly lower equilibrium lines and expand glaciers. Any of these responses would be reflected in alterations of the surface climates on the glaciers.

Sea ice is discussed in chapter 5 as a major element in the oceanic environment. Its effect on terrestrial surface climates of the Arctic Islands has been shown to be pronounced, particularly through lowering of summer temperatures in coastal areas. This influence and others extend into the mainland Arctic and subarctic regions and are presented in the next chapter. The extent of sea-ice is very responsive to climatic variability and change. Perturbations in sea-ice distribution and regime will produce an immediate response in both the surface climates of the coastal zones and the thermal character of the deep-ocean bottom water.

REFERENCES

Addison, P.A. 1977. "Studies on Evaporation and Energy Budget on Truelove Lowland." In C.L. Bliss, ed., *Truelove Lowland, Devon Island, Canada: A High Arctic Ecosystem*, Edmonton: University of Alberta Press, 281–300.

Alt, B.T. 1975. *Energy Balance Climate of Meighen Ice Cap, N.W.T.* Polar Continental Shelf Project Monograph, 2 vols. Ottawa.

Andrews, R.H. 1964. *Meteorology and Heat Balance of the Ablation Area, White Glacier, Canadian Arctic Archipelago – Summer 1960 (Lower Ice Station, 79°26'N, 90°39'W, 208 m).* Axel Heiberg Island Research Reports, Meteorology No. 1. Montreal: McGill University.

Barry, R.G., and Hare, F.K. 1974. "Arctic Climate." In J.D. Ives and R.G. Barry, eds., *Arctic and Alpine Environments*, London: Methuen, 17–54.

Bovis, M.J., and Barry, R.G. 1974. "A Climatological Analysis of North Polar Desert Areas." In T.L. Smiley and J.H. Zumberge, eds., *Polar Desert and Modern Man*, Tucson: University of Arizona Press, 23–31.

Corbet, P.C. 1967. "Terrestrial Microclimate: Amelioration at High Latitudes." *Science* 166: 865–6.

Courtin, G.M., and Labine, C.L. 1977. "Microclimatological Studies on Truelove Lowland." In C.L. Bliss, ed., *Truelove Lowland, Devon Island, Canada: A High Arctic Ecosystem*, Edmonton: University of Alberta Press, 73–106.

Edlund, S.A. 1992. "Climate Change and Its Effects on Canadian Arctic Plant Communities." In M.K. Woo and D.J. Gregor, eds., *Arctic Environment: Past, Present and Future*, Hamilton: McMaster University, 121–37.

Edlund, S.A., and Alt, B.T. 1989. "Regional Congruence of Vegetation and Summer Climate Patterns in the Queen Elizabeth Islands, Northwest Territories, Canada." *Arctic* 42: 3–23.

Edlund, S.A., Woo, M.K., and Young, K.L. 1990. "Climate, Hydrology and Vegetation Patterns, Hot Weather Creek, Ellesmere Island, Arctic Canada." *Nordic Hydrology* 21: 273–86.

Hare, F.K., and Hay, J.E. 1974. "The Climate of Canada and Alaska." In R.A. Bryson and F.K. Hare, eds., *Climates of North America*, Amsterdam: Elsevier, 49–192.

Hare, F.K., and Thomas, M.K. 1979. *Climate Canada.* Toronto: John Wiley & Son.

Havens, J.M. 1964. *Meteorology and Heat Balance of the Accumulation Area, McGill Ice Cap, Canadian Arctic Archipelago – Summer 1960 (Upper Ice Station, 79°41'N, 90°27'W, 1530 m).* Axel Heiberg Island Research Reports, Meteorology No. 2. Montreal: McGill University.

Holmgren, B. 1971. "Climate and Energy Exchange on a Sub-polar Ice Cap in Summer, Arctic Institute of North America Devon Island Expedition 1961–1963, Part E, Radiation Climate." *Report No. 111*, Department of Meteorology, University of Uppsala, Uppsala.

Jackson, C.I. 1958. *Operation Hazen/The Meteorology of Lake Hazen, N.W.T., Part 1: Analysis of the Observations.* Publications in Meteorology No. 15. Department of Meteorology, McGill University, Montreal.

Keeler, C.M. 1964. *Relationship between Climate, Ablation and Run-off on the Sverdrup Glacier, 1963, Devon Island, N.W.T.* Research Paper No. 27. Montreal: Arctic Institute of North America.

Koerner, R.M. 1979. "Accumulation, Ablation, and Oxygen Isotope Variations on the Queen Elizabeth Islands Ice Caps, Canada." *Journal of Glaciology* 22: 25–41.

Maxwell, J.B. 1980. *The Climate of the Canadian Arctic Islands and Adjacent Waters.* Vol. 1. Ottawa: Ministry of Supply and Services.

– 1981. "Climatic Regions of the Canadian Arctic Islands." *Arctic* 34: 225–40.

Ohmura, A. 1981. *Climate and Energy Balance of Arctic Tundra, Axel Heiberg, Island, Canadian Arctic Archipelago, Spring and Summer 1969, 1970 and 1972.* Zürcher Geographische Schriften, No. 3. Zurich: Fachverein Verlag.

– 1982. "Evaporation from the Surface of the Arctic Tundra on Axel Heiberg Island." *Water Resources Research* 18: 291–300.

– 1984. "On the Cause of 'Fram'-Type Seasonal Change in Diurnal Amplitude of Air Temperature in Polar Regions." *Journal of Climatology* 4: 325–38.

Ohmura, A., Kasser, P., and Funk, M. 1992. "Climate at the Equilibrium Line of Glaciers." *Journal of Glaciology* 38: 397–411.

Orvig, S. 1954. "Glacier-Meteorological Observations on Ice Caps in Baffin Island." *Geografiska Annaler* 36: 197–318.

Péwé, T.L. 1974. "Geomorphic Processes in Polar Deserts." In T.L. Smiley and J.H. Zumberge, eds., *Polar Desert and Modern Man*, Tucson: University of Arizona Press, 33–52.

Sagar, R.B. 1966. "Glaciological and Climatological Studies on the Barnes Ice Cap, 1962–64." *Geographical Bulletin* 8: 3–47.

Vowinckel, E. and Orvig, S. 1966. "Energy Balance of the Arctic. 5. The Heat Budget over the Arctic Ocean." *Archives météorologie, géophysique, bioklimatique*, B 14: 303–25.

Ward, W.H. and Orvig, S. 1953. "The Glaciological Studies of the Baffin Island Expedition, 1950; Part 4: The Heat Exchange at the Surface of the Barnes Ice Cap during the Ablation Period." *Journal of Glaciology* 2: 158–68.

Woo, M.K. 1983. "Hydrology of a Drainage Basin in the Canadian High Arctic." *Annals of the Association of American Geographers* 73: 577–96.

Woo, M.K., and Dubreuil, M.-A. 1985. "Empirical Relationship between Dust Content and Arctic Snow Albedo." *Cold Regions Science and Technology* 10: 125–32.

Woo, M.K., Edlund, S.A., and Young, K.L. 1991. "Occurrence of Early Snow-free Zones on Fosheim Peninsula, Ellesmere Island, Northwest Territories." In *Current Research Part B*, Geological Survey of Canada Paper No. 91–1B, 9–14.

Woo, M.K., Heron, R., Marsh, P., and Steer, P. 1983. "Comparison of Weather Station Snowfall with Winter Snow Accumulation in High Arctic Basins." *Atmosphere-Ocean* 21: 312–25.

Woo, M.K., and Marsh, P. 1990. "Response of Soil Moisture Change to Hydrological Processes in a Continuous Permafrost Environment." *Nordic Hydrology* 21: 235–52.

Woo, M.K., Marsh, P., and Steer, P. 1983. "Basin Water Balance in a Continuous Permafrost Environment." *Proceedings Fourth International Conference on Permafrost*, Washington, DC: National Academy Press, 1407–11.

Woo, M.K., and Steer, P. 1979. "Measurement of Trace Rainfall at a High Arctic Site." *Arctic* 32: 80–4.

Woo, M.K., and Young, K.L. 1996. "Summer Solar Radiation in the Canadian High Arctic." *Arctic* 49: 170–80.

Woo, M.K., Young, K.L., and Edlund, S.A. 1990. "1989 Observations of Soil, Vegetation, and Microclimate Effects on Slope Hydrology, Hot Weather Creek, Ellesmere Island, Northwest Territories." In *Current Research, Part D*, Geological Survey of Canada Paper No. 90–1D, 85–93.

CHAPTER NINE

The Low Arctic and Subarctic

WAYNE R. ROUSE, RICHARD L. BELLO,
AND PETER M. LAFLEUR

INTRODUCTION

Extending from the southern Arctic Islands to the close-crowned boreal forest, the low arctic and subarctic region is underlain largely by continuous and discontinuous permafrost and is traversed by the arctic treeline (Figure 9.1). It is bounded on the west by the Western Cordillera and has its eastern limit in western Labrador. The region is non-mountainous and consists of the broad landscape categories of tundra, open subarctic forest, and northern lakes (Figure 9.2 and 9.3). Each category contains several types of terrain, distinguished by vegetation, soil type, and wetness factors.

The vegetation and terrain of tundra are quite varied, ranging from expanses of dwarf birch-willow scrub, through upland sedge meadows and lichen heath, to lowland, wet-sedge meadows with abundant frost features, which give a hummock-hollow character to the landscape. The tundra is underlain largely by continuous permafrost. Summer active-layer thawing extends anywhere from 0.4 to 1.5 m in depth, depending primarily on whether the tundra is wet (ice-rich) or dry (ice-poor) and secondarily on insulation effects from organic soils and from the winter snowpack. The ice content of permafrost soils is crucial to the magnitude of ground heat flux (as is discussed below). The edge of the treeline trends from the coast of the Arctic Ocean in northwest Yukon southeastward until it intersects with northern James Bay south of Cape Henrietta Maria. Treeline in northern Quebec–Ungava is further north and extends to the montane regions of Labrador, where it responds to altitudinal influences.

Like tundra, vegetation in the open subarctic forest is highly varied. In most vegetative classifications (for example, Hare and Ritchie, 1972; Ritchie, 1987) it includes two zones – to the north, a zone often referred to as forest-tundra, where individuals or small groves of trees dot a largely tundra-like surface, and to the south, the open woodland zone, also called open forest or taiga, which consists of continuous open-crown (< 50 per cent areal coverage) forest. The open forest is the larger of these two zones and corresponds roughly to the zone of discontinuous permafrost in Figure 9.1. Most of the information in this chapter refers to the open forest zone (Figure 9.2c), which we refer to below as "subarctic open forest" or simply "subarctic forest."

The three most common tree species, all with transcontinental ranges, are black spruce (*Picea mariana*), white spruce (*Picea glauca*), and tamarack (*Larix laricina*).

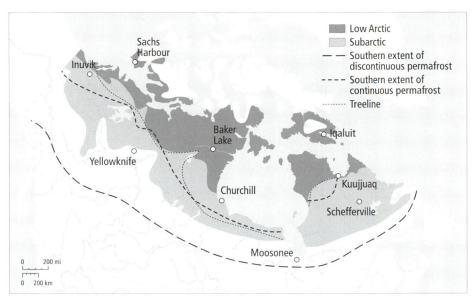

Figure 9.1
Geographical regions in the low arctic and subarctic regions.
Source: Ecoregions Working Group (1989).

There is considerable local and regional variation in the abundance of each species because of site factors (exposure, soil, and wetness) and because of historical climate–fire–regeneration relationships. In the broadest terms, black spruce tends to dominate in eastern Canada and white spruce in western Canada, while tamarack occurs and dominates on wetter soils in some regions. The understorey in the open forest varies with location and often consists of a variety of shrubs, such as willow, blueberry, birch, Labrador tea, alder, other vascular species, and several varieties of moss and lichen. Particularly important are the extensive ground covers of lichen (*Cladina stellaris, C. mitis*, and *C. rangiferina*), which give rise to the characteristic "open lichen woodlands" in central Quebec–Labrador, the Hudson Bay Lowland, and Keewatin.

Lakes and ponds in this region are defined as any surface consisting primarily of standing freshwater during any portion of the year. A subdivision often denotes ponds as freezing to the bottom in winter, whereas lakes do not. In this region this definition means generally that lakes are more than 1.5 m in depth. However, in terms of their surface climate, the two are similar, and the terms are interchangeable. Freshwater bodies, in the low Arctic and subarctic, range in size from the smallest pond to Great Bear and Great Slave lakes, which are the world's eighth- and tenth-largest lakes, respectively. The discussion in this chapter deals only with small lakes and ponds that do not exceed a few square kilometres in area and a few metres in depth (Figure 9.2a). Larger lakes are dealt with in chapter 6, above. In some regions of the north with significant physical relief, drainage is highly channelized, and the proportion of lakes is probably less than 1 per cent. In contrast, areas such as the Hudson and James Bay Lowland have gentle relief and suffer severe impediments to vertical drainage owing to the presence of permafrost and/or impermeable mineral sediments. Bello and Smith

Figure 9.2
Landscapes near treeline, showing (a) wetland sedge fen after snowmelt; (b) upland sedge-lichen heath; (c) open subarctic forest; and (d) shallow tundra lake.

(1990), for example, estimate the areal extent of lakes to be as high as 41 per cent of the landscape in the region south of Churchill, Manitoba (Figure 9.3a). Figure 9.3b illustrates a landscape in which the areal extent of lakes lies between the above extremes. Even within the same physiographic unit, estimates range from 50 per cent to as little as 3 per cent. In flat wetland terrain, it is often difficult to distinguish among swamp or fen and lake because there are no precise boundaries to the water body.

Hudson Bay and James Bay are major features of the low Arctic and subarctic zone of central Canada, and they exert a strong regional influence on surface climates. These inland seas represent a southward extension of the polar ocean and are especially important to the climate in summer (Rouse 1991). Their ice cover persists along the southwestern shore of Hudson Bay until late July (Danielson 1969).

A number of parameters illustrate the nature of the regional physical climate. Annual global solar radiation, $K\downarrow$, decreases from about 140 to 125 W m^{-2} between the southern and northern boundaries, a reduction of less than 15 per cent, but annual net radiation, Q^*, decreases from 40 to 25 W m^{-2}, or close to 40 per cent. This drop in Q^* is almost entirely a result of the longevity of the high-albedo snowcover, which increases with latitude. On an annual basis, the air temperature is sub-freezing, with the southern and northern boundaries lying within the $-3°C$ isotherm and the $-10°C$ isotherm, respectively. Precipitation is highly variable from south to north and from

Figure 9.3
Aerial views of lowland and wetlands. (a) Northern Hudson Bay Lowland,
with frost-created peat polygons in foreground and lakes, ponds, and
bogs in the central area. (b) Wetland environment of the Mackenzie River
delta region, showing open subarctic forest, wetland tundra, lakes, and
rivers.

east to west. In very general terms, it is largest in the southeast, where it can exceed
500 mm, and smallest in the north and northwest, at 200 mm or less. Snowfall shows
similar variation, from more than 2 m in the southeast to 1 m or less in the north and
northwest. The winter snowpack lasts from an average of more than six months in the
southernmost regions to nine months in the north (Findlay 1978). In the south, snow
persists into middle or late May, whereas in the north it lasts on average until after
the summer solstice. Growing seasons are characteristically short. In the vicinity of
Churchill, Man. (Figure 9.1), for example, the period of active growth for sedges on
wet tundra lasts from thirty to thirty-five days, and willow, birch, and Labrador tea
reach full leaf thirteen days after initial emergence (Blanken and Rouse, 1994).

   Though this chapter encompasses a large geographical region, it includes only a
limited number of surface climate studies. For each ecosystem it outlines current

knowledge of the radiation balance and energy balance. It examines differences between tundra and forest in surface climate. It develops the complementary relation of dry uplands, low wetlands, and shallow lakes. It discusses the impact of climate variability and the usefulness of such variability as an indicator of the possible effects of climate change, and it describes the important role of advection, which operates at macro-, meso-, and micro-scales and can often dictate much of the surface climate. The discussion focuses on the summer season, when most of the research has been pursued.

TUNDRA

Seasonal Patterns

A typical seasonal pattern for the energy balance and temperature for upland tundra in the Hudson Bay Lowland near Churchill, Man., is illustrated in Figure 9.4. Midwinter is a five-month period of minimal net radiation, very small fluxes of conductive and convective heat, and a very cold atmosphere. By late winter, from mid-April through May, even though the daylight period is long and solar radiation is plentiful, net radiation is still very weak because of the large surface albedo of the snow; there is a small heat gain by the frozen ground and a small sensible heat flux into the atmosphere. The spring period, in June, experiences the largest net radiation, the greatest heat flux into the ground, the largest sensible heat flux, and a latent heat flux equivalent to that of the summer period. Temperatures during this high-sun season remain low, however, averaging about 7°C. Summer is distinguished from spring by less net radiation because of the shorter daylight period, a smaller but still sizeable ground heat flux, higher temperatures, and a smaller Bowen ratio, because of reduced sensible heat flux. The autumn season sees a return to small net radiation and small convective heat fluxes. During this period, from late August to early October, there is little ground heat flux. The energy balance in early winter consists almost entirely of a large loss of ground heat, which goes into warming the atmosphere. Net radiation and the latent heat flux are negligible during this period, which lasts till the onset of midwinter, usually by mid-December.

It is clear that all significant energy fluxes occur in little more than four months, from early June to early October, with the noteworthy exception of the heat released by the freezing soils in early winter. Because most research on surface climates has been concentrated in spring, summer, and autumn, the rest of the discussion of the surface climate of tundra pertains to this period.

Summer Radiation and Energy Balance

Net radiation over tundra shows one of the largest seasonal step changes of any terrestrial surface, because of the dramatic winter-to-summer change in surface albedo. In winter the albedo of snow normally exceeds 0.80 (chapter 4), whereas for most tundra surfaces in summer it ranges from 0.08 to 0.18 (Petzold and Rencz 1975; Rouse and Bello 1983). The drop in albedo leads to dramatic changes in $Q*$ because final snowmelt in tundra areas coincides with the high-sun season in June.

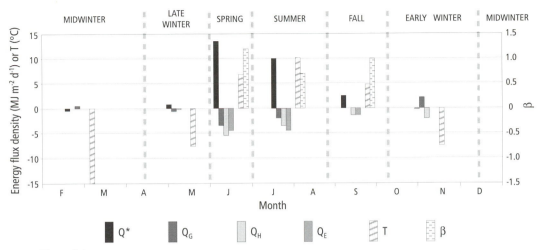

Figure 9.4
Seasonal course of temperature and energy balance components at a high subarctic site.
Source: Rouse (1982).

Because tundra is usually underlain by permafrost, the ground heat flux is large relative to surfaces in non-permafrost terrain, particularly in ice-rich soils, where Q_G can comprise up to 18 per cent of Q^* over the summer period (Halliwell and Rouse 1987), and commonly constitutes at least 10 per cent. This large Q_G is due to the very steep temperature gradients between the ground surface and the frost table and to the large thermal conductivities in the wet, active layer of the melting permafrost.

The partitioning of the available energy, $Q^* - Q_G$, into Q_E and Q_H depends on the soil's moisture supply, atmospheric demand, and surface and aerodynamic resistances to evapotranspiration. These factors are conveniently incorporated in the Penman-Monteith combination equation (equation 2.18), which is used below to describe the evapotranspiration processes in tundra. In addition, the Bowen ratio is employed to describe the division of available energy into the sensible and latent heat fluxes. This analysis assumes a basic division between tundra that is wet most of the summer and that which is dry.

Although flatness of the terrain is a general characteristic, low-arctic and high sub-arctic tundra is as heterogeneous a surface environment as most others in this book. The energy and water balances of lowland wet and upland dry tundra are distinctive. Often a change in relief of a metre or less is all that is necessary to create this division (Figure 9.2a). Upland, well-drained areas may be underlain by beach sands and grav-els, esker or kame deposits, glacial tills, bedrock, or thick layers of peat. Lowland wet areas commonly are underlain by peat deposits, which may vary from a few cen-timetres to more than a metre in thickness. These organic soils in turn have frequently developed on impervious marine or lacustrine clays and silts.

Evapotranspiration is a major component of the summer water balance in wet tundra and usually equals or exceeds precipitation input (Figure 9.5). The three tundra locations in Figure 9.5 lie to the west of Baker Lake, NWT (T1), east of Churchill, Man. (T2); and in the vicinity of Schefferville, Que. (T3). Evaporation rates are of

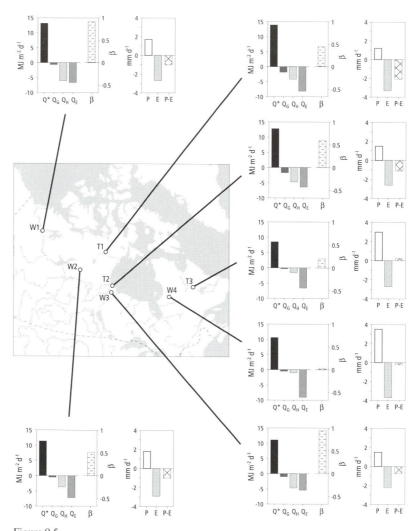

Figure 9.5
Summer energy and water balance of tundra (T) and forest (W) for different geographical regions.
Sources: Fitzjarrald and Moore (1994); Lafleur (1992); Pugsley (1970); Roulet and Woo (1986); Rouse and Bello (1985); Rouse et al. (1977); Singh and Taillefer (1986); and Wright (1981).

fairly similar magnitude, but precipitation increases substantially towards the east. Although the evapotranspiration rates in summer are large, there is still strong sensible heating of the atmosphere. Thus, even over very wet tundra such as that at site T2, the average daily Bowen ratios rarely fall below 0.4 (Rouse, Carlson, and Weick, 1992). This circumstance arises because of the hummocky nature of most wet tundra sites. The frost hummocks and ridges, which lie above water table, are dry and warm compared to the wet areas between; their surface temperatures can be as much as 20 Celsius degrees higher than those in the hollows (Halliwell, Rouse, and Weick, 1991).

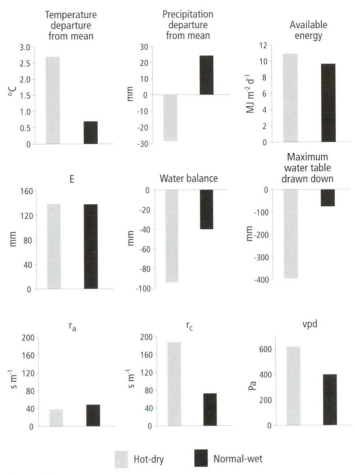

Figure 9.6
Comparative energy and water balances and surface and atmosphere controls
of a hummocky wetland sedge-fen during a hot-dry and a normal-wet growing
season (June through August).
Source: Rouse, Carlson, and Weick (1992).

Figure 9.6 shows the energy and water balances of a hummocky, wetland sedge
fen for two different growing seasons, in terms of the relation of temperature and precipitation to long-term average conditions. One season was hot and dry (H-D) and
the other was normal and wet (N-W). The H-D season's temperature was 2.2°C above
average, and its precipitation was 41 per cent below average, whereas the N-W season had precipitation 28 per cent greater than average. In the H-D season, all standing
water in the hollows disappeared during most of the period. Several observations
emerge. The latent heat flux and hence evapotranspiration were of similar magnitude
in both years. Net available energy was about the same in both years because of similar solar radiation, ground heat flux, mean wind speeds, and surface roughness. The
hot, dry atmosphere, with its large vapour pressure deficit, is able to maintain a substantial latent heat flux in spite of the large moisture deficit in the H-D year. Although

the peat soils undergo surface drying, because of their large volumetric water capacity (typically 75 to 85 per cent), they remain relatively moist beneath the surface. The very dry atmosphere creates large, vertical vapour pressure gradients between the evaporating layer in the peat and the lower atmosphere, which, when accompanied by strong turbulent mixing over the windy tundra, drive the latent heat flux.

As noted above, in the barren lands it does not take much local relief to create dry tundra. On upland tundra, intense summer drying commonly favours a xerophytic vegetation cover, dominated by lichens, mosses, and drought-resistant sedges. Not only does this surface cover inhibit evapotranspiration, it also provides little decaying biomass to the ground surface. Given small biomass input and aerobic decaying conditions, there is often little development of peat soils, in contrast to wet tundra. This failure leads to reduced moisture capacity in the surface soil layers, which favours xerophytic vegetation.

The energy balance of upland, lichen-sedge tundra developed on raised beach deposits is in contrast with that of adjacent lowland, wet-sedge tundra for the same period (Figure 9.7). At both sites, $Q*$ and Q_G are of comparable magnitude. Bowen ratios for upland tundra are twice as large, because Q_E is 69 per cent smaller and Q_H 49 per cent larger. The large sensible heat flux can be witnessed visually, on warm summer days, in the strongly developed shimmering above upland surfaces, the same phenomenon seen above sand desert surfaces or asphalt highways in summer.

As a result of the smaller evaporation loss from upland tundra, more water drains to the water table after periods of rain. Because the water table slopes downward to the adjacent lowlands, the water drains laterally, to supply the lowland areas. This lateral drainage is a major water supplier, serving to maintain the wetlands and shallow lakes.

In more humid areas with frequent rainfall, upland tundra is often covered in a luxuriant lichen mat. The lichen can store large amounts of precipitation, which then evaporate without the moisture ever having reached the soil (Bello and Arama, 1989). The lichen also acts as a mulch, preventing evaporation from the ground beneath. As a result, the soil surface can remain very wet, and anaerobic conditions can exist. This condition gives a favourable environment for the development of peat (Figure 9.3a), and peat soils can develop to depths of 0.20 m or more because of conditions analogous to those favouring development of upland peat blankets in northwestern Europe and Newfoundland. With frequent rainfall of small magnitude, these upland surfaces are capable of evaporating even more vigorously than in the example shown in Figure 9.7.

Advection exerts a major influence on the surface climates of the higher latitudes. This is evident from the advective effect exerted by Hudson Bay on the Hudson Bay Lowland (Rouse, 1991). On a synoptic scale, Hudson Bay imparts its cold, moist characteristics to the overlying air, so that for roughly half the summer there is a cold, moist onshore flow of air into the Hudson Bay Lowland. The cooling effect is exerted up to 600 km inland and helps explain why the southern boundary of tundra and of continuous permafrost is so far south in the coastal regions of Hudson Bay (Figure 9.1). The influence that Hudson Bay exerts on the surface climate in the adjacent lowland is undoubtedly replicated along the Arctic Ocean's coastal regions, including the Arctic Islands. Similar temperature effects have been documented for the coastal zone of the Beaufort Sea in the vicinity of the Mackenzie Delta (C.R. Burn, pers. comm. 1996).

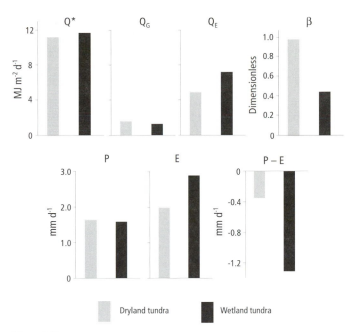

Figure 9.7
Comparative energy and water balances of dryland (sedge-lichen) tundra
and wetland (sedge) tundra under similar climatic conditions.
Sources: Rouse (1984b); Rouse and Bello (1985); Rouse, Carlson, and Weick
(1992).

Advection profoundly affects the surface energy balance. There is an exponential
increase in evapotranspiration rates in from the coast, so that maximum rates are
reached at about 12 km inland (Rouse, 1991). Further inland, the influence of isos-
tatic rebound is such that an older landscape is higher and drier, with more frequent
occurrence of thick peat soils, covered in xerophytic vegetation. Thus there is a
change to a larger proportion of dry, upland tundra and less of wet, lowland tundra,
with accompanying results of the type illustrated in Figure 9.7.

At the micro scale, there is advection between the hummocks and hollows of
wetland tundra. With horizontal temperature differences of up to 20 Celsius degrees
between hummock and hollow (Halliwell, Rouse, and Weick, 1991), there is a strong
horizontal flux of hot, dry air from the hummocks to the wet hollows, which enhances evaporation from the latter. As a result, under hot, dry conditions, the water
standing in wet hollows can disappear very rapidly. Other examples of strong advective influences on lakes and on the subarctic forest are developed below.

Carbon Fluxes

There has not been much study of trace-gas exchanges in Canadian subarctic tundra.
The methane flux from wetlands is documented in Roulet et al. (1994) and Rouse,
Holland, and Moore (1995). These studies indicate that northern wetlands constitute
between 3 and 4 per cent of the global methane source. In tundra there are substantial

emissions from fens, marshes, and ponds, but very small emissions from bogs. Year-to-year variability is large, and a dry, warm year leads to methane emissions that are only one-third those of a wet, cool year. Controls are related to the ability of methane-producing organisms to function vigorously in cool, anaerobic soils, as opposed to their cessation of activity in warm, aerobic soils.

Burton, Rouse, and Boudreau (1996) and Schreader (1995) report on net carbon dioxide (CO_2) from a sedge fen in the northern Hudson Bay Lowland. Both studies found that during the growing season there was a net carbon loss to the atmosphere from the peat soils of the fen. This loss was particularly evident in 1994, which was the warmest and driest summer of the fifty-one–year meteorological record. The implication is that the carbon reservoir of the peatlands, which has built up over 2,200 years, is being depleted. These findings agree with similar results in other tundra areas, such as Alaska (Oechel et al., 1993). The negative carbon balance is related to growing seasons with a strong negative water balance. With warm, well-aerated organic soils, carbon dioxide is evolved from the organic breakdown of peats, which exceeds the rates of photosynthetic uptake by plants, particularly by mosses, which during dry periods go into dormancy. The soil's water balance strongly controls the methane and carbon dioxide components of the carbon balance. With dry conditions, methane release is inhibited and carbon dioxide release enhanced. Under saturated conditions, the pattern is reversed. Thus one of the most important influences of climate change involves the degree to which the water balance is altered. In a $2 \times CO_2$ equilibrium climate, warmer and wetter conditions in wetland tundra are likely to increase methane evolution and favour photosynthesis (CO_2 gain) over respiration (CO_2 loss). Warmer and drier conditions will decrease methane evolution and favour respiration losses over photosynthetic uptakes.

OPEN SUBARCTIC FOREST

The climatic significance of the subarctic forest zone was first examined by Hare and Ritchie (1972). Drawing on earlier works, these authors showed that large gradients in important bioclimate parameters (annual net radiation and length of thaw season) are encountered across the zone of subarctic forest and are linked to strong gradients in vegetative indices such as standing biomass and annual net production. The largest gradients in net radiation occur in spring, when albedo differences among tundra, open forest, and closed-crown boreal forest are greatest. They suggested that these $Q*$ differences lead to greater convective heating over the closed-crown forest, and hence a longer thaw season. However, they cautioned that these broad-scale comparisons were tentative and called for rigorous field investigations of all bioclimate aspects of these zones, including their surface energy exchange. A limited amount of work has been conducted to answer this call.

Radiation Balance

The radiation climate of the subarctic forest is governed by the openness of the canopy, which permits large quantities of light to penetrate to the forest floor. The only mathematical treatment of this problem specifically for subarctic open forest was given by Wilson and Petzold (1973). Though the authors obtained good agreement between model results and field measurements, the model they used was developed for a theoretical distribution of trees, and it is not clear how this fact affects the model's wider application. The input data are mean tree height, mean branch radius, and mean distance between trees. The model output is most sensitive to distance between trees. Subsequent field observations support this early modelling work. There appears to be a close linear relationship between the proportion of above-canopy radiation that reaches the forest floor and canopy closure (that percentage of the ground surface covered by tree crowns) in subarctic forests (Figure 9.8). From the available data, it appears that as much as 50 per cent of the incident solar flux penetrates to the surface for denser canopies, and as much as 80 per cent in sparser canopies. These values contrast with the 1–30 per cent typically found for closed-crown boreal forest (chapter 11).

Given the ease of penetration for solar irradiance, it might be expected that the understorey significantly determines the albedo of these forests. However, despite a large range in albedos of understorey surfaces, from 0.07 (water) to 0.22 (lichen) (Petzold and Rencz, 1975), this does not seem to be the case. Mean summer albedos reported for a number of sites in the subarctic fall near 0.12 (Lafleur, Rouse, and Carlson, 1992). This finding is comparable to the albedos (0.09–0.12) reported for closed-crown, coniferous boreal forest (chapter 11). Albedos of the open forest (0.34–0.42) are three to four times larger in winter than in summer. Clearly, the transition from closed-crown boreal forest, through open forest, to tundra sees changes in summer-time net radiation that are driven largely by receipt of solar radiation. Significant latitudinal differences in net radiation are experienced in winter and spring, mostly because of stepwise albedo changes, supporting the earlier hypothesis of Hare and Ritchie (1972).

Energy Balance and Evapotranspiration

The partitioning of net radiation over subarctic forest into other energy fluxes can vary considerably among sites (Figure 9.5). As with tundra, the proportion of $Q*$ consumed by Q_G in open forest depends on soil type, wetness, and presence or absence of permafrost. During summer, Q_G consumes about 16 per cent of $Q*$ for open forest underlain by permafrost (Rouse, 1984a). Where permafrost is sporadic or absent, midsummer values of $Q_G/Q*$ range from 5 per cent to 9 per cent. During winter, trapping of snow by open forest helps control the ground thermal regime. The deep snowpack that collects in the forest insulates soil from the extreme cold in the atmosphere. However, the tundra environment is windswept, leaving little or no snowcover. Accordingly, soil in the forest experiences a smaller total heat loss and remains warmer than that on the tundra. Midwinter measurements indicate that Q_G accounts for less than 2 per cent of $Q*$ in subarctic forest (Renzetti, 1992).

Summertime energy budgets for subarctic forest, assembled from the available studies, are given in Figure 9.5. Of central interest is the partitioning of energy between Q_H and Q_E, as expressed by the Bowen ratio. The variation in Bowen ratios among open-forest sites seen in Figure 9.5 can be attributed to several factors: local soil wetness, regional climatic features, advective influences (described above in this

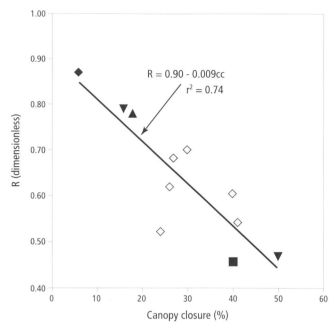

Figure 9.8
Ratio (R) of global solar radiation received at the forest floor to that above
the canopy, as a function of canopy closure for subarctic forests.
Sources: Lafleur and Mantha (1994). Data derived from Lafleur and Adams (1986): upright triangle; Lafleur and Mantha (1994): open diamond;
Petzold (1981): inverted triangle; Slaughter (1983): square; and Wilson
and Petzold (1973): closed diamond.

chapter), and, to some extent, differences in the methods used to calculate energy
fluxes. For example, while sites W3 and W4 were both "wet" (in a relative sense),
with similar canopy structures, the Bowen ratios are markedly different. W4 was in a
continental location and represents only a limited number of "warm" days. In contrast, W3 was conducted over a complete summer near the Hudson Bay coast, where
cold-air advection from the bay strongly influences Bowen ratios (Lafleur, 1992). W2
and W5 represent drier environments, where ground cover between trees is almost exclusively a contiguous lichen mat. Despite similar meteorological conditions at both
sites, the derived Bowen ratios are markedly different, probably in part because of
canopy structure. The trees at W2 were on average twice as tall and somewhat denser
than those at W5. However, different methods were used to derive evaporation. Q_E at
W2 was measured by the soil water balance method, while energy fluxes at W5 were
measured by sophisticated eddy-correlation techniques. Fitzjarrald and Moore (1994)
also describe the influence of synoptic climate conditions on the surface energy balance at W5. In early summer, Q_E responds to topographically controlled wind channelling and frequent synoptic disturbances. Later, a regional shift in air mass brings
winds from the southwest and the warmest and driest time of the year, when daytime
Bowen ratios often reach 2.5–3.0. The study at W1 was not based on field measure-

ments; instead, fluxes are estimates from climatological formulae. Clearly, the energy balance over subarctic open forest can vary greatly from site to site, depending on terrain and climate. The range of Bowen ratios, however, is not markedly different from those presented for closed-crown boreal forest, as discussed in chapter 11. Average water losses for open forest range from 1.1 to 3.6 mm d^{-1}, figures comparable to those reported for tundra (Figure 9.5).

The only simultaneous comparisons of energy balances over tundra and subarctic forest are reported by Lafleur, Rouse, and Carlson (1992). They show that daily Q_E from wetland tundra (T2) usually exceeds Q_E from a nearby forest site (W3). In a single summer season, total Q_E from the tundra was about 35 MJ m^{-2} (equivalent to 14 mm of water) larger than that from the forest, but Q_H from the forest was 70 MJ m^{-2} larger than that from the tundra. Seasonal average Bowen ratios were 0.81 and 0.63 for forest and tundra, respectively. Such differences can affect atmospheric circulation, since sensible heat usually warms only the first few kilometres of the atmosphere, while latent heat (released as condensation at the levels of cloud formation) warms the middle and upper troposphere. This study shows also that the differences in β between tundra and subarctic forest persist during both wet and dry years, but are smallest during dry years. One conclusion from this research is that water losses by evapotranspiration from northern basins dominated by wet tundra are larger than those from basins dominated by open forest, at least where they are exposed to the same regional climate. Advances and retreats of treeline in response to changing climate would then have profound implications for basin hydrology.

The second focus of the study by Lafleur, Rouse, and Carlson (1992) and a later one by Lafleur and Rouse (1995) was the sensitivity of the surface energy balance of subarctic forest and tundra to climatic changes. For this purpose, the highly variable meteorological conditions experienced at the Hudson Bay coastal sites were used as a surrogate for climate change. Bowen ratios for tundra seemed significantly more sensitive to changes in regional airmasses than did those of the forest. This behaviour is captured in Figure 9.9, which shows that the relative magnitude of the Bowen ratio in tundra and forest varies with atmospheric vapour pressure deficit. When the atmosphere is cool and moist, Bowen ratios in tundra and forest differ only marginally. As the air becomes drier (vapour pressure deficit and temperature increase), the tundra and forest Bowen ratios begin to diverge significantly. These studies conclude that this behaviour is the result largely of greater sensitivity of Q_E in wet tundra to changes in temperature and atmospheric dryness. This result is somewhat unexpected, because evapotranspiration from forests, which typically have small aerodynamic resistances, tends to be linked closely to changes in vapour pressure deficit through stomatal control of the plant canopy. However, we hypothesize that latent heat flux from this subarctic-forest site was influenced by other controls, related to canopy architecture and evaporation at the forest floor.

It appears that the physical and physiological controls on Q_E in subarctic forests differ markedly from those in closed-crown boreal forest. Lafleur (1992) shows that the "big leaf" approach of the one-dimensional Penman-Monteith model of evapotranspiration (equation 2.18), which often gives a good description of canopy–atmosphere interactions for closed-crown forest, is inappropriate for open forest. In particular, the assumption that fluxes of latent and sensible heat have the same

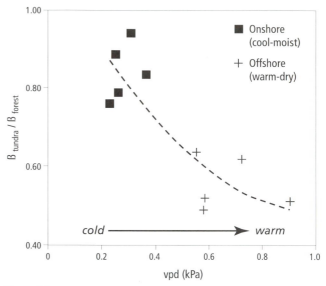

Figure 9.9
Ratio of tundra to subarctic forest Bowen ratios ($\beta_{tundra}/\beta_{forest}$), as a function of vapour pressure deficit. Data represent mean values for each of five summers and are grouped according to wind direction relative to the Hudson Bay coast. "Onshore" are winds originating from Hudson Bay; "offshore" are winds originating from inland. Dotted line represents approximate trend.

source/sink distribution within the canopy is not valid. Lafleur shows that the tree canopy of the open forest contributes a minor fraction (10–30 per cent) of the total latent heat exchange; the larger, remaining amount is derived from sources at the forest floor. The canopy's ineffectiveness in supplying water to the atmosphere is attributed to the small leaf area of these forests and large stomatal resistances at the leaf surface (Lafleur, 1992). Investigation of evapotranspiration at the forest floor for the site suggests that the main water sources are transpiration from the shrub layer, evaporation from pools of open water, and evaporation from non-vascular plant species, such as moss and lichen beds. The contribution from each of these sources depends on its absolute evaporation rate and its areal extent in the forest (Figure 9.10). Absolute evaporation rates vary considerably, with non-transpiring lichen beds having the lowest rate and open water the highest (Lafleur and Schreader, 1994). These results can at least partially explain the large Bowen ratio observed in the study at W5 (Figure 9.5), where the forest floor was covered exclusively with lichen beds.

   As yet, there is little known about the controls on Q_E from sources at the forest floor or about the nature of the turbulence structure within the canopy, which couples the forest floor to the overlying atmosphere. Despite this dearth of information, it is interesting to speculate about the role of changing forest structure on energy partitioning over the open-forest zone. In the present context, infilling of the open forest, which has been observed in northern Quebec (Payette and Filion, 1985), will alter water loss, depending on the present contributions of the forest floor. For sites such as

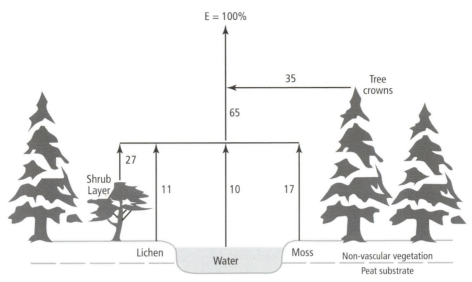

Figure 9.10
Conceptual model of evapotranspiration sources in a subarctic forest. Values represent components of evapotranspiration (E), given as percentage of total water flux (= 100 per cent).
Source: Lafleur and Schreader (1994).

W3, where water loss from the floor is a large proportion of forest Q_E, evapotranspiration might be reduced as forest density increases. However, in the drier sites, such as W5, infilling of the forest could increase summer water losses.

There is virtually no information available at this time on the fluxes of trace gases from subarctic forests.

Clearly, the unique and interesting features of the radiation and energy balances of the subarctic open forest, and those that influence surface water balance, are likely to alter with the make-up of the forest, in response to climatic changes. However, prediction of the effects of climate change on this type of terrain must consider that variables will feed back into climate by changing the energy balance. One must agree with Hare and Ritchie (1972), who describe the interaction between the subarctic forest and climate and state: "This relationship works as an interlocking system rather than as a simple control of vegetation by climate."

SUBARCTIC LAKES AND PONDS

The ice-free period for shallow subarctic and tundra lakes and ponds is very short. The smaller and shallower lakes are the first to freeze in early winter and the first to thaw in spring. Typically, lakes begin freezing in early winter, when air temperatures remain below freezing for several days. On a cold, calm evening in early October, the first skin of ice will form on the water surface. Turbulent mechanical mixing in the lake ceases, as it becomes decoupled from the ambient wind. As winter sets in, daily net radiation becomes increasingly negative, and, as water temperatures drop below 4°C, the colder, least dense water stratifies just below the ice surface. This is the period of accelerated conductive heat loss and ice growth. Ice thickens until net radiation becomes positive in spring and in shallow tundra lakes results in 1.50 to 1.75 m of ice growth, on average. Since 10 mm of fresh snow has the same insulating capability as 0.25 m of ice, large variability in ice thickness results if drifting snow is prevalent. In general, windswept and snow-free lakes such as those in Keewatin experience greater ice growth than those where drifts 0.5 m deep are commonplace, as in northern Quebec.

In spring, primary ice generally melts from the bottom up. Slushing occurs at the ice–water interface in deeper lakes or at the ice–lake sediment interface in shallow ponds. Development of a complete and continuous regional permafrost zone is prevented because the lakes do not freeze to the bottom (taliks). Lake depth is also critical to the mechanisms of ice break-up. For at least a week before the ice starts to melt, snowmelt on tundra progresses. Water that percolates down from the snow reaches the depressions formed by lakes and ponds. At this time, it is not uncommon for all water bodies to take on an aquamarine hue, with meltwater ponding on top of the ice surface. This somewhat unnatural juxtaposition of relative densities is short-lived on deeper lakes, as the lake ice that is bonded to the shoreline weakens and allows the ice mass to float again. All surface water and subsequent meltwater find their way below the ice surface.

In contrast, ice on shallower ponds is bonded to the lake bottom. Meltwater may accumulate to a depth of 0.5 m or more before bonds weaken sufficiently to permit the ice to rise. In this case, restoration of floating ice is known to occur in dramatic fashion, with a single, explosive rise. With removal of surface meltwater, secondary thinning of ice proceeds from the top down. The process involves both convective energy transfer downward from the atmosphere and radiative melt. The latter is particularly distinctive. Solar irradiance is preferentially transmitted downward along the vertical ice-crystal boundaries in black ice. The resultant, localized melt produces equally-spaced hexagonal columns of ice on the lake surface, which are 30 to 70 mm across. These eventually melt to sharpened spikes about 0.3 m long, attached at their bases to an elastic, floating ice slab perhaps 0.1 to 0.2 m thick. The quaking mass of candle ice undulates in symphonic fashion with gusts and lulls in the wind and shifts about the lake as open leads permit. Up to this point, water temperatures remain at 0°C. With arrival of a renewed wind, the ice cover disintegrates and is driven to the downwind end of the lake. Within hours of the onset of mechanical lake mixing, water temperatures rise above 0°C. The last vestiges of ice disappear within hours on smaller ponds and within a couple of days on deeper lakes. In the case of the latter, two to three weeks have elapsed since the open tundra became snow-free.

Until they have completed their freezing cycle in midwinter, lakes and ponds give off substantial amounts of heat during the freezing process. In spring, equivalent amounts of latent heat must be used in thawing, and this requirement delays thaw until after final snowmelt on terrestrial surfaces, as noted above. Lakes more than 2 m deep may remain partially frozen during spring. In midwinter, lakes are dominated by the energy balance of snow, and their energy balance is probably not much different from that of the contiguous tundra. As in the surrounding tundra, for lakes the most dynamic time for energy exchange is the ice-free period – the subject of the subsequent discussion.

At first glance, the summer radiation and energy balance of northern water bodies is invitingly simple. Compared to heterogeneous, vegetated surfaces, water is uniform in specific heat and thermal conductivity and relatively uniform in albedo. The air layer next to the surface is saturated at surface temperature, and the supply of moisture to the atmosphere is unlimited. The ground heat flux into the lake bottom is of relatively minor importance, but the storage flux in the water is often very large, even in relatively shallow ponds. Net available energy is distinctly asynchronous with the diurnal flux of solar radiation – the single most distinguishing feature of lake microclimate, as we see below, for it affects not only the timing of fluxes but also their magnitude and direction.

Radiation Balance

The lake's control on surface net radiation is exerted by albedo and surface temperature. Water has the lowest average albedo of all natural terrestrial surfaces, which results in the highest net shortwave radiation. Northern lakes show a strong dependence of albedo on zenith angle, when the waters are calmest. For small, shallow ponds, calm frequently corresponds to periods of largest solar zenith angle in early morning and late evening. Lake albedo is also affected by suspended sediment and water depth. In shallow ponds the entire water column participates in absorption and scattering of solar irradiance. This effect can be quite variable. With an albedo of 30 per cent, bottom calcareous sediments near shorelines decrease net solar radiation locally. Often northern ponds contain considerable organic matter washed from surrounding peat banks. As bottom sediment, this material has an albedo of close to 4 per cent. It is slightly more absorptive than water itself and would have minimal effects on net solar radiation. In shallow ponds if the bottom has a low albedo it is effective as a heat source for both ground heat flux and lake warming from below.

The longwave flux emitted from the lake is coupled to absolute surface temperature through the Stefan-Boltzmann law. Because of their shallowness and resultant thorough mixing, northern lakes' anomalies in surface temperature should be comparable to similar mid-latitude lakes, with maximum anomalies of 3 Celsius degrees or less (Ledrew and Reid, 1982). At normal summer temperatures, this corresponds to maximum differences in the radiation flux of about 12 W m^{-2} between shallow and deep areas. This variation probably need not inhibit estimation of mean net longwave loss from the lake, but it may make it difficult to obtain accurate diurnal estimates of surface vapour pressure.

On average, over a moderate-sized lake during the ice-free period net radiation comprises 65 per cent of incoming solar radiation. There is little to suggest that differences in albedo or outgoing longwave radiation should significantly affect the spatial pattern of net radiation on small northern lakes. This is fortunate, since most radiation studies have depended on measurements from a single instrument location on their lake.

Energy Balance and Evaporation

A large portion of the net radiation of a lake goes into warming the water during daytime (Figure 9.11). The most important warming process is penetration of solar

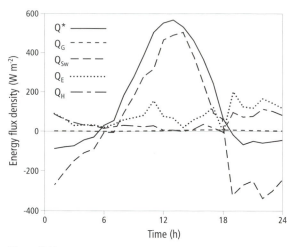

Figure 9.11
Energy balance of a shallow tundra lake. Q^* is net radiation, Q_G heat flux across lake bottom, Q_{SW} heat storage in water, Q_E latent heat flux, and Q_H sensible heat flux.

radiation beneath the surface. Several metres of water participate simultaneously in the radiation exchange. This feature is not shared by a forest canopy, for example, where conduction progressively diminishes as the tree core warms up. Another factor favouring substantial heat storage is the high heat capacity of water, which retards a lake's temperature increase in response to a large input of solar radiation. Convective mixing is also important. Once ice melts and water temperature exceeds 4°C, the warmest water is least dense and tends to stratify near the lake's surface. Strong wind-driven vertical mixing distributes this warm water throughout the lake's total depth and encourages heat storage. This thorough mixing leaves vertical temperature gradients (thermoclines) poorly developed or absent in shallow tundra lakes. Exceptions occur only rarely, during daytime, under very calm conditions. In addition, where a dark-bottomed lake strongly absorbs solar radiation, a second radiatively-active surface is established from which local buoyant convection enhances the mixing process. Field experiments indicate that temperature differences between surface and bottom waters in shallow tundra lakes in summer rarely exceed one Celsius degree (Stewart and Rouse, 1976; Bello and Smith, 1990).

During daylight periods of positive net radiation, there are many hours when lake warming consumes all the energy and there is no net available energy for evaporation or sensible heat transfer into the atmosphere. However, at night, lake cooling typically releases sufficient energy to promote strong fluxes of latent and sensible heat. Thus where lakes comprise a major part of the landscape, there is a large source for nocturnal evaporation and atmospheric heating, and the atmospheric fluxes are decoupled from the diurnal net radiation regime.

As pond depth increases, the magnitude of temperature change decreases linearly for a given heat input (Figure 9.12). Thus the magnitude of daily heat storage in

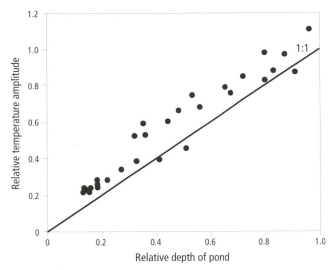

Figure 9.12
Pairwise comparison of diurnal temperature changes in eight tundra ponds. Vertical axis is T_d/T_s (dimensionless), and horizontal axis z_s/z_d (dimensionless), where T and z are daily temperature range and mean pond depth, respectively, and s and d are shallower lake and deeper lake, respectively.

tundra lakes and ponds is probably almost independent of lake depth, and so lake heat storage can potentially be modelled for isothermal tundra lakes in the absence of knowledge about the depth of individual lakes.

An unpublished study of the ground heat flux across the bottoms of ponds that average 1.3-m in depth shows the following. The bottoms were frozen at the beginning of the measurement period. The maximum fluxes of ground heat represent about 6 per cent of net radiation, which is considerably less than the maxima reported for vegetated tundra terrains, as discussed above. Despite saturated sediments and high thermal conductivities, vertical temperature gradients across the bottom sediments are never large enough to promote heat flux as large as that for permafrost terrain. However, gradients are greater than those for large lakes in temperate latitudes, and ground heat fluxes are relatively more substantial. The comparatively larger gradients are a response to the thorough mixing of the ponds, the presence of permafrost and of continuous, cold bottom-soil temperatures, and radiative warming of the top layer of the bottom sediments.

Evaporation from the shallow, smaller northern lakes is influenced strongly by the nature of the surrounding terrain and the process of advection. In general, northern lakes and ponds are not large enough to maintain their own, characteristic climatic identity. During the early hours of a typical sunny day, most net radiation is consumed in lake warming. However, at the same time latent heat flux is substantial. All the energy driving evaporation is derived from the downward-directed flux of sensible heat. The temperature inversion that promotes downward-directed Q_H over the lake occurs whenever warmer air from the heated, upwind terrain is advected over the colder lake waters. Thus lake evaporation is coupled to the upwind supply of radiant energy through the intermediary of the upwind flux of sensible heat.

The combination model, as presented by Priestley and Taylor in equation (2.20), provides a framework within which to examine the relative roles of energy supply and atmospheric demand on lake evaporation. The amount by which the α_P coefficient is larger than 1.00 represents the contribution of atmospheric demand through advection, which is over and above the input from available radiant energy. Figure 9.13 shows the hourly readers of α_P during midsummer at the centre of an oval, tundra lake approximately 500 m long and 300 m wide and 1.3 m deep. The surrounding vegetation is predominantly dryland lichen heath. The day-to-day and diurnal variation of α_P is large. There is a pronounced relationship between the magnitude of α_P and sky conditions; α_P is largest under sunny or partially cloudy skies, when sensible heat flux from the dry, upwind surface is largest, and as a result advection of hot, dry air over the lake is greatest. In contrast, α_P is smallest during overcast periods associated with rain, when sensible heat flux from the upwind surface is smallest. The maximum values are among the largest reported for any terrestrial surface, and the variability demonstrates the strong complementarity between lake and land evaporation.

Scale is very important in explaining surface climates of northern lakes. It is the "smallness" of ponds that presents unique conditions that are relatively unimportant in larger water bodies. Not only does advection influence lake and pond evaporation from day to day, but its influence varies with distance from the leading edge as well. Evaporation is greatest at the upwind margin, where advected air from the upwind surface is hottest and driest, and it decreases exponentially downwind (Bello and Smith, 1990). Thus, with all other factors being constant, smaller ponds evaporate more than larger lakes.

CONCLUSIONS

Variation in surface climate across the subarctic and low Arctic is shown to be a function of local surface and vegetation features, advective influences at several scales, and climatic gradients induced by latitude and the distribution of land and sea. Vascular vegetation seems to be less a control on energy and water balances in the region than it is in more southerly landscapes, largely because of the small leaf areas and biomass of vascular plants. However, non-vascular vegetation, particularly lichens, are a key element in the functioning of the water balances of tundra and open forest. While net radiation varies little between terrain types during snow-free periods, the partitioning of energy is controlled by surface wetness and the presence of permafrost. Advective effects are shown to be significant at all spatial scales. The complementary relationship between water bodies and surrounding tundra is particularly important. However, its effects on the convective boundary layer have not been fully investigated. Large-scale advection from the polar seas strongly influences energy partitioning for many kilometres inland. Sea ice enhances these effects. Overlying all the above features are the climatic gradients imposed by latitude. The longevity of snow-cover in spring is a crucial feature of this region, and it is potentially sensitive to

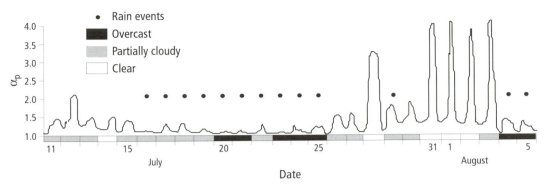

Figure 9.13
Priestley-Taylor evaporation coefficient (α_P) (dimensionless) over a shallow tundra lake during midsummer. Rain events are denoted by •.
Source: Bello and Smith (1990).

climatic changes. For example, snowmelt one month earlier in spring would increase net radiation during the growing season by about 18 per cent for tundra and about 22 per cent for lakes (Rouse, Carlson, and Weick, 1992).

Several questions about surface climates in the low Arctic and the subarctic remain. Many concern the influence of climate change on these systems. It is known that not all terrain types will respond in the same magnitude to an imposed change in temperature (Lafleur and Rouse, 1995). As well, even small shifts in the precipitation regime could enhance or offset precipitation and landscape response to a temperature change. The feed-backs arising between the subarctic landscape and the atmosphere are still poorly understood – particularly those among vegetation and climate, the carbon balance, and permafrost. For example, since peat soils of this region represent a major store of carbon, there will almost certainly be changes in the net flux of carbon across the surface–atmosphere interface if temperature and precipitation regimes are altered. The response of the carbon of many subarctic surfaces to climate change is still not understood. In another example, many climate models conceive of the albedo changes expected with significant northward advances in treeline when climate warms as a significant bioclimate feed-back. The resulting changes in energy partitioning and on the surface water balance are less well studied.

REFERENCES

Bello, R.L., and Arama, A. 1989. "The Water Balance of Lichen Canopies." *Climatological Bulletin* 23: 74–8.

Bello, R.L., and Smith, J.D. 1990. "The Effect of Weather Variability on the Energy Balance of a Lake in the Hudson Bay Lowlands, Canada." *Arctic and Alpine Research* 22: 98–107.

Blanken, P.D., and Rouse, W.R. 1994. "Modelling Evaporation from a High Subarctic Willow-Birch Forest." *International Journal of Climatology* 15: 97–106.

Burton, K.L., Rouse, W.R., and Boudreau, L.D. 1996. "Factors Affecting the Summer Carbon Budget of Subarctic Wetland Tundra." *Climate Research* 6: 203–13.

Danielson, E.W. 1969. *The Surface Heat Budget of Hudson Bay.* Marine Science Manuscript Report 9. Montreal: McGill University.

Ecoregions Working Group. 1989. *Ecoclimatic Regions of Canada, First Approximation.* Ecological Land Classification Series No. 23. Sustainable Development Branch, Canadian Wildlife Service, Environment Canada, Ottawa.

Findlay, B.F. 1978. "Dates of Formation and Loss of Snow Cover." In *Hydrological Atlas of Canada*, Ottawa: Department of Fisheries and Environment, Plate 10.

Fitzjarrald, D.R., and Moore, K.E. 1994. "Growing Season Boundary Layer Climate and Surface Exchanges in a Subarctic Lichen Woodland." *Journal of Geophysical Research* 99: 1899–1917.

Halliwell, D.H., and Rouse, W.R. 1987. "Soil Heat Flux in Permafrost: Characteristics and Accuracy of Measurement." *Journal of Climatology* 7: 571–84.

Halliwell, D.H., Rouse, W.R., and Weick, E.J. 1991. "Surface Energy Balance and Ground Heat Flux in Organic Permafrost Terrain under Variable Moisture Conditions." In *Permafrost: Fifth Canadian Proceedings*, Nordicana No. 54, Quebec City: Université Laval, 223–9.

Hare, F.K., and Ritchie, J.C. 1972. "The Boreal Bioclimates." *Geographical Review* 62: 333–65.

Lafleur, P.M. 1992. "Energy Balance and Evaporation from a Subarctic Open Forest." *Agricultural and Forest Meteorology* 52: 163–75.

Lafleur, P.M., and Adams, W.P. 1986. "The Radiation Budget of a Subarctic Woodland Canopy." *Arctic* 39: 172–6.

Lafleur, P.M., and Mantha, B.K. 1994. "Global Radiation within Subarctic Forest Canopies." *Canadian Journal of Forest Research* 24: 1062–6.

Lafleur, P.M., and Rouse, W.R. 1995. "Energy Partitioning at Treeline Forest and Tundra Sites and Its Sensitivity to Climate Change." *Atmosphere-Ocean* 33: 121–33.

Lafleur, P.M., Rouse, W.R., and Carlson, D.W. 1992. "Energy Balance Differences and Hydrologic Impacts across the Northern Treeline." *International Journal of Climatology* 12: 193–204.

Lafleur, P.M., and Schreader, C.P. 1994. "Water Loss from the Floor of a Subarctic Forest." *Arctic and Alpine Research* 26: 152–8.

Ledrew, E.F., and Reid, P.D. 1982. "The Significance of Surface Temperature Patterns on the Energy Balance of a Small Lake in the Canadian Shield." *Atmosphere-Ocean* 20: 101–15.

Oechel, W.C., Hastings, S.J., Vourlitis, G., Riechers, G., and Grulke, N. 1993. "Recent Change of Arctic Tundra Ecosystems from a Net Carbon Dioxide Sink to a Source." *Nature* 361: 520–3.

Payette, S., and Filion, L. 1985. "White Spruce Expansion at the Tree Line and Recent Climate Change." *Canadian Journal of Forest Research* 15: 241–51.

Petzold, D.E. 1981. "The Radiation Balance of Melting Snow in Open Boreal Forest." *Arctic and Alpine Research* 13: 287–93.

Petzold, D.E., and Rencz, A.N. 1975. "The Albedo of Selected Subarctic Surfaces." *Arctic and Alpine Research* 7: 393–8.

Pugsley, W.I. 1970. *The Surface Energy Budget of Central Canada.* Publication in Meteorology No. 96, Department of Meteorology, McGill University, Montreal.

Renzetti, A.V. 1992. "Wintertime Energy Balance of a Subarctic Forest." MSc thesis, Trent University.

Ritchie, J.C. 1987. *Post Glacial Vegetation of Canada.* New York: Cambridge University Press.

Roulet, N.T., Ritter, J., Jano, A., Kelly, C.A., Klinger, L., Moore, T.R., Protz, R., and Rouse, W.R. 1994. "Role of the Hudson Bay Lowland as a Source of Atmospheric Methane." *Journal of Geophysical Research* 99(D1): 1439–54.

Roulet, N.T., and Woo, M.K. 1986. "Wetland and Lake Evaporation in the Low Arctic." *Arctic and Alpine Research* 18: 195–200.

Rouse, W.R. 1982. "Microclimate of Low Arctic Tundra and Forest at Churchill, Manitoba." In *Fourth Canadian Permafrost Conference, Edmonton*, Ottawa: National Research Council, 68–80.

– 1984a. "Microclimate at Arctic Treeline. 2: Soil Microclimate of Tundra and Forest." *Water Resources Research* 20: 67-73.

– 1984b. "Microclimate at Arctic Treeline. 3: The Effects of Regional Advection on the Surface Energy Balance of Upland Tundra." *Water Resources Research* 20: 74–8.

– 1991. "Impacts of Hudson Bay on the Terrestrial Climate of the Hudson Bay Lowlands." *Arctic and Alpine Research* 23: 24–30.

Rouse, W.R., and Bello, R.L. 1983. "The Radiation Balance of Typical Terrain Units in the Low Arctic." *Annals of the American Association of Geographers* 73: 538–49.

– 1985. "The Potential for Climatic Modification in the Hudson Bay Lowlands through the Influence of the Ocean on the Energy Balance." *Atmosphere-Ocean* 23: 375–92.

Rouse, W.R., Carlson, D.W., and Weick, E.J. 1992. "Impacts of Summer Warming on the Energy and Water Balance of Wet Tundra." *Climatic Change* 22: 305–26.

Rouse, W.R., Holland, S., and Moore, T.R. 1995. "Variability in Methane Emissions from Wetlands at Northern Treeline near Churchill, Manitoba, Canada." *Arctic and Alpine Research* 27: 146–56.

Rouse, W.R., Mills, P.F., and Stewart, R.B. 1977. "Evaporation in High Latitudes." *Water Resources Research* 13: 909–14.

Schreader, C.P. 1995. "Carbon Budget Exchange over a Subarctic Wetland." MSc thesis, McMaster University.

Singh, B., and Taillefer, R. 1986. "The Effect of Synoptic-Scale Advection on the Performance of the Priestley-Taylor Evaporation Formula." *Boundary-Layer Meteorology* 36: 267–82.

Slaughter, C.W. 1983. "Summer Shortwave Radiation at a Subarctic Forest Site." *Canadian Journal of Forest Research* 13: 740–6.

Stewart, R.B., and Rouse, W.R. 1976. "A Simple Method for Determining the Evaporation from Shallow Lakes and Ponds." *Water Resources Research* 12: 623–8.

Wilson, R.G., and Petzold, D.E. 1973. "A Solar Radiation Model for Sub-arctic Woodlands." *Journal of Applied Meteorology* 12: 1259–66.

Wright, R.K. 1981. *The Water Balance of a Lichen Tundra Underlain by Permafrost.* McGill Subarctic Research Paper No. 33, Centre for Northern Studies, McGill University, Montreal.

CHAPTER TEN

Alpine Environments

IAN R. SAUNDERS, D. SCOTT MUNRO,
AND W.G. BAILEY

INTRODUCTION

The Cordillera of Canada exhibits a physiographic complexity unmatched by most other natural environments considered in this book. The landscape displays a multiplicity of slope angles and azimuths and is made even more complex by the diversity of surface types. Not only is the variability readily apparent at the micro-scale, but there are latitudinal changes too, for the alpine zone decreases in altitude as one moves further north. In addition, the effects of continentality produce longitudinal gradations. One of the outstanding characteristics of the Cordillera therefore is the tremendous spatial variability of the surface, which consequently produces a wide range of surface microclimates. At a broader scale, the influence of the western mountains on Canada's climates extends well beyond the geographical limits of the Cordillera itself. The continentality effect is intensified by the mountain barriers, and large rain shadow areas exist to the east of the Cordillera.

The great relief of the Cordilleran landscape dictates a range of surface types in close proximity. Valley bottoms are generally forested or, in the case of south-central British Columbia, dry grasslands. If the term alpine is defined as the zone that lies above the treeline, then the climates considered in this chapter are confined to those of the highest mountain ranges of the country. There we find a cold and windy environment, and one that lies above a significant proportion of the lower atmosphere and so is exposed to the regional airflow. The alpine areas of western Canada encompass a wide variety of surface types, with meadow, tundra, fellfield, bare rock, snow, and glacier ice forming the key elements (Figures 10.1 to 10.4). With such diversity of surfaces, defining the climatology becomes problematic, since local-scale advection must clearly play a strong role.

Alpine surfaces are characterized by poorly developed soils, sparse vegetation adapted to the extreme conditions, and cryonival/periglacial geomorphic processes and landforms in evidence. Alpine plants occur near the bounds of existence, and subtle changes in geology, soil development, water-supply, slope, and aspect conspire to produce highly variable communities. Areas just above the treeline often display a continuation of the meadow ecotone common in subalpine open forests. Vegetation is relatively abundant and diverse and grows to approximately 0.5 m high. In the higher and more exposed tundra locations, vegetation cover is restricted to approximately

Figure 10.1
A winter view of the Cascade Mountains in southern British Columbia, showing the topographic complexity characteristic of the Western Cordillera. The snow covers all the types of alpine surface, so that bare rock, glacier ice, moraines, tundra, and meadow are largely indistinguishable in the absence of ground inspection. The treeline is clearly defined in most areas, with montane coniferous forest in the valleys. The low-angle sunlight emphasizes the strong role of topography in determining the radiation balance at any one point on the ground, with topographic shading and reflection clearly evident.

Figure 10.2
View north from Peyto Peak in the Rocky Mountains of Alberta in October. Topographic shading effects are clearly evident. A myriad of surface types are shown, including dry bare rock surfaces (rock faces, talus slopes, and moraines), vegetated tundra (with some development of patterned ground indicating the effects of ground-frost action), and snow and glacier ice.

Figure 10.3
Patterned ground on a fellfield tundra surface at Scout Mountain, BC. The sparse vegetation and poor soil development are characteristic of all alpine tundra environments. A significant proportion of the surface consists of bare rocks, which drain quickly and, when dry, contribute nothing to the evaporative flux.

Figure 10.4
Researchers standing on bare ice in the ablation zone of the Llewellyn Glacier, BC, in August. The micro-relief of the surface of the bare ice of the ablation zone is about 0.5 m. A light and discontinuous layer of aeolian dust covers the entire surface. The dark strip running across the glacier in the middle distance is a medial moraine.

0.1 m high, contains fewer species of vascular plants and more lichens and mosses, and is spatially discontinuous. Where significant percentages of rock intrude, "fell-field" is the term commonly used and is often synonymous with periglacial patterned ground (Figure 10.3).

The non-vegetated parts of the alpine zone comprise three main components: rock, snow, and ice. Rock surfaces vary from smooth, glacially polished slabs to rough, physically weathered blockfields ("felsenmeer") and talus. Seasonal and perennial snow and glacier ice are characteristic features in most of the Cordilleran alpine environments and display a wide range of surface characteristics.

In western Canada, the principal alpine areas lie in the mountains encompassed by the major ranges depicted in Figure 10.5. The general distribution of the alpine zone is influenced by the combined effects of latitude and continentality. At 50°N, timber-line occurs at an elevation of ~1,600 m on the Pacific coast and rises to ~2,300 m in the Front Ranges of the Rocky Mountains. The corresponding timberlines at 60°N are ~600 m and ~1,500 m, giving a northward decline of roughly 100 m per degree of latitude. Superimposed on this general distribution are locally operative conditions such as lithology, slope steepness, and aspect, which help determine the treeline at any particular location.

The state of knowledge of physical climatology in the alpine environments of western Canada is almost wholly derived from only a few research sites (Figure 10.5). In general, these are topographically simple glacier basins and valley or mountaintop tundra sites. All the climatology studies of the Cordillera have had to overcome the logistical difficulties of access to alpine terrain and maintenance of instrumentation under inclement weather conditions. The relative paucity of data is therefore not surprising.

The network of weather stations in the Cordillera operated by the provincial and federal governments almost entirely excludes the true alpine zone. A disproportionate fraction of alpine climate measurements are made in valleys, which, although accessible, are not representative of adjacent alpine terrain. The importance of glaciers to the water resources of Canada has been recognized by the greater amount of data collected from snow and ice surfaces, as compared to tundra surfaces.

WIND AND MOMENTUM

The wind field in mountainous terrain is more complicated than that over land with low relief. Two major factors affect surface climates. First, mountains modify the regional airflow. At large spatial scales, mountain ranges affect the jetstreams of the upper troposphere and nearer to the surface impose the chinook effect on the westerly flow of maritime air. At the local scale, airflow is channelled in valleys. Somewhere between these two extremes we also see physically modified airflow in the genesis of lee waves and rotors downwind of mountain barriers. Second, spatial differences in thermal properties of the surface generate local-scale winds. In the absence of strong regional airflows, slope and valley winds are common in mountainous environments.

During the day, radiative heating of slopes warms the surface air layer, which then rises. This process generates upslope anabatic winds and a compensating downward

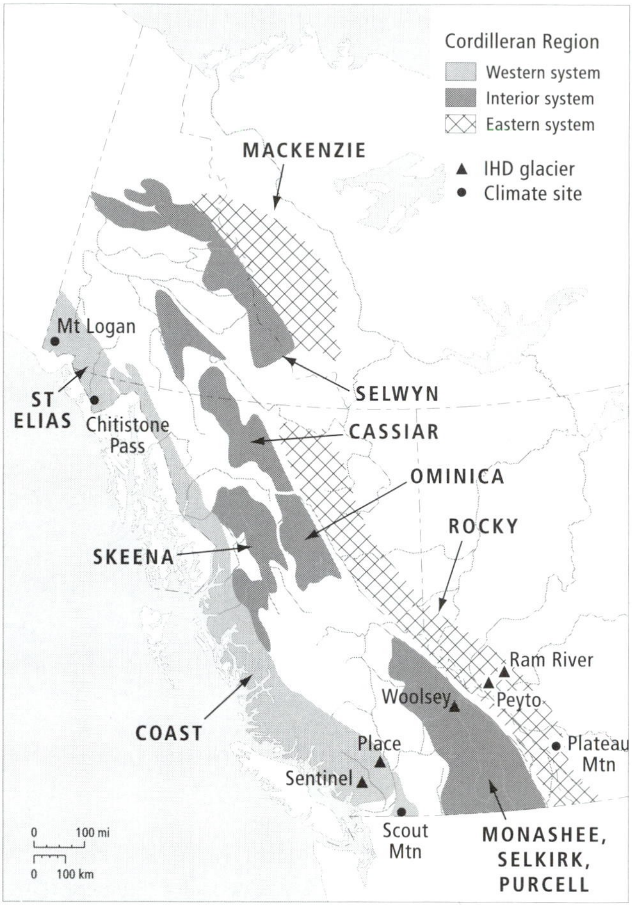

Figure 10.5
Generalized physiographic divisions of the Canadian Cordillera, showing
major mountain ranges (upper-case labels), International Hydrologic
Decade (IHD) glaciers, and other experimental sites mentioned in the text.

flow in the centre of the valley. At the same time a general up-valley airflow (the valley wind) develops, with a compensating down-valley flow (anti-valley wind) above the ridge level (Figure 10.6a). At night, the circulation reverses as a result of the surface cooling induced by the negative net radiation, and downslope katabatic airflow develops (Figure 10.6b). In glacierized basins, the permanently cold surface leads to katabatic winds, which do not show diurnal reversals in direction.

Alpine surfaces are typically very windswept. This results from their being at high altitude and so exposed to the airflow in the free atmosphere. Furthermore, the absence of large types of vegetation allows for higher wind speeds in the boundary layer. At the micro-scale, most alpine surfaces are relatively smooth compared to most other natural surfaces, and surface roughness is therefore insufficient to promote much drag or forced convection. Offsetting the lack of large-scale surface roughness elements in

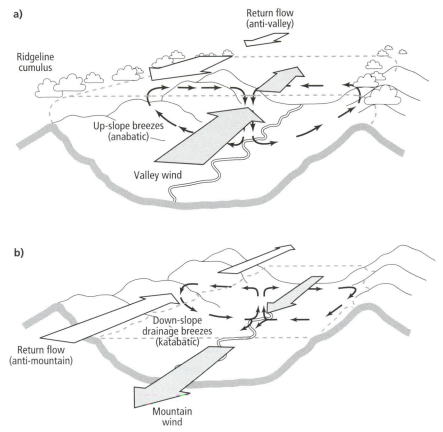

Figure 10.6
Thermally generated winds in mountain terrain. (a) Daytime: up-slope, anabatic winds generated
by surface heating, with a valley wind within the valley confines and an anti-valley wind above;
(b) night-time: down-slope, katabatic winds generated by surface cooling, with a mountain wind
within the valley and an anti-mountain wind above.
Source: Oke (1987).

the surface boundary layer is the turbulence induced in the planetary boundary layer
by the topographic complexity of the mountains themselves. Additionally, radiative
surface heating on bare rock and tundra surfaces can be extreme as a result of the
shorter atmospheric path lengths. This leads to the decreased stability of the boundary
layer.

For snow surfaces, momentum transfer is affected in several ways, producing
marked seasonality in turbulent transfer. A snowcover disallows surface heating above
the freezing point and so reduces free convection, and the smoothing effect of a snowpack on topographic irregularities reduces forced convection. As a result, momentum
transfer is relatively small in winter. As the summer progresses, snow and ice surfaces
commonly develop relief, which promotes momentum transfer. Snow develops "sun-
cups" and other micro-relief features. In extreme cases névé penitentes may develop on

snow or ice surfaces – these are pinnacle-like features, typically no more than a metre high, that ablate out of the snow/ice and are thought to result from strong evaporation or sublimation. The internal structure and flow patterns of alpine glaciers often result in the ice developing a longitudinal "grain," so that the ice surface's micro-relief consists of low ridges aligned in the flow direction. Momentum transfer is therefore dependent on wind direction, since roughness lengths are smaller for airflow along the glacier (such as a typical katabatic wind) than for crosswinds (Munro, 1989).

Since the depth of the boundary layer depends on fetch the heterogeneity of alpine surfaces presents logistical and theoretical problems for the physical climatologist. Unless the surface approximates an "infinite homogeneous plane," boundary-layer theory becomes difficult to apply, even to the point of being invalid altogether. In fact, the theory may apply to only the first metre of air above melting glacier ice, provided that a suitable flow structure for the application of log-linear profiles is established (Munro and Davies, 1977).

Such a structure can be observed in summer, when warm air overlies a glacier. This situation gives rise to a persistent temperature inversion, induced by the cooling of air in contact with the surface. The air flows down the glacier in response to gravitational acceleration, thus creating a katabatic flow known as the glacier wind. To the degree that anticyclonic weather conditions prevail, as is generally the case on fine summer days, the glacier wind dominates the airflow regime. However, glacier size is also important. Peyto Glacier, some 15 km^2 in area, is known to produce strong glacier winds, but weaker glacier winds have been experienced over the much smaller Place Glacier, where up-slope, anabatic winds from the valley below can influence the flow regime.

THE RADIATION BALANCE

The topographic complexity of the Cordillera helps ensure that the radiation balance in the alpine environment is accordingly complex, but there are few data to confirm this fact. Most of the measured radiation balances in the alpine environments of western Canada are derived from a few sets of experimental data. Field measurement requires that several site criteria be met. Unless the purpose of the research is to examine the effects of surrounding terrain on the radiation balance, the ideal measurement site is one with surface homogeneity and a minimum of obstructions to the atmospheric hemisphere. It is usually difficult to satisfy both requirements in the rugged terrain of western Canada's mountain ranges.

Atmospheric Controls on Incoming Radiation

It is to be anticipated that global solar radiation increases with height, since the atmospheric path lengths become shorter. Measurements of $K\downarrow$ in a variety of mountainous regions around the world confirm this hypothesis and show that $K\downarrow$ increases with altitude in both cloudless and cloudy weather. Several studies have also measured instantaneous values of $K\downarrow$ that exceed extraterrestrial radiation, usually when clouds produce some form of focusing effect on the solar beam (Marcus and Brazel, 1974; Bowers, 1988; Saunders, 1990).

Since alpine environments have 25 to 50 per cent less atmosphere overhead when compared to lowlands, the ratio of diffuse to global solar radiation in cloudless conditions also changes at higher elevations. Saunders (1990) measured a value of 0.16 at 2,350 m on Scout Mountain. At 2,480 m on Plateau Mountain, Huo (1991) noted values in the range of from 0.08 to 0.11, which should be compared to typical low-altitude values of about 0.25.

Despite the clear seasonality of the synoptic weather patterns that dominate the Cordillera (see chapter 1) and the attendant changes in cloud amounts, Saunders (1990) observed that the mean monthly atmospheric transmissivity changed little from winter to summer. Even during summer anticyclonic conditions, when the region experiences generally cloudless skies, thermally generated cumulus cloud develops over the mountains, decreasing transmissivity (Figure 10.7a). Indeed, one of the signatures of alpine radiation balances is the diurnal growth and decay of locally generated cloud cover. The effect may also extend beyond the physical limits of the mountains – orographically generated convective storms are common downwind of the interior mountain ranges during periods of hot weather.

The effects of altitude on atmospheric longwave radiation are less well documented, especially for Canadian mountains. Regardless of cloud cover, a decrease in $L\downarrow$ should be expected in alpine environments, since at higher altitudes the temperature is lower and there is less atmospheric mass overhead. Measurements made at alpine sites in the Cordillera demonstrate that, in the absence of strong topographic effects, the pattern of atmospheric longwave radiation there is similar to that non-alpine environments: relatively constant $L\downarrow$ is observed at hourly, daily, and seasonal time-scales (Figure 10.7). The tendency for cloudy-sky $L\downarrow$ to exceed cloudless-sky $L\downarrow$ remains a basic tenet (Figure 10.7a). Brazel (1974), Saunders (1990), and Huo (1991) have tested several published empirical $L\downarrow$ models in the Cordillera and find their performance generally satisfactory. The latter two authors also found that application of Stefan-Boltzmann's law, using air temperature and an assumed effective atmospheric emissivity of 0.70, worked as well as the published empirical models.

Characteristic of alpine radiation inputs are the modifications to the individual fluxes induced by surrounding terrain. At any given point on the surface, mountains may decrease the global solar radiation by shading that point (as observed by Saunders and Bailey, 1996), or they may increase it by reflecting radiation to that point. Given that the alpine environment is snow-covered much of the time, slope reflections may affect local inputs of shortwave radiation. Shading of direct-beam radiation by higher ground can easily be included in radiation models, but reflections from the same ground are less readily calculated. Topography may substantially affect the input of longwave radiation as well, though extensive research has yet to be directed towards this problem.

Some idea of the importance of topography, as opposed to cloud cover, can be gained by viewing a sequence of $K\downarrow$ and $L\downarrow$ measurements at Place Glacier and comparing the cloudy periods of the first and third days with the cloudless-sky measurements in between (Figure 10.7c). Global radiation is lower than expected for a cloudless sky when mountains block the sun, an effect that can readily be seen on the second day, soon after sunrise. Clouds reduce $K\downarrow$ whenever they obstruct the sun, and they cause substantial increases in $L\downarrow$. Mountains do not in this case obscure enough sky to affect the $L\downarrow$ measurements discernibly.

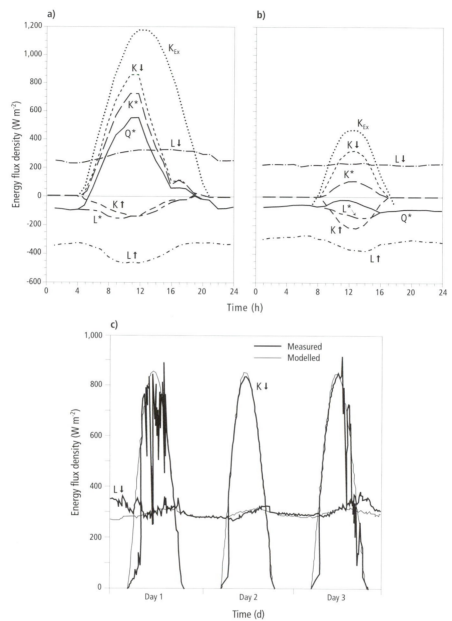

Figure 10.7

Hourly radiation balances for different alpine surfaces. (a) Summer tundra, Scout Mountain. The day began with cloudless skies, but a complete, orographically generated cumulus cover had developed by 1400 h. The afternoon cloud limits $K\downarrow$ but enhances $L\downarrow$, so both K^* and L^* tend towards zero values. (b) Winter tundra, Scout Mountain. Absorbed solar radiation is small on this cloudless day due to the low global solar radiation levels of the season and the high albedo of the snowcover, and Q^* is negative for the whole day. This is a typical winter radiation balance for a cloudless sky, regardless of the type of surface that underlies the snowcover. (c) Summer glacier, Place Glacier. Three-day sequence shows a partly cloudy day preceding a cloudless day, which is followed by a day with afternoon cloud. Topographic shading causes $K\downarrow$ measurements to be much less than modelled $K\downarrow$ for short periods after dawn. Large deviations of $K\downarrow$ and $L\downarrow$ from modelled values are the result of cloud.

The ability to model $K\downarrow$ in complex topography is invaluable for three reasons: the measurement network is inadequate (Hay and Suckling, 1979); most differences in the radiation balance between slopes and horizontal sites are generated by solar, not longwave, radiation (Huo, 1991); and $K\downarrow$ shapes some geomorphic, hydrologic, and ecologic processes. Early work on modelling of $K\downarrow$ in the Cordillera focused on relatively simple models driven by standard meteorological data. Recently, Huo (1991) successfully applied several physically based $K\downarrow$ models to horizontal and sloping terrain at Plateau Mountain and accounted for topographic effects in $L\downarrow$ models as well.

Surface Controls on Outgoing Radiation

Albedo exerts a strong control on alpine radiation balances at all temporal scales. It is high when snow covers the ground (Figure 10.7b) and lower when rocks, tundra, or glacier ice are uncovered (Figure 10.7a). The heaviest snowfalls in the Cordillera occur in the alpine zones of the Coast Mountains of British Columbia, where seasonal snowpacks several metres deep are common and the snow-free season may last only two or three months. Conversely, snowpacks are thinner in the drier interior ranges. They melt away earlier than the coastal snow and are more susceptible to wind transport and removal, thus increasing spatial and temporal variability.

Snow and ice possess a dynamic range of structural and radiative properties that give rise to a wide range of albedo values (Figure 10.8). Snow albedo depends on age, wavelength of radiation, zenith angle, grain size, water content, cloud cover, and surface micro-relief ("sastrugi"). Generally, snow albedo is between 0.5 and 0.9, with the lower end of the range applying to old snow and glacier firn, while glacier ice typically reflects from 0.2 to 0.5 of the global solar radiation. Sediment accumulation on and within the ice lowers the albedo.

Zenith angle dependence is well illustrated in the hourly patterns of Figure 10.8, which also show the dramatic fall in albedo that results from the melting away of winter snowcover. Seasonally, the drama is enacted over the entire glacier basin, accentuated slightly by the accumulation of wind-blown sediment, as the snowline migrates upward across the glacier surface. Thus, even though global solar radiation decreases following the summer solstice, $K*$ increases for a while because of the darkening of the glacier basin as snow melts away to reveal bare ice (Munro, 1991).

In extreme cases, alpine glaciers carry enough supraglacial sediment that they effectively behave like bare soil or rock, and some glaciers in the northern ranges even possess forest growth. More typically, variations in sediment content across the glacier surface cause enough albedo change to give rise to such subjective terms as "clean ice" and "dirty ice" (Figure 10.8), though what seems to be clean on one glacier may well appear dirty on another. Additionally, glaciers vary in surface topography, displaying, for example, smooth, planar surfaces, which may reflect like mirrors, and heavily crevassed icefalls, which trap solar radiation effectively.

In the absence of snow, tundra albedo is relatively constant – 0.16–0.17 at Scout Mountain (Saunders and Bailey, 1996); 0.17–0.19 at Plateau Mountain (Bailey, Weick, and Bowers, 1989; Huo, 1991); and 0.19 at Chitistone Pass (Aufdemberge, 1974). Rainfall appears to have little effect on the albedo of vegetated tundra, beyond a

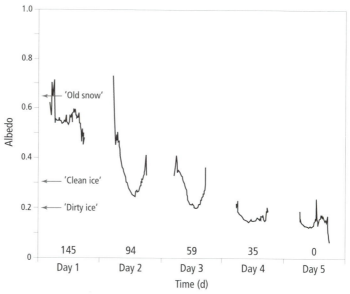

Figure 10.8
Five-day sequence of albedo changes that attend melting of snowcover from the ice surface of Place Glacier. The gaps in the record occur at night. The lowest values for albedo (dimensionless) on each day tend to be recorded around noon, depending on solar zenith angle. Mean daily snow thickness (mm) is given for each day. Typical α values for old snow and ice are given.

slight darkening of the surface. However, rock surfaces may show a greater range of albedos, depending on the lithologies involved.

The Stefan-Boltzmann law governs outgoing longwave radiation primarily through surface temperature, since surface emissivities are close to unity for most alpine surfaces. $L\uparrow$ from melting snow and ice, however, is constrained to no more than 316 W m^{-2} because surface temperature cannot rise above 0°C. Consequently, net longwave radiation may approach zero over glaciers in summer, implying the potential for glacier ice to have higher $Q*$ values than adjacent tundra, despite the generally higher albedo of ice (Aufdemberge, 1974).

Net Radiation

Net radiation is generally observed to decrease with increasing altitude. The decrease in $K*$ promoted by the persistence of snowcover plays a major seasonal role, but topography is important as well. Such effects explain the differences in daily total net radiation between snow and ice sites on Peyto Glacier (Table 10.1). There, despite higher $K\downarrow$ totals from a less-obstructed sky at the snow site, the higher albedo more than compensates, making $K*$ there much less than that over ice. Though some decrease in the magnitude of $L*$ at the ice site is to be expected because $L\downarrow$ increases at lower elevations, most of the increase is contributed probably by mountains overlooking the site.

Table 10.1
Average daily total radiation components for ice and snow on Peyto Glacier

	Ice	Snow
$K{\downarrow}$ (MJ m^{-2} d^{-1})	15.91	18.37
Albedo, α	0.36	0.67
$K*$ (MJ m^{-2} d^{-1})	10.24	6.09
$L*$ (MJ m^{-2} d^{-1})	−1.62	−2.63
$Q*$ (MJ m^{-2} d^{-1})	8.63	3.46

The annual trend of net radiation displays marked seasonality, which varies in character by surface type (Figure 10.9). The principal control is persistence of seasonal snowcover. Winter snowcover suppresses transfers of radiation regardless of atmospheric conditions; $K*$ is therefore small, and so at these times $Q*$ is dominated by $L*$. Thus we observe generally small, negative values of net radiation throughout winter. Net radiation does not become positive until snowcover becomes discontinuous. This process may take as little as a few days for exposed tundra and rock, where wind has kept snow accumulation to a minimum, or as long as a few months, for the deeper snow found on glaciers and in sheltered subalpine meadows.

In summer, on rock, tundra, and meadow surfaces, albedo is typically small and almost constant, and high inputs of solar radiation result in a large, positive value of net radiation. Atmospheric transmissivity is the dominant control, and so $Q*$ depends primarily on cloud cover. Similar conditions exist for glacier ice. Sporadic summer storms can upset this pattern at any time in the Cordillera by replenishing the snowcover and causing a step change in albedo. Perennial snow maintains a relatively high albedo, which keeps $Q*$ small.

Exploiting the strong dependence of net radiation on global solar radiation, empirical relations, of the form $Q* = f(K{\downarrow})$, have been derived from a wide variety of locations around the world. Bailey, Weick, and Bowers (1989), Huo (1991), and Saunders and Bailey (1994) have presented such relations for Canadian alpine tundra surfaces. Similar results have also been derived from other alpine areas. Empirical relations between $K{\downarrow}$ and $Q*$ for snow-free alpine tundra surfaces correspond very closely to those of Davies (1967) and confirm measurements (mostly European) which report that increased solar radiation loading at high altitudes is offset by greater net losses of longwave radiation. Since solar radiation strongly controls $Q*$, which in turn affects melt, many snow/ice studies have attempted to determine relations between solar radiation and ablation. Use of the form $Q* = f(K*)$ instead of $Q* = f(K{\downarrow})$ is a more logical approach, since it accounts for albedo changes. Munro and Young (1982) successfully employed this method at Peyto Glacier.

ENERGY BALANCE

Storage in the vegetation canopy and energy used in photosynthesis are both negligible in the energy balance of the alpine environment. Generally, heat input from rain is

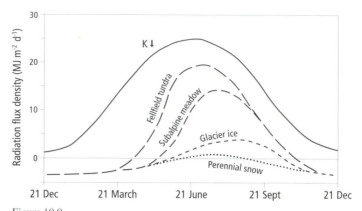

Figure 10.9
Net radiation for selected alpine surfaces at a mid-latitude location: highly
generalized annual trend. Idealized annual trend of $K\downarrow$ also is given.

also negligible in seasonal energy balances, although it may be important during individual storms. Therefore, if advective influences are absent, the simplified version of
the energy balance (equation 2.9) is appropriate for alpine tundra, meadow, and rock
surfaces. For snow and ice surfaces, heat storage changes in the snowpack and the
energy changes associated with internal freezing and thawing need to be considered.

All the methods of turbulent flux evaluation discussed in chapter 2 have at some
time been employed in alpine studies. Unfortunately, one of the greater difficulties
in determining turbulent fluxes in the alpine environment involves finding a measurement site with sufficient fetch and surface homogeneity. The problem is usually
smaller for glacier surfaces than it is for others in the alpine zone.

For modelling sensible heat in the tundra environment, Bowers (1988) and Saunders (1990) have used the energy balance–Bowen ratio, aerodynamics eddy correlation, and Ohm's law approaches with general success. Of particular interest is the
promising performance of the Ohm's law approach (right-hand side equation 2.12),
which has simple data requirements and has potential as an operational method. However, the precise specification of surface temperature is problematic. Use of satellite
thermal imagery is an obvious research avenue to explore.

On glacier surfaces, Munro (1989; 1990) has employed both the eddy-correlation
and the bulk-aerodynamic approaches effectively, though the former is not suitable
for routine, long-term use. That is best handled by the bulk aerodynamic method,
which has been extensively employed in glacier energy balance studies because the
instrumentation required is relatively simple to deploy and is robust.

Methods tested in alpine tundra for evaporation modelling include the aerodynamic equation, equilibrium evaporation, the Priestley-Taylor approach, and treating
Q_E as a residual in the energy balance equation. The Priestley-Taylor approach (equation 2.20) has potential operational value in alpine tundra, but reliable data on soil
moisture need to be generated. The relationship between α_p and soil moisture is not
sharply defined (Saunders and Bailey, 1994). Data collated from the Cordillera nei-
ther support nor reject the validity of $\alpha_p = 1.26$ as representative of potential evaporation in alpine tundra (Saunders and Bailey, 1994). Further investigations from wetter

alpine tundras, such as those found in the BC Coast Mountains, would be of value in defining the behaviour of α_p. The specification of surface moisture availability is rendered problematic by the presence of patterned ground. The alpine data do, however, show reasonable agreement with a similar trend found in arctic tundra by Marsh, Rouse, and Woo (1981) and are suggestive of characteristic links between arctic and alpine tundras. Assuming that $\alpha_p = 1.00$ (equilibrium evaporation) for well-drained tundra has proven useful in the subarctic (Stewart and Rouse, 1976). In the alpine environment, Saunders and Bailey (1990) found that calculated equilibrium evaporation agreed with measured Q_E on a wet day, but not for dry conditions.

Problems of surface specification also inhibit use of the combination model (equation 2.18). Though the model has been used to derive surface resistances for alpine tundra, insufficient data have been accumulated for researchers to determine r_c reliably for predictive purposes, and there are also concerns about the model's validity for a surface that lacks continuous vegetation cover.

Snow and Ice Surfaces

One of the principal factors that determines the behaviour of the energy balance is availability of surplus energy to initiate melting. The seasonal transition to a melt regime generally occurs in the spring as a result of increasing $K\downarrow$, decreasing albedo, and warmer air. However, melting may occur even in midwinter at any of the mountains in western Canada during the advection of warm air, though the southern coastal ranges are the most susceptible.

In winter, lack of strong solar radiation forcing leaves energy fluxes very small, displaying only conservative diurnal changes. Turbulent transfer is driven by negative net radiation, which promotes a temperature inversion at the surface and so heat fluxes are directed downward (Figure 10.10). The ground heat flux acts as a net source of energy in early winter as it transfers heat stored in deeper layers. For much of the snow season, however, Q_G is small, as are changes in the snowpack's energy content.

Winter energy balances of thin snowpacks in the tundra ecotone need not be so very different from those of a snow-free surface, even in winter. At Scout Mountain, Saunders and Bailey (1996) commonly observed that in the diurnal pattern of the winter energy balance daytime net gains in radiation were offset by losses in both Q_H and Q_E. The near-permanent exposure of vegetation and rocks above the thin, windswept snowpack caused the surface to behave simply as a colder, whiter variant of the summer surface, rather than as a thick mountain snowpack more typical of the subalpine and coastal alpine zones.

Though the very high mountains in some of the northern ranges experience a winter-like energy balance regime for the whole year, spring and summer witness a change in the energy balances of snow and ice in most of the Cordillera. At this time, sufficient surplus energy becomes available for melting to occur. Most studies show that Q^* is the dominant source of energy during melt, but the roles of Q_H and Q_E appear to vary widely among studies. Some experiments show both turbulent fluxes to be insignificant, others indicate that only sensible heat is an important turbulent flux, and yet others show that both Q_H and Q_E are major sources of energy to the melt process. The last-named cases arise in maritime locations, where persistent heavy cloud

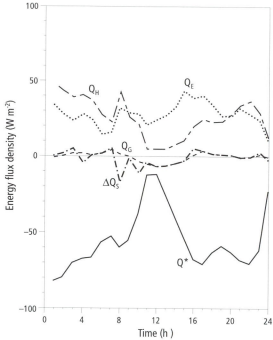

Figure 10.10
Hourly energy balance for snow-covered tundra at Scout
Mountain in winter. Negative $Q*$ forces the turbulent fluxes
(Q_H, Q_E) to become sources of energy for the surface, while
Q_G is almost negligible for the whole day. Compare the
magnitude of the fluxes to those for summer in Figure 10.12.

cover produces much condensation at the surface. During conditions of heavy cloud
cover and warm maritime air masses, turbulent fluxes may replace $Q*$ as the dominant heat source for a mountain snowpack. The relative contributions to the melt process of $Q*$, Q_H, and Q_E also differ between snow and glacier ice. Since the latter has a
lower albedo and absorbs more radiation than snow, it has a higher contribution from
$Q*$ and less from the turbulent fluxes (Munro, 1990). One of the consequences is that
the source of the energy for melting a glacier can change during summer as the snowline migrates and the lower glacier changes from a snow to an ice surface.

Most sites in the Cordillera could well follow the summer energy balance patterns
observed at Peyto Glacier. Regardless of whether one deals with ice (Figure 10.11a)
or snow (Figure 10.11b), Q_E appears to be a small term, arising from evaporation on
some occasions, condensation on others. Q_H is clearly the dominant turbulent transfer
at both sites, persistently supplying heat to the surface. The main contrasts occur
in the $Q*$ values for the two sites. While $Q*$ is arguably the more important energy
source for melting snow during daylight hours (Figure 10.11b), it is unquestionably
the most significant for melting ice (Figure 10.11a) because ice absorbs roughly twice
as much solar radiation as snow does. At night, the patterns take on a more familiar
form, in which negative $Q*$ is compensated by one, or more, of the other energy
balance terms.

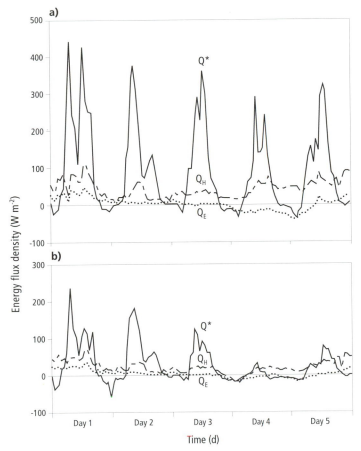

Figure 10.11
Surface energy balances of Peyto Glacier: matched five-day sequences from
(a) the ice surface and (b) the snow surface. The relatively low magnitudes of all
the energy terms over snow result from the higher albedo, lower temperatures,
and weaker winds, compared to those over the ice.

Average daily totals place $Q*$ and Q_H in better perspective (Table 10.2). Net radiation supplies twice the energy to Q_M that Q_H does over ice, but only about 1.4 times as much over the snow surface. This is not to diminish the importance of $Q*$ to the melting of snow. As $Q*$ over snow falls to half its value over ice, so too does the magnitude of Q_M. This underlines the role of surface albedo in the melting process.

Turbulent fluxes directed towards the ice surface provide a source of energy in addition to $Q*$ and so greatly enhance capacity for melting. At the same time heat is being drawn from the surface air layer, which, as a consequence, becomes colder and denser, so forming the characteristic katabatic drainage flows discussed above.

Tundra Surfaces

Since surface energy balance regimes are superimposed on the regional climate, the maritime mountains bordering the Pacific coast tend to display energy balance regimes

Table 10.2

Average daily total heat balance components for three glacier sites on Peyto Glacier over ten-day periods

	Site no. 1	Site no. 2	Site no. 3
Surface	Ice	Ice	Snow
Elevation (m)	2,000	2,300	2,500
$Q*$ (MJ m^{-2} d^{-1})	13.49	8.63	3.46
Q_M (MJ m^{-2} d^{-1})	−17.57	−12.94	−6.30
Q_H (MJ m^{-2} d^{-1})	3.14	4.20	2.54
Q_E (MJ m^{-2} d^{-1})	0.94	0.11	0.30
Bowen ratio, β	3.3	38.2	8.5

dominated by wet-surface conditions (Brazel, 1970; Aufdemberge, 1974) and drier, moisture-limiting regimes in the more continental ranges inland (Bowers and Bailey, 1989; Saunders and Bailey, 1994), but there are seasonal differences too. Delineation between the energy- and moisture-limiting regimes at Scout Mountain was defined by Saunders (1990), using equations (2.21) and (2.22). For a wet spring season, $\Omega \rightarrow 1.0$, indicating a decoupled, energy-limiting evaporation regime, whereas $\Omega \rightarrow 0.0$ during a drier spring-summer season, indicating that evaporation was then coupled and moisture-limiting.

Alpine tundra energy balances respond to energy and moisture supplies in broadly similar ways as other short vegetation surfaces do: Q_G is small, $Q_E/Q*$ is high when surfaces are wet, and $Q_H/Q*$ is high when surfaces are dry. Latent heat fluxes directed towards the surface (condensation) are common when clouds cloak the mountains.

Thermal properties of tundra soils show great spatial variability as a result of the distribution of vegetation and soil moisture and the presence of patterned ground, rendering Q_G difficult to determine. Measurements made with buried soil heat flux plates indicate that Q_G is typically of the order of 10 per cent of $Q*$ (Bowers and Bailey, 1989; Saunders, 1990).

In response to intense radiative heating, alpine tundra surface temperatures may increase from below 0°C at night to more than 40°C on cloudless summer days. One consequence is that alpine vegetation takes a long time to get re-established following disturbance, since the temperature requirements for germination are met only infrequently. Air temperature does not show such a great diurnal range because it is more readily mixed with air in the free atmosphere. As a result, very large vertical temperature gradients are common in the lowest metre of the surface boundary layer.

One might expect to see altitudinal gradients in evaporation regimes, because alpine meadows just above treeline tend to be wetter than tundra higher up the mountain, which in turn is drier than the barren rock surfaces that characterize the very highest tracts of land. Insufficient data prevent extensive quantitative assessment of this pattern in the Western Cordillera, though qualitative observations support the idea.

The alpine tundra energy balances shown in Figures 10.12a and 10.12b encapsulate several of the salient features of this environment. Ground heat flux tends to mimic $Q*$, but with values of about 10–15 per cent the magnitude. On a wet day, Q_E exceeds Q_H,

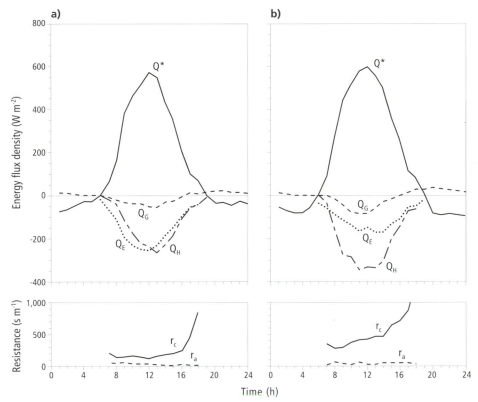

Figure 10.12
Tundra energy balances and aerodynamic (r_a) and surface resistances (r_c) for (a) a wet surface and (b) a dry surface on cloudless summer days at Scout Mountain.

since recent precipitation has increased the moisture availability and the evaporation regime is an energy-limiting one (Figure 10.12a). Surface resistances are generally small. But most alpine tundra environments possess significant proportions of barren rock, which dry quickly and thereafter contribute nothing to evaporation. As a result, Q_E no longer greatly exceeds Q_H, even though the soil and vegetation may be well watered. As a general rule, therefore, $Q_E \approx Q_H$ for wet tundra surfaces, and $Q_E < Q_H$ for dry.

Desiccation of alpine tundra surfaces is rapid. Without fresh precipitation input, evaporation quickly declines as a result of a combination of factors: the coarse soil texture allows rapid drainage, which limits soil moisture; the shallow root system limits the potential soil moisture reservoir; transfer of moisture from the surface to the atmosphere by transpiration is restricted by the high proportion of non-vascular plants; and the bare rock surfaces store no water and so contribute nothing to the evaporative flux once surface drying is complete.

Soon (often within a day or two) after precipitation receipt, the tundra's energy balance is characterized by a regime in which sensible heat is the dominant means of turbulent heat transfer and surface resistances are high because of moisture stress as vascular plants attempt to prevent further loss of moisture (Bowers, 1988; Bowers and Bailey, 1989; Bailey, Saunders, and Bowers, 1990; Saunders and Bailey, 1996). Diurnally, r_c commonly displays morning minima and afternoon maxima, showing the effects of dew availability and afternoon heating of the surface (Figure 10.12b). This moisture-limiting energy balance regime persists until precipitation again supplies water to the surface.

The basic characteristics of summer energy transfers in alpine environments are encapsulated in the summary diagram shown in Figure 10.13. One of the key determinants of the energy balance is clearly whether or not the surface is snow/ice; if it is, the melting process dominates energy exchange by forcing the turbulent fluxes to become sources. Non-vegetated surfaces such as rock walls and fresh glacial moraines have high Bowen ratios. With increasing vegetation cover, fellfield and especially subalpine meadows allow transpiration to play an increasing role in energy and mass transfer. Note that Figure 10.13 sacrifices realism (for the sake of clarity) by ignoring the effects of advection and topography. In reality, advection would occur probably in the vicinity of the boundaries between each surface type, particularly where snow/ice is adjacent to unfrozen surfaces.

Physiological and ecological similarities between arctic and alpine tundras suggest that they may create comparable climatic regimes. The climates of the alpine tundras studied in the Cordillera broadly agree with those of arctic and subarctic tundras (for example, Stewart and Rouse, 1976; Rouse, Mills, and Stewart, 1977; Marsh, Rouse, and Woo, 1981). Apart from the effects of latitude, alpine climates tend to show greater temporal and spatial variability because of the effects of topographic complexity on such factors as the wind field, radiation loading, and cloud generation.

HYDROLOGY AND THE WATER BALANCE

Precipitation

The prevailing westerly airflow, which brings cyclonic precipitation to the Cordillera, controls general spatial and temporal distributions, but on the ground local-scale orographic effects determine precipitation amounts at any given point. The orographic influence on precipitation manifests itself in a variety of ways, and results in a precipitation distribution which is more complex in time and space than that in any other natural environment in Canada. The form of precipitation changes with elevation as dictated by the environmental lapse rate: the rain that falls at lower elevations is in the form of snow higher up. Mountains in maritime regions receive 2,000–4,000 mm of precipitation annually, decreasing to less than 1,000 mm for continental ranges. Snow forms much of the precipitation input throughout the Cordillera. The percentage of annual precipitation that falls as snow increases with latitude and altitude, such that the highest parts of the St Elias Mountains and the Mackenzie Mountains receive only negligible amounts of rain.

Topographic effects on airflow produce rain shadow and wind shadow effects. Rain shadows occur to the east of each of the major mountain ranges in western Canada, and the same effect is also observed in the lee of individual large mountain massifs. Offsetting these effects at the basin scale are wind shadows, where precipitation accumulates in drifts in the lee of obstructions. At the micro-scale, lee-side drifts are significant sources of water for tundra vegetation.

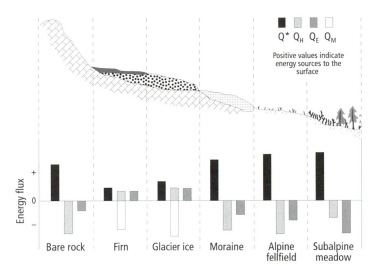

Figure 10.13
Generalized energy balance diagram for alpine surfaces in summer. Length of
bars indicates magnitude of the energy flux. Positive fluxes indicate sources of
energy for the surface, while negative fluxes are energy losses. The effects
of topography and advection between different surface types have been
ignored.

Up to a certain point, precipitation increases with elevation as a consequence
of orographic cloud (resulting from forced uplift or thermally generated convection).
However, precipitation cannot increase indefinitely with elevation because of moisture starvation during orographic uplift as available water is precipitated lower down
and increased exposure to winds that blow precipitation off the mountain. Decreases
in precipitation at higher elevations have been measured on Mount Logan (summit
elevation 5,959 m), Yukon. Maximum precipitation occurs in the 1,500–3,000–m
zone, although heavy precipitation can still occur at the highest elevations. Different
precipitation–elevation relations are observed on the windward and leeward sides of
the mountain, but relations among precipitation and topography, exposure, and continentality are indistinct (Marcus and Ragle, 1970). The variability in precipitation
on Mount Logan is probably echoed throughout all major mountain massifs in the
Cordillera.

Precipitation is usually assumed to be the sole input of water into the basin
system, with interbasin transfers being either ignored or considered negligible.
Errors in measurement of precipitation therefore make it more difficult to determine
the complete water balance, even if the other terms can be accurately measured.
Accuracy of precipitation measurement is thus crucial, and yet representative estimates from alpine environments are nearly impossible to obtain. Interbasin and
intrabasin transfers of water mass can occur in the alpine basins of the Cordillera
through the actions of groundwater flow (especially prevalent in the porous strata
of the Rocky Mountains) and by removal of snow from ridge crests by wind or by
avalanches. The controls on the last phenomenon vary: in winter, snowfall amounts
and rates are a major factor in initiating avalanches, and so the synoptic weather
is key. In spring, after snow has metamorphosed to a more stable form, wet-snow

avalanches are the most common, and these are controlled by surface snowmelt rates, as governed by the surface energy balance on that slope, which may be synoptically or locally controlled.

Evapotranspiration

The key elements of the evaporation regime have been discussed in the above section on the energy balance. The moisture reservoirs available for evapotranspiration are the vegetated parts of tundra, snow, ice, and free water in lakes and rivers and over-land flow derived from glaciers and snowbanks. The substantial amounts of bare rock in the alpine environment contribute little to the evaporative loss from the surface, and so the amount of evaporation is restricted at all times when the surface is not freshly wettened by precipitation or meltwater. Since water typically drains or evaporates from the tundra surface very rapidly following rainfall, it is the frequency, rather than the magnitude, of rainfall events that governs surface energy partitioning in alpine tundra.

Daily evaporation losses from snow and ice surfaces tend to be small over extended time periods and offset to various degrees by condensation gains. At Peyto Glacier, in the continental regime of the Rocky Mountains, losses may be sufficiently large almost to balance the gains, whereas at Place Glacier, in the more humid Coast Mountains, condensation sufficiently exceeds evaporation to bring almost 1 MJ $m^{-2}\,d^{-1}$ of heat to the glacier surface (Table 10.2). Attempts to generalize such results must be viewed cautiously, and the lower the glacier elevation, the more likely it is to condense moisture from the atmosphere. All the Bowen ratios are large, as one would expect in a snow-and-ice situation, but the smallest of the three is from the more maritime regime of Place Glacier.

Surface Storage

A variety of water storage locations are found in the alpine environment, including soil moisture, lakes, glaciers, and both seasonal and perennial snowpacks. Glaciers and snowpacks serve as buffers between input of precipitation and output of runoff, operating at time-scales of months to millennia. Most lakes in alpine environments are relatively small and so do not greatly influence the water balance. The magnitude of soil storage is typically very small in alpine environments. Slopes are often steep, promoting rapid runoff. Soils are poorly developed, with limited biomass, and once field capacity is reached water quickly drains away, unless prevented from doing so by frozen ground. As noted in the energy balance discussion, the surface–atmosphere exchange of water is small, once the surface has dried out.

Storage of water in glaciers is a major factor in the alpine water balance. Glacier mass balances respond to the prevailing climate and so change volumes of stored ice. In addition, liquid water is stored in many glaciers, either in subglacial chambers or in the accumulation-zone snowpack. The Coast Mountains and the Rocky Mountains contain many alpine glaciers in their southern sections and extensive ice-fields and large valley glaciers further north. Östrem (1973) determined that the effects of location and altitude are paramount in the mass balance of alpine glaciers. The glaciation level in the Coast Mountains rises from southwest to northeast across the mountains, possibly in response to the decreasing snowfall from maritime to interior regions, and so reflects the influence of the prevailing, southwesterly storm tracks. Additionally, relations exist between glacier mass balance and transient snowline elevation, which may help scientists discover the health of glaciers by remote sensing methods.

Young (1990) lists Canadian glaciers for which mass balance records are available. The most focused research on controls of glacier mass balances in the Cordillera, initiated during the International Hydrological Decade (1965–74), have been at Peyto, Place, and Sentinel Glaciers. Letréguilly (1988) determined that the mass balance of the maritime Sentinel Glacier is affected primarily by winter precipitation, whereas that of the continental Peyto Glacier responds to summer temperature. Glacier mass balances show at least partial dependence on synoptic-scale weather patterns (Yarnal, 1984). Winter precipitation not only determines accumulation of mass but also plays a role in summer ablation: heavy winter snowfall produces thicker snowcover, which delays onset of icemelt in the ablation zone. In general, most of the monitored glaciers in the Cordillera have experienced negative mass balances during the past two to three decades.

Drainage and Runoff

Only in a minority of basins in the permeable sedimentary strata of the Rocky Mountains is drainage significant, and so runoff is the major output term in the alpine water balance equation in the Cordillera. Alpine hydrographs are characterized by a peak produced by spring snowmelt. Summer runoff may be maintained by glacier ablation or decline to low flows in non-glacierized basins. In either case, flows are supplemented by precipitation. Since the melting process depends strongly on the surface energy balance, alpine hydrographs commonly reflect the dominant role of energy rather than precipitation. Actual runoff amounts from alpine environments are typically greater than those from adjacent lowlands because of the increased precipitation and lower evaporation/drainage losses. Differentiation of snowmelt, glacier melt, and baseflow in an alpine basin can be accomplished by hydrochemical analysis of runoff. Winter runoff is small or zero.

Since evapotranspiration, surface storage changes, and drainage are all small, runoff amounts are strongly tied to precipitation, but with some exceptions pertaining to glacierized basins. First, runoff routing in glaciers becomes faster as summer progresses, because internal conduits in the ice enlarge. Second, a heavy winter snowpack insulates the glacier, delaying melt and runoff. Conversely, small amounts of accumulation lead to greater ablation of glacier ice. Thus the total volume of runoff from glacier ablation may show less-than-expected change from year to year despite differences in precipitation. As a consequence, variations in annual runoff from alpine basins depend on the fraction that is glacierized as much as on the prevailing climate.

Catastrophic outbursts of stored water from within, or dammed by, glaciers (jökulhlaups) are also characteristic of many alpine basins in the Canadian Cordillera. Although relatively common, they have seldom been effectively documented because most occur in remote, ungauged basins. Common mechanisms of flood release include: thermal enlargement of subglacial drainage tunnels by lake water, retreat or lowering of the ice dam, and lifting of the ice dam by hydraulic pressure. In addition, lakes ponded by recessional moraines are prone to collapse and catastrophic release of water. Recent jökulhlaups have been documented in the Coast Mountains for Ape, Summit, and Tulsequah lakes and for Nostetuko and Klattasine creeks.

CONCLUSIONS

The importance of the Cordillera to the climates of Canada cannot be understated. Although physically confined to the extreme west of the country, it helps define energy and water exchanges for a much larger area. As noted in chapter 1, it interrupts the mid-latitude westerlies to allow arctic-air incursions into the Prairie regions and can generate cyclones that affect central and eastern Canada.

It is readily apparent that the present state of knowledge falls short of complete understanding of surface climates of alpine environments in the Cordillera. The present database considers physical climates of snow, glacier-ice, and tundra surfaces. A major proportion of glacier research is derived from only a few glaciers, and most of the study of tundra is from only two sites, Plateau Mountain and Scout Mountain. Ironically, nearly all efforts focus on horizontal (or very low angle) surfaces, despite the rugged terrain and extreme relief that characterize much of the region.

For the scientist, the western mountains present major logistical problems, which have inhibited extensive collection of data, and the physiographic complexity of the landscape prevents easy interpretation of the results. Although we still have much to learn about this environment, and some basic challenges remain, an outline of the essential character of radiation, energy, and water exchanges is now in hand and represents a marked advance in the state of knowledge of Canada's mountain environments.

For an environment whose single most obvious attribute is its physiographic heterogeneity, there are some clear directions for future research. Radiation and energy balances of sloping surfaces are probably the highest priority. Research from other mountain ranges has suggested that effects of topographic complexity are offset by the mitigating influence of orographic cloud cover, but confirmation is needed. Sloping-surface radiation balances have facilitated modelling of slope wind circulations, which is of value in air pollution studies in mountain valleys. In Canadian mountains, high quality research from sloping terrain is at a premium and confined largely to radiation balance work at Plateau Mountain (Huo, 1991; Huo and Bailey, 1992).

The spatially complex array of surface types in the alpine environment makes advective heat transfers key factors in basin-scale climates. Few studies address this concern, although it is clear that the edge effects presented in chapter 3 must abound throughout the alpine zone. The very nature of alpine tundra surfaces, for example, begs the question of the sources of heat and vapour there. Patterned ground inherently contains micro-scale variation: the spatially discrete nature of the vegetation, rocks, and snow that characterize fellfield tundra make assumptions of spatial homogeneity unrealistic. When the stone polygons are dry, they contribute no, or very little, water to the evaporative flux. Thus spatial distributions of the source of the sensible and latent heat fluxes are not the same, and micro-scale lateral advection must be occurring. Although the high wind speeds typical of the alpine tundra environment ensure that the surface boundary layer is well mixed, it would be of experimental interest to evaluate the micro-scale contributions of energy of the vegetated and non-vegetated parts of the surface.

REFERENCES

Aufdemberge, T.P. 1974. "Energy-balance Studies over Glacier and Tundra Surfaces, Chitistone Pass, Alaska, Summer 1969." In V.C. Bushnell and M.G. Marcus, eds., *Icefield Ranges Research Project, Scientific Results*, Montreal: Arctic Institute of North America, 4: 63–79.

Bailey, W.G., Saunders, I.R., and Bowers, J.D. 1990. "Atmospheric and Surface Control on Evaporation from Alpine Tundra in the Canadian Cordillera." In H. Lang and A. Musy, eds., *Hydrology in Mountainous Regions*, International Association of Hydrological Sciences, Publication 193, Wallingford, England, 45–52.

Bailey, W.G., Weick, E.J., and Bowers, J.D. 1989. "The Radiation Balance of Alpine Tundra, Plateau Mountain, Alberta, Canada." *Arctic and Alpine Research* 21: 126–34.

Bowers, J.D. 1988. "Surface Radiation and Energy Balances at a Mid-latitude Alpine Tundra Site during the Summer." MSc thesis, Simon Fraser University, Burnaby, BC.

Bowers, J.D., and Bailey, W.G. 1989. "Summer Energy Balance Regimes for Alpine Tundra, Plateau Mountain, Alberta, Canada." *Arctic and Alpine Research* 21: 135–43.

Brazel, A.J. 1970. "Surface Heat Exchange at Chitistone Pass, Alaska." *Proceedings, Association of American Geographers* 2: 26–30.

– 1974. "Comparison of Estimated and Observed Solar Radiation and Counterradiation at Chitistone Pass, Alaska." In V.C. Bushnell and M.G. Marcus, eds., *Icefield Ranges Research Project, Scientific Results*, Montreal: Arctic Institute of North America, 4: 49–62.

Davies, J.A. 1967. "A Note on the Relationship between Net Radiation and Solar Radiation." *Quarterly Journal of the Royal Meteorological Society* 93: 109–15.

Hay, J.E., and Suckling, P.W. 1979. "An Assessment of the Networks for Measuring and Modelling Solar Radiation in British Columbia and Adjacent Areas of Western Canada." *Canadian Geographer* 23: 222–38.

Huo, Z. 1991. "Measurement and Modelling of the Radiation Balance of Alpine Tundra, Plateau Mountain, Canada." MSc thesis, Simon Fraser University, Burnaby, BC.

Huo, Z., and Bailey, W.G. 1992. "Evaluation of Models for Estimation of Net Radiation for Alpine Sloping Surfaces." *Acta Meteorologica Sinica* 6: 189–97.

Letréguilly, A. 1988. "Relation between the Mass Balance of Western Canadian Mountain Glaciers and Meteorological Data." *Journal of Glaciology* 34: 11–15.

Marcus, M.G., and Brazel, A.J. 1974. "Solar Radiation Measurements at 5365 metres, Mt. Logan, Yukon." In V.C. Bushnell and M.G. Marcus, eds., *Icefield Ranges Research Project, Scientific Results*, Montreal: Arctic Institute of North America, 4: 117–19.

Marcus, M.G., and Ragle, R.H. 1970. "Snow Accumulation in the Icefield Ranges, St. Elias Mountains, Yukon." *Arctic and Alpine Research* 2: 277–92.

Marsh, P., Rouse, W.R., and Woo, M-K. 1981. "Evaporation at a High Arctic Site." *Journal of Applied Meteorology* 20: 713–16.

Munro, D.S. 1989. "Surface Roughness and Bulk Heat Transfer on a Glacier: Comparison with Eddy Correlation." *Journal of Glaciology* 35: 343–8.

– 1990. "Comparison of Melt Energy Computations and Ablatometer Measurements on Melting Ice and Snow." *Arctic and Alpine Research* 22: 153–62.

– 1991. "A Surface Energy Exchange Model of Glacier Melt and Net Mass Balance." *International Journal of Climatology* 11: 689–700.

Munro, D.S., and Davies, J.A. 1977. "An Experimental Study of the Glacier Boundary Layer over Melting Ice." *Journal of Glaciology* 18: 89–99.

Munro, D.S., and Young, G.J. 1982. "An Operational Net Shortwave Radiation Model for Glacier Basins." *Water Resources Research* 18: 220–30.

Oke, T.R. 1987. *Boundary Layer Climates*. 2nd edn. London: Routledge.

Östrem, G. 1973. "The Transient Snowline and Glacier Mass Balance in Southern British Columbia and Alberta, Canada." *Geografiska Annaler* 55A: 93–106.

Rouse, W.R., Mills, P.F., and Stewart, R.B. 1977. "Evaporation in High Latitudes." *Water Resources Research* 13: 909–14.

Saunders, I.R. 1990. "Radiation and Energy Budgets of Alpine Tundra, Scout Mountain, British Columbia, Canada." PhD thesis, Simon Fraser University, Burnaby, BC.

Saunders, I.R., and Bailey, W.G. 1990. "Evaluation of Evaporation Models for Alpine Tundra, British Columbia, Canada." In H. Lang and A. Musy, eds., *Hydrology in Mountainous Regions*, International Association of Hydrological Sciences, Publication 193. Wallingford: England, 71–8.

– 1994. "Radiation and Energy Budgets of Alpine Tundra Environments of North America." *Progress in Physical Geography* 18: 517–38.

– 1996. "The Physical Climatology of Alpine Tundra, Scout Mountain, British Columbia, Canada." *Mountain Research and Development* 16: 51–64.

Stewart, R.B., and Rouse, W.R. 1976. "Simple Models for Calculating Evaporation from Dry and Wet Surfaces." *Arctic and Alpine Research* 8: 263–74.

Yarnal, B. 1984. "Relationships between Synoptic-scale Atmospheric Circulation and Glacier Mass Balance in South-western Canada during the International Hydrological Decade, 1965–74." *Journal of Glaciology* 30: 188–98.

Young, G.J. 1990. "Glacier Hydrology." In T.D. Prowse and C.S.L. Ommanney, eds., *Northern Hydrology*, NHRI Science Report 1, Saskatoon: National Institute of Hydrology, 135–62.

CHAPTER ELEVEN

Forest Environments

J. HARRY MCCAUGHEY, BRIAN D. AMIRO,
ALEXANDER W. ROBERTSON, AND
DAVID L. SPITTLEHOUSE

INTRODUCTION

The forests of Canada can be divided into eight distinct regions that reflect the country's different climatic regimes (Figure 11.1). The uniqueness of the climate of forests compared to most other vegetated surfaces is largely a result of their lower albedo, lower stomatal conductance, and higher aerodynamic roughness. The last-named induces efficient mixing in the boundary layer above forests, which has characteristically small temperature gradients (< 0.10 °C m^{-1}) and humidity gradients (< 10 Pa m^{-1}).

Macroclimatic Controls

The forest regions and their boundaries are related to macroclimatic controls – usually measures of available radiant energy and aridity. However, exact cause and effect can seldom be proved between a single climatic variable – for example, temperature (Larsen, 1980) – or a particular mix of variables and the species mix or the boundaries of forest regions. Hare and Ritchie (1972) found that in the northern boreal region, net radiation was the variable that most differentiated zones. In western Canada, the northern boundary of tundra and open forest – the arctic treeline – is correlated strongly with an annual net radiation of 750 MJ m^{-2}, and the same boundary in Labrador-Ungava is 800 MJ m^{-2}. The transition from open to closed-crown forest – the northern forest line – varies from 1,170 MJ m^{-2} in western Canada to 1,300 MJ m^{-2} in Labrador-Ungava. The closed-crown forest persists southward until annual net radiation of 1,460 MJ m^{-2} is reached (Larsen, 1989). The BC forest regions have been divided into fourteen zones and their climatic regions defined (Meidinger and Pojar, 1991).

Microclimatic Controls

At the local level, species occurrence and extent of cover are affected by small changes in relief, soil type, and soil drainage, which in turn influence the microclimate. For instance, in the boreal region, aspen and jack pine favour dry, upland sites, and black spruce tends to be found in low, poorly drained sites.

Figure 11.1

Forest regions of Canada.

Source: Rowe (1972).

Forest region

f	Boreal – predominantly forest
g	Boreal – forest and grassland
b	Boreal – forest and barren
s	Subalpine
m	Montane
c	Coast
o	Columbia
d	Deciduous
gt	Great Lakes–St Lawrence
a	Acadian
gr	Grassland
t	Tundra

0   200 mi

0   200 km

The significance of climatic control at the local scale is most clearly seen in seedling survival (Spittlehouse and Stathers, 1990). Other studies demonstrate the possibility of people's maximizing microclimatic resources. McMinn and Herring (1989) document the benefits of some site preparation practices involving scarification to alleviate low soil temperatures. The changes in the site microclimate of BC clearcuts imposed by prescribed burning are considerable (Silversides, Taylor, and Hawkes, 1989) and include increased absorption of solar radiation as a result of lowered albedo, elevated surface and shallow-soil temperatures, a greater diurnal range of air temperature in the first two metres of the atmosphere, and increased levels of light at the surface.

ENERGY AND GAS EXCHANGE
IN THE FOREST

Radiation Balance

Figure 11.2 shows the diurnal variation in the components of the surface radiation balance for the (20-m) mixed forest at Petawawa National Forestry Institute (PNFI), Chalk River, Ont., for two days with contrasting radiative forcing: 11 and 14 August, 1987. 11 August was cool and clear, with low wind speeds (< 2.5 m s^{-1}); the canopy was dry, and the soil moisture very low (6 per cent by volume). 14 August had a similar low wind speed regime but was overcast for the whole day; 1 mm of rain was recorded prior to 0600 h, and moisture persisted on the canopy until late morning.

Albedo and Solar Radiation Penetration within Canopies

Low albedo values are typical of forests because most of the radiation is below the top of the canopy before it is reflected, and therefore efficient trapping ensues. Albedo values of 15 per cent to 18 per cent are typical of temperate deciduous forest in full foliage (Shuttleworth, 1989), and this value drops by approximately 1 to 2 per cent when the leaves fall. Coniferous forests have lower albedos; an average value of 12 per cent is a typical midday value (Jarvis, James, and Landsberg, 1976). For the mixed forest at PNFI, McCaughey (1987) found the following: daily mean albedo varies from 12 per cent to 15 per cent with the canopy in full leaf in summer; winter albedo drops to an average of 10 per cent in the absence of snow; following snow, albedo increases to 50 per cent, but quickly decays to an average of 20 per cent while snow remains on the forest floor; there is no systematic relation between snow depth in the canopy and the albedo; there are well-marked transition periods in the spring and autumn as the canopy adjusts to leaf-up and leaf-fall.

Solar radiation decreases as it passes through the canopy and spectral composition changes. Photosynthetically active radiation is lost most strongly towards the top of the canopy, and at the canopy base the infrared component dominates. There is considerable spatial variation in the flux on the forest floor as a result of the clumping that characterizes most forest vegetation. Penetration of solar radiation for temperate forests varies from as high as 30 per cent to as low as 1 per cent (Shuttleworth, 1989).

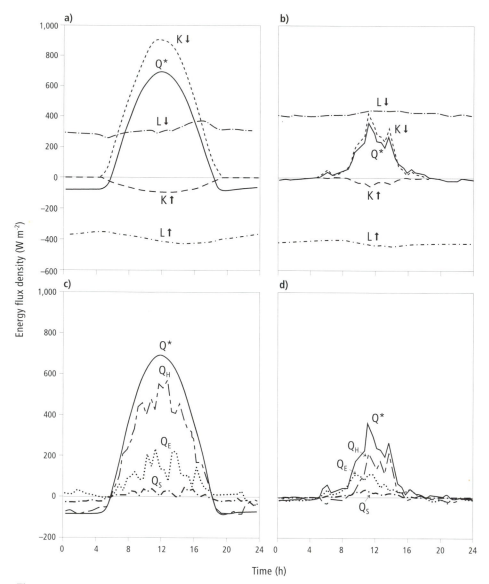

Figure 11.2
Diurnal radiation and energy balance regimes for a mixed forest at the Petawawa National Forestry Institute (PNFI), Chalk River, Ont. Radiation balance for (a) 11 August 1987, a clear day when the canopy was dry; and (b) for 14 August, an overcast day. Energy balance for (c) 11 August and (d) 14 August, when the canopy was wet from rain before 0600.

Net Radiation

The surface directly controls net radiation through albedo and surface temperature. In energy balance work, net radiation is a fundamental input variable and is routinely measured with a single radiometer. There is a decrease in net radiation with depth in a forest, and most of the change occurs in the top third of the canopy. At the ground, spatial variability of the net radiation field can be substantial (Denmead, 1984).

The degree of control by the forest on the net radiation regime is most clearly seen when the forest is removed by clearcutting or burning. McCaughey (1981) reported that, at Montmorency, Que., following clearcutting, net radiation decreased by between 10 per cent and 20 per cent on a daily average basis. The range depends on surface dryness. For wet surface conditions, the change tends towards the lower end; when the surface is dry, the opposite happens. After four years, there was no significant change in the impact because the site's revegetation was slow enough not to have significantly altered the surface controls.

Energy Balance

The energy balance shown for the mixed forest at PNFI for two contrasting days in 1987 (Figure 11.2) demonstrates the differences possible on a short-term basis depending on cloud amount and whether the canopy is wet or dry. 1987 was an extremely dry year, and by August this forest was under severe water stress. With cloudless conditions and low soil moisture content (Figure 11.2c), Q_H dominates the daytime regime (the average daytime Bowen ratio was 3.3). There is a sensible heat transfer to the canopy at night which, together with heat emitted from storage, sustains a small positive latent heat flux. When the canopy is wetted by an early rain and the sky is overcast, the energy input to the system is drastically reduced and the evaporation regime changes (Figure 11.2d). Following the rain, Q_E dominates the energy balance until just before noon. After noon, the relative sizes of the convective fluxes return to the pattern found for a dry canopy (Figure 11.2c).

Ground Heat Flux and Biomass Storage

For forests, the minor energy balance terms – especially heat storage in the phytomass, Q_{Sp}, and canopy air, Q_{Sa} – can collectively constitute a significant component of the overall energy balance because of the physical dimension of the canopy, as compared to agricultural and lower, smoother natural sites. Part of the distinctive nature of the climate of forests is a result of this storage. Many studies of forest energy balance oversimplify treatment of the canopy storage by assuming either that soil heat flux is always a good proxy for total storage or that it is the primary storage term. If the canopy is open, with low stem density that allows high transmission of radiation to the forest floor, then the Q_G term is pre-eminent. However, Q_{Sp} and Q_{Sa} dominate for the majority of stands, especially during the night, at the transition points around sunrise and sunset, after periods of rain and in the early morning. For the tall (20-m), mixed canopy at PNFI $Q_{Sp} + Q_{Sa}$ was 50 per cent of Q^* on clear nights, 20 per cent of Q^* in the early morning following sunrise, and less than 10 per cent of Q^* in the afternoon (McCaughey and Saxton, 1988). For daily or daytime average conditions, the size of $Q_{Sp} + Q_{Sa}$ seldom exceeds 5 per cent to 10 per cent of Q^* (McCaughey, 1985). The clear conclusion is that the minor energy balance terms should be estimated as completely as possible, especially for tall forests and for calculations of short-term energy balance.

Sensible and Latent Heat Flux

The latent heat flux from dry forests is generally less than that from short vegetation because of the lower surface (or canopy) conductance, and it typically ranges between 55 per cent and 80 per cent of the potential evapotranspiration rate (Shuttleworth and Calder, 1979; Calder, 1990). It is insensitive to short-term changes in net radiation because transpiration is coupled closely to the atmospheric vapour pressure deficit. This linkage is demonstrated for a Douglas-fir canopy for a day when there was a precipitous drop in net radiation at midday as a result of cloud build-up (Figure 11.3). In the morning the Bowen ratio was 2.3, and in the afternoon it fell to 0.4. Sensible heat flux decreases markedly, following the decrease in net radiation, but latent heat flux remains essentially unchanged.

For wet canopy conditions, evaporation of intercepted moisture from forests is rapid, and the rate exceeds available energy (and the potential rate) because aerodynamic conductance is extremely high and the wet leaves maintain an extensive wet "surface" to interact with the air. The elevated evaporation rate is maintained by energy drawn from the air, resulting in a negative flux of sensible heat. The exact mechanisms of supply for this excess energy over extended periods is still not entirely understood. Elevated evaporation rates from wet canopies are not quite as elevated for forests in the interior of continents (for example, at Manaus, Brazil) as they are for forests at continental edges (such as Plynlimon, Wales), where meso-scale advection can occur over extended periods (Shuttleworth, 1989).

Understorey Vegetation

The vast majority of work on forest energy balance has been concerned with the whole canopy's response. The energy balance of the soil and understorey has received little attention. Denmead (1984) reported soil evaporation under forests to be variable and at times large, with values ranging from 10 per cent to 27 per cent of the total canopy evapotranspiration. In the coastal Douglas-fir forests of British Columbia, there is usually a well-developed understorey of salal whose latent heat flux is often substantial, with values as high as 50 per cent of the stand's daily evapotranspiration (Kelliher, Black, and Price, 1986; Black and Kelliher, 1989).

Coupling of Forest Evapotranspiration to
the Planetary Boundary Layer

Coupling refers to vegetation interaction with, and response to, changes in the above-canopy environment (Jarvis and McNaughton, 1986). Well-coupled systems respond more to the vapour pressure deficit of the overlying air than to net radiation. This can be illustrated in terms of equation (2.21), in which the decoupling coefficient (Ω) varies from 0 to 1 and depends on the ratio of the surface and aerodynamic conductances. The first term on the right-hand side of the equation represents the response to radiation expressed as the equilibrium evapotranspiration rate (E_{eq}). The second term is the response to the vapour pressure deficit of the overlying air.

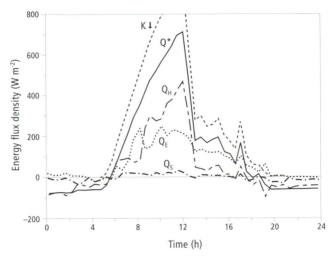

Figure 11.3
Energy balance regime for a Douglas-fir canopy in British Columbia.
Rapid decrease in solar radiation between 1200 and 1300 h is accompa-
nied by a similar decrease in Q_H but relatively little change in Q_E.
Source: Price and Black (1989).

Forests are rough surfaces with high aerodynamic conductances and relatively low
surface conductances. This results in values for Ω of 0.2 to 0.3 (McCaughey, 1989;
Price and Black, 1991), and the trees are well coupled to the vapour pressure deficit of
the overlying air. Therefore, because of efficient mixing of the air, the stomata are im-
mersed in air that is almost identical to that above the canopy. Short canopies, such as
agricultural crops, clearcuts, and the understorey of a forest, have $\Omega > 0.5$ and are
strongly coupled (responsive) to net radiation (Jarvis and McNaughton, 1986; Black
and Kelliher, 1989; McCaughey, 1989; Price and Black, 1991). In this case, air in the
canopy is not well mixed with air above the canopy.

Carbon Fluxes

Carbon dioxide (CO_2) and methane (CH_4) are the main carbon gases absorbed and released by forest ecosystems. The size of the sources and sinks for these gases varies
with the age of the forest, its location, and the time of year. Photosynthesis and respiration are the major sinks and sources, respectively, of CO_2. Forest fires or prescribed
burning to enhance regeneration release an array of aerosols and gases. CO_2 and carbon particles are a major component of the smoke, along with carbon monoxide and
CH_4 (Cofer et al., 1990). Hydrocarbons such as naphthalene, phenathene, and pyrene
are released in much smaller quantities.

Other hydrocarbons – for example, such volatile organics as isoprenes and mono-
terpenes – are released by forests, but in much smaller amounts than CO_2 or CH_4.
Some are released as part of the defence mechanisms of trees. They display diurnal
and seasonal emission patterns and impart the smell typical of a warm sunny day in
the forest. They can have localized effects on air quality, as in haze.

The large leaf area of a forest implies a large daytime sink for CO_2. However, this is counteracted by release of CO_2 through respiration of the vegetation at night and by the decomposition of dead wood and other detritus by organisms. The CO_2 concentration in forests shows a large diurnal variation during summer. Poor mixing of air below closed forest canopies and the CO_2 released through respiration result in nighttime concentrations of 450 to 600 μl l^{-1} below the crowns. Photosynthetic uptake of CO_2 reduces concentrations in the tree canopy to 2 to 4 μl l^{-1} less than above-canopy values (about 350 μl l^{-1}) during the day. The daytime reduction of CO_2 concentration in the canopy is less than that in agricultural crops because the crowns are well coupled to regional air flow. Concentrations close to the ground remain high.

CO_2 fluxes into and out of forests are affected by air temperature, intensity of photosynthetically active radiation, and soil water content. Most of Canada's forests are snow-covered during winter. Leaves either have been lost or are quiescent, and soil temperatures are low, resulting in only a small respiratory loss of CO_2. However, coastal forests, particularly in southwestern British Columbia, may have a small net uptake of CO_2 on sunny days through late autumn, winter, and early spring.

Daytime fluxes of CO_2 to temperate forests (trees and understorey vegetation) during summer range from 0.3 to 1 mg CO_2 m^{-2} s^{-1} (Jarvis, James, and Landsberg, 1976; Desjardins et al., 1985; Price and Black, 1990; 1991). There is also a respiratory flux of CO_2 from the soil surface that is taken up by the trees. This flux was estimated at about 0.1 mg CO_2 m^{-2} s^{-1} for a BC Douglas-fir forest while the net CO_2 flux during the daytime was equivalent to photosynthesis rates of 0.05 to 0.2 mg CO_2 m^{-2} of leaf s^{-1} (Price and Black, 1990; 1991).

Net twenty-four–hour fluxes include nighttime respiration from the forest (trees, understorey, and soil). This release was about 0.2 mg CO_2 m^{-2} s^{-1} for the forest studied by Price and Black (1991) and ranged from −15 to +20 g CO_2 m^{-2} d^{-1}, depending on weather. The maximum value is about half that of rapidly growing corn. A net loss occurred on days with high solar radiation and high air temperature (Figure 11.4). Stomatal opening and CO_2 uptake were reduced because of high transpiration demands, high leaf water stress, and increased photorespiration and internal cycling of carbon dioxide. Reduction in reserves of soil moisture late in summer resulted in stomatal closure and less photosynthetic uptake of CO_2. This can translate into significant limitations to annual tree growth (Spittlehouse, 1985).

Forest fires and prescribed burning of logging residues return carbon to the air. Amounts depend on the temperature of the fire but average about 120 t CO_2 ha^{-1} per burn. Prescribed fires usually consume the finer materials that would otherwise decay within a few years after harvest.

Forest ecosystems differ widely in rates of accumulation of carbon. For example, coastal Douglas-fir stands between 20 and 40 years of age have a maximum rate of accumulation of 15 t carbon ha^{-1} y^{-1}, while the rate for interior stands of lodgepole pine is about 3 t carbon ha^{-1} y^{-1} (one tonne of carbon is equivalent to 3.67 tonnes of CO_2). The different rates of accumulation and differences in disturbance regime, such as wild fires, produce large variations in the amount of and proportion of trees, dead wood, and soil at maturity. Old-growth coastal stands of Douglas fir (Figure 11.5) contain about 600 t carbon ha^{-1} – four times that of the lodgepole-pine stands – and

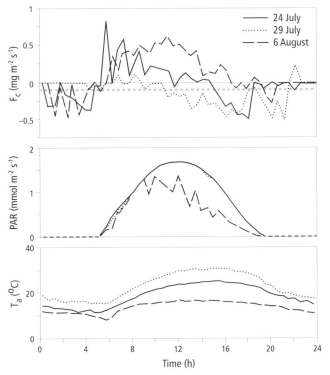

Figure 11.4
Diurnal net fluxes of carbon dioxide (F_C) in a Douglas-fir stand for three summer days: 24 and 29 July and 6 August. The horizontal, dotted line on the F_C graph indicates the contribution of soil respiratory CO_2 to F_C. The diurnal courses of the photosynthetically-active radiation (PAR) and air temperature (T) are also shown. PARs for 24 and 29 July are almost identical. Evapotranspiration, stomatal conductance, and vapour pressure deficit for the same days are shown in Figure 11.12.
Source: Price and Black (1990).

less than 60 per cent is in above-ground material (Harmon, Ferrell, and Franklin, 1990; Kurtz and Apps, 1992).

In old-growth forests with large amounts of dead wood, much of the annual carbon gain by photosynthesis is offset by release of CO_2 through decay and plant respiration. Disturbing this balance, by harvesting the forest, releases CO_2 to the air by increasing decay rates, through wood processing, and by decay of wood products. Depending on the forest ecosystem, it can take eighty to 250 years for carbon storage to return to pre-harvest levels on the site (Figure 11.5). Harvesting rotations of less than the recovery age of the forest result in net release of CO_2 to the atmosphere (Harmon, Ferrell, and Franklin, 1990; Kurtz and Apps, 1992).

Much less is known about fluxes of methane into and out of forests. Temperate, aerobic forest soils act as a sink for methane from the atmosphere, absorbing 2 to 3 mg CH_4 m^{-2} d^{-1}, whereas anaerobic soils (wetlands) are a source of methane. In contrast, the fluxes of methane from rice paddies can be ten times as great.

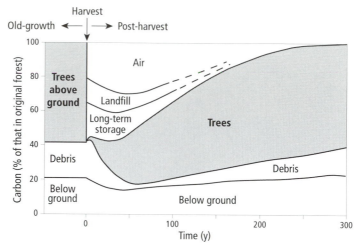

Figure 11.5
Location history after harvest for carbon originally in an old-growth Douglas-fir forest in the Pacific northwest. Total amount of carbon before harvest is 600 t ha^{-1}. At harvest (year 0), the carbon becomes on-site debris, long-term wood products (for example, in buildings), paper, mill residues, and fuel. Carbon in fuel is rapidly released to the air. On-site debris and below-ground carbon are lost through decay. Paper, mill residues, and wood products enter landfills, where they decay slowly. Regrowth of trees returns carbon to the site.
Sources: Data in Harmon, Ferrell, and Franklin (1990) and B.C. Ministry of Forests (unpublished data).

The concentration of methane above the forest is usually about $2\ \mu l\ l^{-1}$. Burning forests release methane, increasing above-canopy concentrations to 3 to $7\ \mu l\ l^{-1}$ (Cofer et al., 1990). Emission rates are about 0.3 t CH_4 ha^{-1} per burn, less than 1 per cent of the emission rate of CO_2.

Atmospheric Turbulence within Forest Canopies

Forest canopies have some unique properties influencing mass and energy transport within them. The large scale of the trees increases surface roughness, resulting in greater mixing near the top of the canopy, compared to shorter plant canopies. Flow regime within forests is affected by the structure of the tree canopy and the community of shrubs and other plants under the trees. Therefore different types of forest may have different characteristics of air flow. Hutchison and Hicks (1985) describe many of the basic principles of momentum exchange and atmospheric turbulence within forests in general.

There have been a number of analyses of atmospheric turbulence within Canadian forests. Extensive studies include those in a deciduous forest near Camp Borden, Ont. (Shaw, Paw, and Gao, 1989; Leclerc et al., 1991), in the boreal forest near Pinawa, Man. (Amiro, 1990), and in a Douglas-fir forest on Vancouver Island (Lee et al., 1991). Others have been conducted within forests near Petawawa, Ont. (McBean, 1968;

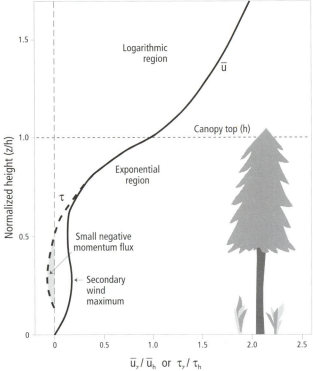

Figure 11.6
Typical profiles of mean wind ($\bar{u}$) and momentum flux (τ) within a dense
spruce forest. The data at each height ($\bar{u}_z$, τ_z) are normalized to those
measured at the canopy top ($\bar{u}_h$, τ_h).
Source: Adapted from Amiro (1990).

Martin, 1971), near Banff, Alta (McBean 1968), and near Elmira, Ont. (Shaw, Silversides, and Thurtell, 1974) and above forests in British Columbia (Spittlehouse and
Black, 1979) and in New Brunswick (Bourque, Arp, and Dickison, 1989). Described
next are some of the main features of these forests.

Profiles of Wind and Temperature

The three-dimensional wind field is described in chapter 2. Mean horizontal velocity,
$\bar{u}$, is known to decrease with height in a logarithmic fashion under neutral atmospheric stability conditions (Figure 2.5b). This logarithmic decrease dictates that, at
some height, mean wind speed approaches zero. For a forest canopy, this height is an
effective displacement height, d. However, $\bar{u}$ does not approach zero at the canopy
top at heights less than d, and momentum penetrates the canopy, resulting in windflow within the forest. The velocity profile within the forest is observed to be exponential in the vicinity of the canopy top but again becomes logarithmic near the
ground (Figure 11.6). Often, a secondary velocity maximum is observed deep within
the forest canopy (Amiro, 1990; Lee et al., 1991). This phenomenon could be caused

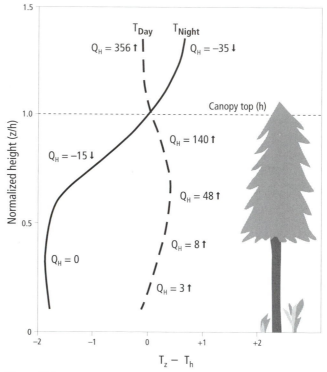

Figure 11.7
Typical temperature profiles within forests during day and night. $T_z - T_h$ is the temperature difference between height z and the canopy top, h. During the day, there is a lapse above the forest and an inversion within, contributing to negative buoyancy. Upward (counter-gradient) heat fluxes (Q_H – arrows indicate direction, units W m^{-2}) are often observed here. At night, there may be a strong inversion across the top of the canopy, with close-to-isothermal conditions below.
Source: Adapted from Amiro (1990).

partly by localized wind penetration through clearings in the forest and through the dead-branch space. However, Shaw (1977) has shown that the secondary maximum can be explained when all aspects of turbulent transport within canopies are considered. The ratio of wind speeds within the forest to those above the forest depends on atmospheric stability, with the nighttime ratio being on average about 75 per cent of the daytime ratio (Martin, 1971).

The penetration of momentum into the forest from above is affected by the atmospheric stability. Radiant energy is absorbed most strongly in the crown space, as discussed above, resulting in vertical profiles of temperature that vary with radiation exchange. The crown space experiences the temperature maximum in daytime, and the temperature minimum at night (Figure 11.7), as a result of gain or loss, respectively, of net radiant energy. During the day, these temperature gradients produce unstable atmospheric conditions (positive buoyancy) within the upper crown space and above the canopy, but stable conditions (negative buoyancy) below the crown. The reverse is true at night, although the weaker radiation exchange often results in near-isothermal conditions within the trunk space. During the daytime, negative buoyancy within the forest suppresses mixing because mechanical forces must fight the temperature inversion, inhibiting vertical transport of energy and mass.

The flux of momentum or shear stress (τ), equation (2.3), displays a profile similar to that of mean wind speed ($\bar{u}$) (Figure 11.6). It is approximately logarithmic near the canopy top and becomes exponential within the canopy. Deep within the canopy, the momentum flux may even be negative, almost always corresponding to the secondary wind maximum. Thus momentum flux is not counter-gradient in this region, although the apparent source of momentum near the ground must be accounted for (Amiro, 1990; Lee et al., 1991). This negative flux of momentum is always small and little affects momentum fluxes through forests.

Another phenomenon within forests consists of counter-gradient fluxes of sensible heat (Q_H) within the bottom part of the canopy (Amiro, 1990; Lee et al., 1991), as illustrated in Figure 11.7, where the apparent heat flux does not follow the mean temperature gradient at all heights. This phenomenon is probably caused by intermittent, large-scale eddies which transport heat through the canopy, over distances larger than the mean local gradient. Therefore the assumption that eddy fluxes are dependent on transport along mean gradients is inappropriate within the canopy. This conclusion probably applies also to transport of other scalar quantities, such as water vapour and CO_2. In order to obtain correct values of the fluxes it is then necessary to employ alternative modelling concepts, or they must be measured directly.

The intensity of turbulence is much greater within the forest than in the free atmosphere. Although mean horizontal wind conditions might be light, there can still be substantial energy in vertical motions within the canopy.

Coherent turbulent structures occur within forests and have a pattern unlike the "random" type of fluctuations often associated with turbulence (Figure 11.8). These structures can be related to physical causes. They are most noticeable as characteristic ramp patterns in time traces of quantities such as temperature (Gao, Shaw, and Paw, 1989; Shaw, Paw, and Gao, 1989). Figure 11.8 illustrates a typical type of ramp pattern in temperature: the ramp moves downward through the canopy as it is advected laterally (Shaw, Paw, and Gao, 1989). These structures often dominate transport through canopies, sometimes being responsible for 75 per cent of the total flux of momentum and heat (Gao, Shaw, and Paw, 1989). They also help to explain some of the phenomena observed in the fluxes of momentum and heat.

Atmospheric waves can also be observed within forests (Figure 11.8). These are generated by the shear, or drag, at the top of the canopy during very specific atmospheric stability conditions, usually at night or in early morning (Amiro and Johnston, 1991). Their presence is noticeable within the canopy as a wave structure with a period of about 35 s. The wave is less obvious in the top part of the canopy, where it is masked by turbulence in the background flow. The role of waves in turbulent transport of energy and mass is not well understood and depends largely on whether the period of the wave is in phase with a scalar quantity, such as temperature.

The identifiable nature of coherent structures and wave patterns facilitates the experimental approach towards study of turbulent motions within forests. Future research on

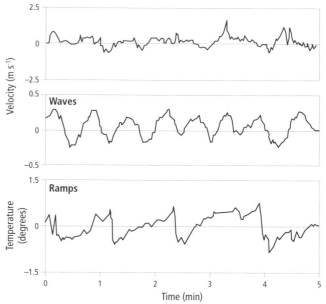

Figure 11.8
Examples of time-series data of wind speed (u) and temperature (T)
within forests at a height of 0.6 h. The top trace shows fluctuations
of u without coherent structures; the middle, waves; and the bottom,
ramps in temperature within a deciduous forest.
Sources: (Top and middle) Amiro and Johnston (1991); (bottom) Gao,
Shaw, and Paw (1989).

these structures will probably help us to understand the physical forces governing turbulence within forest canopies.

THE HYDROLOGICAL BALANCE

Atmospheric processes and plant factors result in the water balance of forests being
significantly different from that of short vegetation, such as agricultural crops and
grasslands. An extensive review of the influence of forests on the hydrologic regime
is presented in Hetherington (1987). This section focuses on the seasonal course of
precipitation, interception, evapotranspiration, and soil water storage and on modelling forest evapotranspiration.

Precipitation Input

Probably the greatest difference in the hydrological balance between forests and short
vegetation lies in the interception of precipitation. A much greater fraction of precipitation is intercepted by a forest than by a short canopy, because of the large surface
area of foliage, the canopy structure of forests, and interactions with the atmospheric
planetary boundary layer, as noted above. Interception can be divided into two processes: precipitation (snow and rain) and fog drip.

Precipitation either is intercepted by foliage or falls directly to the forest floor as throughfall. Intercepted precipitation may remain on the canopy, be lost to the air (through evaporation of water or sublimation of snow), or fall to the forest floor (be blown off, drain down stems, or drip from the canopy). Conifers intercept more water (snow plus rain) than hardwoods, since they maintain their leaves throughout the year (Hetherington, 1987).

The amount of snow intercepted depends on canopy density, whether the snow is dry or wet, the amount already on the canopy, and meteorological conditions (Calder, 1990; Schmidt and Gluns, 1991). The large trees in the BC coastal forests intercept up to 50 per cent of a snowfall (up to 16 mm snow–water equivalent). Shorter, less dense forests in the BC interior and in other parts of Canada and leafless, hardwood forests intercept substantially less snow (Hetherington, 1987). As snowload on the branches increases, they bend and the snow may fall off. The fate of the remaining snow depends on air temperature, wind speed, and solar radiation. This snow can sublimate, melt and evaporate, melt and drip onto the snowpack below, or be blown off (Calder, 1990; Schmidt and Gluns, 1991). Over the winter, 20 to 40 per cent of the snowfall on conifer forests is intercepted and returned to the atmosphere (Figure 11.9). Small openings in the forests act as snow traps and may have up to 40 per cent more snow than large openings (Hetherington, 1987; Toews and Gluns, 1986).

Intercepted rainfall can remain on the canopy, drain to the ground, or evaporate. The throughfall pattern shows high spatial variability. There can be significant evaporation of intercepted water while rain is falling, leaving room for further interception. This loss of water to the air is a consequence of the high aerodynamic conductance of a forest and large-scale advective enhancement. Thus light showers are almost fully intercepted and little rain reaches the ground, while only 10 to 30 per cent of the rainfall in heavy rainstorms suffers the same fate. Interception increases with forest canopy density. During a year, 20 to 40 per cent of rainfall does not reach the forest floor (Spittlehouse, 1985; Calder, 1990).

Fog drip is the result of an interception process that produces a gain in water for the forest. It is a localized feature of coastal and mountaintop environments. Fog consists of moisture droplets suspended in air. As air moves through the forest canopy, these droplets collide with and adhere to foliage. As the drops coalesce, they run off the leaves and drip to the forest floor. Wind speed and radiant energy are usually low during foggy conditions, and the deposition rate is much greater than the evaporation rate. It is not known how significant this gain in water is in Canadian forests.

Water Storage and Runoff

Water is stored in forests as snow on the ground or water in the soil; the amount depends on the season and soil factors. In many parts of Canada, snow is a major water store, and recharge of the soil's moisture reserves occurs only when snow melts.

Canopy density and topography are the site factors most influencing snowmelt and accumulation of snow (Figure 11.9). The depth of snow increases as canopy density decreases, and leafless hardwood canopies act as a porous trap for snow (Hetherington, 1987). The depth of the snowpack and snow density are quite variable in forests containing evergreen trees. With less interception and intercepted snow sloughing off

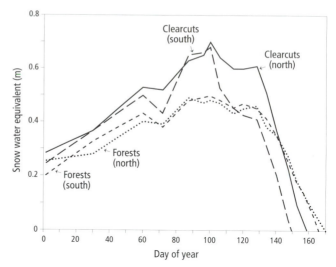

Figure 11.9
Comparison of snow accumulation and melt in clearcuts and forests. The melt rate is faster, and the snow is lost sooner in the clearcuts. South-facing clearcuts have a higher rate of melt than those on north-facing slopes.
Source: Toews and Gluns (1986).

the crown, greater depths occur between the trees. There is often an increase in snow accumulation at the edge of a clearing. Snow depth usually increases with elevation. Aspect affects snow accumulation through its relation to the direction of the prevailing weather systems.

Snowmelt depends on the energy balance of the snowpack (see chapter 4). Canopy density influences the amount of radiant and turbulent energy reaching the snow surface by shading the surface and reducing wind speed. Snowpacks under leafless hardwoods and in large clearings usually melt faster than those under a coniferous forest (Toews and Gluns, 1986; Hetherington, 1987). South-facing slopes have the highest radiation loads in the spring and are usually the first to lose all their snow (Figure 11.9). Where snowcover is thin, late developing, or intermittent, the soil may freeze, and melted snow will run off rather than infiltrate into the soil.

The ability of a soil to store water depends on soil texture, stone content, depth of soil, and slope position. Coarse-textured, stony soils can hold only half the moisture of a stone-free loam of similar depth. Fine-textured and organic soils can hold a lot of water and are often poorly drained and waterlogged. The tops of slopes usually have shallow soils and hold little water. Soils near the base of slopes receive water drained from higher up the slope.

Coarse-textured soils drain rapidly when wet and can show rapid response to rainfall. The old-growth BC forests are permeated by channels left when roots die and decay. These channels act as pathways during heavy rainfall, causing rapid below-ground runoff and response of streams (Hetherington, 1987).

Runoff from forests depends mainly on the annual temperature and precipitation regimes. Coastal watersheds in British Columbia and part of the east coast of Canada have air temperatures above zero most of the year. The runoff hydrograph usually shows a peak in winter because of heavy winter rains (Hetherington, 1987). The contribution of snowmelt to runoff depends on the elevation of the watershed. Runoff decreases through summer and increases again as autumn rains recharge the soils. In forested watersheds in the rest of Canada, the peak runoff is dominated by late winter and spring snowmelt. Flows then decrease during summer and may or may not increase in autumn, depending on the rainfall regime.

Removing the forest cover can significantly affect the hydrological balance. Reduction in interception of precipitation, changes in the rate of snowmelt (Figure 11.9), and reduction in evapotranspiration result in an increase in the soil's water content, drainage, and stream flow. A cover of moss and/or understorey plants can protect the soil from damage by erosion. Disturbance of this cover, significant compaction of the soil during harvesting, and poor road management can lead to an increase in surface runoff, erosion, and landslides. Usually, 20 to 30 per cent of a watershed's forest cover must be removed before significant changes in stream flow are seen (Hetherington, 1987). The rate of regrowth of vegetation controls the length of time it takes the hydrological balance to recover to pre-harvest conditions. Recovery occurs relatively quickly in moist environments such as coastal British Columbia and eastern Canada. However, it can take thirty years or more in dry environments such as the southern BC interior.

Seasonal and Diurnal Course of Evapotranspiration

Evapotranspiration in Canadian forests varies seasonally. It is small during winter for most forests, since the energy available to melt and evaporate or sublimate snow is low (except in chinook conditions), the soil surface is covered with snow, and trees are dormant. An exception occurs at lower elevations on the west coast of Canada, where transpiration from conifers and soil evaporation can take place, though rates are less than 1 mm d^{-1}, even on sunny days. As noted above, 20 to 30 per cent of winter precipitation in coniferous forests is lost through evaporation or sublimation of intercepted precipitation.

Figure 11.10 shows the change in daily evapotranspiration and net radiation from spring through autumn for a successional boreal-forest ecosystem dominated by deciduous species. Evapotranspiration follows daily and seasonal trends in net radiation. The fraction of net radiation used to evaporate water changes with leaf area. On days with no intercepted rain on the foliage, 40 to 50 per cent of net radiation is used to evaporate water in spring, 70 per cent when the canopy is fully developed, and about 50 per cent in the autumn as the leaves senesce. A significant fraction of the evapotranspiration in spring and autumn is from the understorey and soil surface. These data are for a wet site, and it is unlikely that transpiration is limited by lack of soil water. Conifer forests display only small changes in leaf area during the year and show less seasonal variation in the fraction of net radiation used in transpiration. The potential contribution of understorey to these fluxes for a thirty–year-old Douglas-fir forest with an understorey of salal is shown in Figure 11.11. Rainy periods presented in Figures 11.10 and 11.11 have evapotranspiration rates close to the water equivalent of the net radiation.

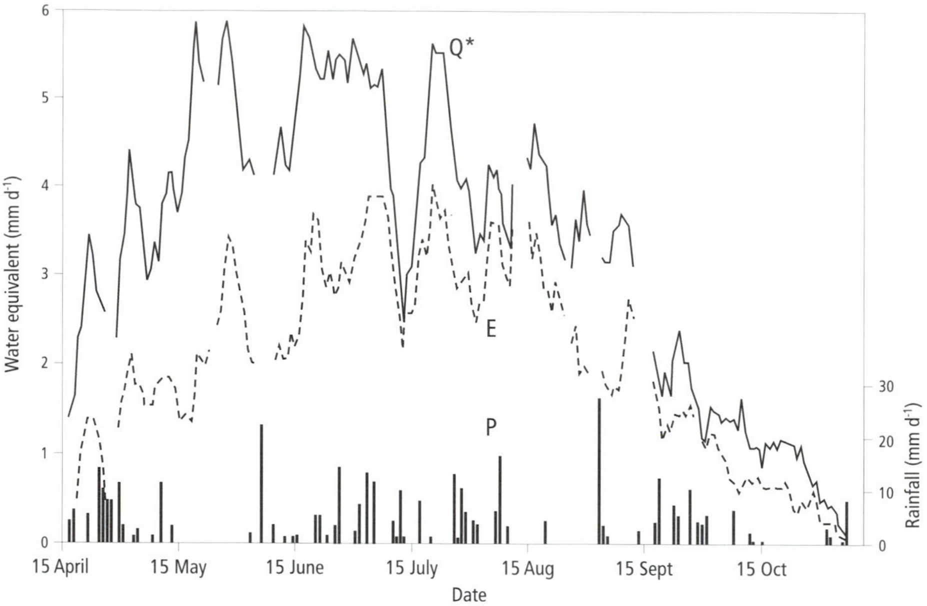

Figure 11.10

Daily net radiation (Q^*) and evapotranspiration (E) from a partially closed aspen-alder-willow stand with grass and sedge understorey at Pinawa, Man. Data are presented as a five-day moving average for clarity. Daily rainfall (P) is also presented.

Source: Amiro and Wuschke (1987).

Transpiration does not continue to increase indefinitely with leaf area. Leaves lower in the canopy and understorey are shaded and in a cooler environment, and there is a limit on the energy available to evaporate water. Transpiration tends to peak at leaf area indices of 5 to 6. A similar situation occurs with agricultural crops (Spittlehouse, 1989). Thus, when soil water is not limiting, forests transpire at the same maximum rate for much of their lifetime.

Seasonal variation in soil water status is important. Depletion of water reserves results in the development of water stress in the trees. There is a reduction in transpiration as the stomata do not open as much in the morning and close more rapidly than under moist soil conditions. Significant reduction in transpiration is seen when the soil's water potential drops below about −0.2 MPa (Spittlehouse, 1989). The seasonal rainfall pattern and soil depth control the length and intensity of drying (Spittlehouse, 1985). The lower panel of Figure 11.11 shows how drying the soil resulted in transpiration dropping from an average of 3.5 mm d^{-1} to about 0.5 mm d^{-1} during summer for a Douglas-fir forest.

On a diurnal basis, evapotranspiration is affected by available energy, the dryness of the air, and stomatal conductance. Stomatal conductance usually decreases as the air temperature and vapour pressure deficit increase above some critical value, and as the soil dries. Maximum stomatal conductances range from 2 to 6 mm s^{-1} (Iacobelli and McCaughey, 1993; McCaughey and Iacobelli, 1994).

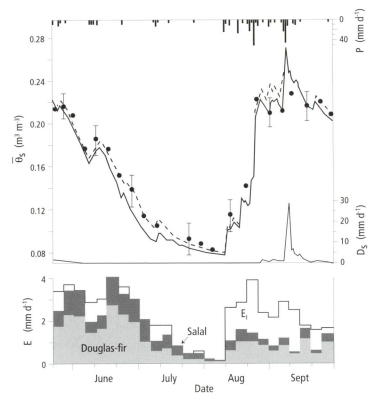

Figure 11.11

Comparison of modelled (solid and dashed lines) and measured (•) mean root-zone water content ($\bar{\theta}_s$) for a thirty–year-old stand of Douglas fir. The solid line is for a water balance model, using a layered Penman-Monteith equation and hourly weather data. Modelled values of transpiration (E) from the Douglas fir and salal, and intercepted water that is evaporated (E_I), are shown as five-day averages in the lower panel. Also shown is modelled daily drainage (D_s) and measured daily rainfall (P). The dashed line represents modelled $\bar{\theta}_s$ with evapotranspiration calculated using the Priestley-Taylor and soil-limiting evapotranspiration equations.

It is often convenient to consider the forest canopy as a single, large leaf. This giant "leaf" has a canopy or surface conductance that incorporates soil surface, understorey, and tree factors. The combination of these factors is not linear, making calculation of canopy conductance difficult in certain situations (Raupach and Finnigan, 1988). Maximum canopy conductances for forests, obtained by calculation from above-canopy measurements of evapotranspiration, range from 5 to 20 mm s^{-1}, and deciduous trees tend to have slightly higher canopy conductances than conifers.

Transpiration often peaks by late morning (Figure 11.12). Water loss during the daytime from the leaves usually precedes uptake by the roots. Consequently, water is lost from leaves and twigs, and the water potential in the tree decreases. On days with a high vapour pressure deficit in the afternoon, demand may be greater than the plant

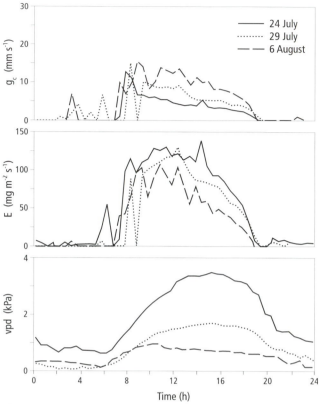

Figure 11.12
Diurnal course of evapotranspiration (E) and canopy conductance (g_c) from a young Douglas-fir stand in coastal British Columbia for three days (24 and 29 July and 6 August) with different vapour pressure deficit (*vpd*) regimes. Fluxes of carbon dioxide for these days are shown in Figure 11.4.
Source: Price and Black (1990).

can meet, and stomata start to close, lowering the transpiration rate (Price and Black, 1990). This negative feedback tends to dampen the range of variation in transpiration, as can be seen in Figure 11.12.

Modelling Forest Evapotranspiration

Examples of a range of approaches used to model forest evapotranspiration can be found in Kelliher, Black, and Price (1986), Black et al. (1989), Spittlehouse (1989), and Calder (1990). There are two classes of models, depending on whether plant factors are explicit (stomatal conductance and leaf area index are inputs) or implicit (with some form of evapotranspiration coefficient needed). There is also differentiation in terms of time step (for example, hourly versus daily or monthly) and scale (for example, leaf or canopy versus forest). Models with plant factors explicit usually use the shorter time step and the leaf or canopy as the unit of vegetation. It is usually necessary to combine the evapotranspiration model with a soil water balance model, where interception and soil drainage are calculated.

Models with time steps of an hour or less generally use some form of the Penman-Monteith equation (2.18). A surface conductance or stomatal conductance (and leaf area index) are determined as a function of atmospheric and soil conditions (Kelliher, Black, and Price, 1986; Adams, Black, and Fleming, 1991). Intercepted water is allowed to evaporate in the model by setting the plant conductance to zero and by calculating a leaf wetness factor. Predictions from a two-layered (trees and understorey) Penman-Monteith model (similar to that of Kelliher, Black, and Price, 1986) in a forest water balance model are shown in Figure 11.11. The lower panel shows the partitioning of evapotranspiration among the salal understorey (leaf area index of 3), Douglas fir (leaf area index of 5), and intercepted rainfall. Drying soil restricted transpiration during late July and early August. The upper panel of Figure 11.11 indicates that modelled (solid-line) and measured (dots) water storage in the root zone agree quite well. Significant drainage does not occur until after late summer rains rewet the soil.

Daily transpiration rates from dry foliage are usually well correlated with daily net radiation when the soil is moist. Consequently, formulae such as the Priestley-Taylor equation (2.20) can model transpiration for a forest stand on a daily basis as long as the appropriate evapotranspiration coefficient (α_p) is used. During the snow-free part of the year, on days without intercepted water on the foliage, this coefficient is in the order of 0.8 for conifers, 1.0 for deciduous and mixed woods, and 1.2 for many agricultural crops and clearcuts (McCaughey, 1989; Spittlehouse, 1989; Price and Black, 1991). Lower values have been obtained for sparsely vegetated clearcuts where a surface layer of litter is suppressing soil evaporation (Adams, Black, and Fleming, 1991). A larger value of α_p is required to evaporate intercepted water. When soil water is limiting (soil water potentials drier than -0.2 MPa), transpiration is usually calculated as a function of the water content of the root zone. The dashed line in Figure 11.11 shows the course of soil water storage as determined by a forest water balance model (Spittlehouse, 1985) that uses this approach to calculate evapotranspiration.

EFFECTS OF POLLUTION AND CLIMATE CHANGE

Wind Flow in Forested Terrain

The orientation and degree of deformation of trees growing in the open (crown flagging and leaning trunks) often represent vectors of surface wind patterns and therefore are indicators of the effects of wind on landscape ecology. On sites experiencing little wind, where average wind speed is less than 3 m s^{-1}, trees growing in the open have symmetrical tree crowns and vertical boles. Conversely, trees growing in windy locations, where average wind speed is greater than 5 m s^{-1}, have asymmetrical crowns. Therefore, since crown asymmetry and stem inclination are proportional to mean wind speed, it is possible to estimate wind direction and average wind speed of near-surface wind flow patterns for particular seasons of the year. Szeicz, Petzold, and Wilson (1979) developed a unique approach for predicting surface wind speeds, based on the relative, vertically projected stand density of trees in the subarctic forest near Schefferville, Quebec.

Graphical techniques suitable for on-site estimation of mean wind direction and wind speed vary from simple profiles of crown deformation – the Barsch and Griggs-Putnam indices and Yoshino's grading system – to empirical equations based on the combined measurements of crown deformation and stem deflection (Robertson, 1987a; b).

Estimates of prevailing wind direction derived from tree deformation may not coincide with those obtained statistically from wind vane measurements at standard meteorological stations because certain types of trees respond differently to wind throughout the year. Also, in many localized areas, the prevailing wind changes throughout the day because of the diurnal reversals of mountain and coastal breezes. In these cases, flagging of trees is more likely to represent cold, down-slope mountain winds and onshore ocean winds than warmer up-slope or offshore winds. Tree deformation may also reflect localized features in surface winds, such as jet effects and lee vortices created by the orientation of valleys, mountain ridges, forests, lakes, and coastlines.

Disturbance is a continuous process controlling forest heterogeneity on many temporal and spatial scales. From the perspective of landscape ecology, there is no general theory to predict the effects of spatial heterogeneity on disturbance propagation. However, new concepts such as percolation theory combined with fractal geometry (Turner et al., 1989) have been used to show how minor perturbation in an ecosystem can be propagated to the landscape level. For example, small-scale exogenic disturbances, such as trees damaged by non-pyrogenic lightning strikes, or small gaps in a forest created by wind, are often the epicentres for large-scale, endogenic disturbances, such as bark-beetle epidemics. Atmospheric circulation also provides mass transit for insects, such as the spruce budworm and the hemlock looper, that cause major changes to the forest landscape.

Landscape heterogeneity influences the direction and rate of spread of disturbances. If disturbance begins to diffuse within a forest type, landscape heterogeneity should retard the process. However, if the disturbance moves between forest types, heterogeneity should enhance the spread. Wind, of course, is a dominant factor influencing landscape heterogeneity. It affects such physical processes as carbon and nitrogen cycling, energy budgets, and rate of spread of fire.

Within the most stable forests, wind accounts for a certain amount of attrition, as dead and weakened trees are thrown or snapped by wind as part of a natural recycling process. The sporadic nature of forest turnover is largely a function of variations in wind regimes acting on the above-ground portions of stands. Indeed, most litterfall occurs during very windy days, particularly in coniferous forest in winter, when wind loading combines with glaze and snow loading to break off tender shoots.

Wind loading can exert bending stresses sufficient to cause mortality of fine-root mass. In winter when the ground is frozen, the rootplates are held firm by soil frost and hence would be subject to low flexural stress during windy days. During short, mild periods in winter, when only the soil near the surface thaws, the finer roots bound in the subsurface soil frost may be torn away from parts of the rootplate in the unfrozen surface layer of soil. Young to middle-aged trees respond to flexural stresses by shifting their growth downward to the root mass, at the expense of height growth and longevity. Physiologically, older trees are less tolerant of flexural stresses and become susceptible to dieback and wind throw. Forests growing in the windiest climates in Canada withstand hurricane-force winds. In Newfoundland, for example, catastrophic wind throw is rare and occurs only in forests weakened by insect epidemics and in old-growth stands exposed by clearing. In contrast, Quebec and Ontario and also the benign climate of coastal British Columbia have a comparatively low mean wind speed. Consequently, forests there are much more susceptible to catastrophic damage. In the Lac Suel region of northwestern Ontario, there is an unusual pattern of wind throw along a 160-km transect, thought to be the result of short-lived, violent downbursts from low-level jet streams (Flannigan, Lynham, and Ward, 1989).

From a physiognomic standpoint, wind-induced tree deformation is a form of biotic and abiotic streamlining that reduces the drag coefficient of an isolated tree and forest canopy. Streamlining of crowns and stems, resulting from constant exposure to wind, enables individual trees and groups of trees to withstand strong winds, including hurricanes.

Krummholz is a type of stunted forest characterized by an impenetrable mass of foliage, twisted branches, and short, stout trunks, with main roots often exposed above the ground. In cool, windy regions of the Hudson Bay Lowlands and the high plateaux in the Coast and Rocky Mountain chains, wherever there are insufficient heat units to allow for maturation of seed, it is common to find clonal krummholz of spruce and fir that regenerate by epicormic shoots (layering). In some areas, such as southern Labrador, coniferous forests (black spruce and balsam fir) may have created microclimates that permit seed development. However, removing these forests by cutting or fire increases the fetch and invokes a degree of exposure that prevents seed development, and the tree cover reverts to clonal krummholz.

Where the climate is particularly harsh, as in the subarctic and subalpine forest-tundra ecotones, distinctive wind-shaped forests are formed like "hedges" and aligned parallel to the direction of the prevailing wind. In contrast, snow-barrier "ribbons" and "lenses" intercept wind and snow more or less normal to the prevailing winter wind direction.

In 1985, a peculiar type of wind-shaped forest – a wave forest – extending over approximately 150 km^2 was discovered on a coastal plain near Port Saunders in western Newfoundland. Wave forests are distinguished by bands or arcs of dead tree strips that move across the landscape in wave-like formations (Figure 11.13). The general hypothesis is that wave forests evolved from gaps, but the precise aeromechanical features that create and maintain them are unknown. It is suggested, however, that the influence on surface wind flow of helical roll vortices and Honami winds generated by the ocean-land interface may be the principal features controlling the Spirity Cove wave forest in Newfoundland. This type of forest has been known for centuries in Japan as Shimagare (forest with the dead tree-strips), and more recently in New England as wave regeneration. Since then it has been observed throughout much of eastern Canada. The only known case of a wave krummholz, regenerating entirely by epicormic shoots, was found at L'Anse au Clair, southern Labrador, in 1988. It is of interest that wave forests and the wave krummholz are monospecific (*Abies balsamea*) in North America but conspecific (*Abies veitchii* and *A. mariesii*) in Japan.

Figure 11.13
Wave forest at Spirity Cove, Nfld, showing the wave-like dead tree strips, which move across
the landscape (from bottom left to top right in the photograph) in 55-to-60-year cycles.

The wave forests in the eastern United States and Japan are found in subalpine fir
forests, whereas those in eastern Canada, notably in Newfoundland, are found mostly
near sea level.

There are many similarities between natural wave forests and dieback on forested
edges of plantations, as well as primaeval forests within harvest areas and along clearings. The isolation of forest patches in clearcut areas increases the fetch and hence
susceptibility to dieback and wind throw. In mature and old-growth forest ecosystems,
wave-like dieback increases landscape heterogeneity. In severe cases, dieback along
forest edges of old-growth coniferous forests can spread at rates of 5–10 metres per
year under moderately windy conditions and several times that rate during storms.

Acid Precipitation

There is an increasing acidity of precipitation over much of North America. Acid precipitation is two to sixteen times higher in industrial belts than in sparsely populated
regions. Also, acid deposition by fog on vegetation has a lower pH (2.9) than deposition by rain. Despite widespread concern about the effects of acid rain on forests,
it has not been possible to link acid pollution to forest damage in a quantitative way
(Borman, 1985).

It is difficult to establish a direct quantitative link between forests and acid rain
because forests constantly react to multiple stresses (Table 11.1). For example, causes
other than acid precipitation have been ascribed to the forest decline in eastern North
America. The decline of hardwood forests in southeastern Quebec has been attributed to a warming trend punctuated by periods of cold, snowless winters, followed by

Table 11.1
Stresses on the forest ecosystem

Factors	Long-distance pollutant transport	Effects of air pollution on forests
Primary factors	*Gaseous*	*Direct effects*
Climate factors	Sulphur compounds	Nutrient leaching from foliage,
Drought	Nitrogen compounds	ozone damage, and decreased
Excess moisture	Photo-oxidants	photosynthetic efficiency
Increased soil temperature	Hydrocarbons	Ozone damage and increased
Climate change	Pollutant mixtures	foliar leaching
Winter damage		Sulphur-dioxide damage
Wind and ice storms	*Particulates*	Foliar fertilization and
Insects and disease	Acid (S and N compounds)	increased winter injury
Defoliating	Heavy metals	Foliar fertilization and altered
Sucking bark beetle	Hydrocarbons	nutrient allocation
Fungal pathogens of shoots		Growth-altering substances
and needles	*Solutions*	
Root rots	Oxidants	*Possible indirect effects*
Stand conditions	Acid rain/fog	Nutrient leaching from soil
Overmaturity	Hydrocarbons	Aluminum mobility and toxicity
Logging		Heavy metal toxicity
Combined interaction of soil,		
climate, and trees		
Air pollutants		
Secondary factors		
Nematodes		
Viruses		
Mycoplasms		
Fastidious xylem-limited bacteria		

Source: Hall and Addison (1991).

summer droughts that cause severe root damage. However, replacement of red spruce (*Picea rubens*) by maple (*Acer* spp.) and birch (*Betula* spp.) on lower and middle slopes of the Appalachian Mountains casts some doubt on climatic warming as the principal cause of hardwood decline.

Wind/Forest Relationships in a Changing Climate

The relationships between changing surface wind patterns and forest were reviewed by Robertson (1991). The ability of general circulation models (GCMs) to predict changes in bioclimatic distribution is rather imprecise, especially with regard to shifts in surface wind patterns. Nevertheless, GCMs indicate that an incremental increase in CO_2 concentration would increase the potential energy of the lower atmosphere. This would increase surface winds over the oceans and produce less frequent but more intense meteorological events (windstorms, heavy rainfall) and prolonged droughts.

Shifts in the subarctic forest-tundra ecotone in response to a warming trend are measured in scales ranging from a few hundred metres per century on mineral soils to several kilometres per century wherever drying peatlands are present. However, it is incorrect to assume that northward warming will result in a northward shift in forests. When wind is factored into predictions of forest migration, it has notable effects on drying and removal of snowcover. It is readily understood why there has been an expansion of the tundra southward in the Labrador-Ungava region; it is a result of increased forest fires and environmental stresses inhibiting forest regeneration. Further, human disturbances tend to mask natural responses to climate change. In fact, in harsh climates, human disturbances such as forest harvesting and clearing may invoke short-term climate changes resulting in conversion to shrub or heath barrens. For example, clearing of forests in southern Newfoundland and southern Labrador by fire, and by constant logging since the start of European settlement, has created large barrens that are extending northward because of the cooling effects of onshore, prevailing, south-westerly winds blowing over the cold Labrador Current. In fact, Newfoundland is also one of the few places in the world where the progression from tundra to forest is reversed; it goes northward. It is probably the only region where even the caribou migrate north for the winter.

THE BOREAS PROJECT

There remain large gaps in our understanding of interactions between the boreal forest biome and the global climate system. We need to know more about the sensitivity of the biome to changes in climate, such as the summer warming and drying projected by several GCMs in the event of a doubling of CO_2; about whether the boreal forests constitute a major sink for the carbon placed in the atmosphere by the burning of fossil fuels; and whether the biophysical responses of the surface (such as changes in albedo, roughness, and surface conductance) will produce significant feedbacks on the climate system itself.

These scientific issues spawned the Boreal Ecosystem-Atmosphere Study (BOREAS) – a multi-institutional, international field project, which incorporates the largest coordinated set of surface climate studies ever conducted in Canada (Sellers et al., 1995). The two main study areas, in central Sakatchewan and near Thompson, Man., represent the southern and northern ecotones of the boreal, respectively, and scientists have collected data on surface fluxes and surface biophysics at individual sites in each. In addition, there was a network of mesoscale meteorological stations across the ecosystem, including the zone between both study areas, where surface climate and radiosonde data were collected. An extensive set of remote sensing data, together with these mesonet data, will be central to the modelling of surface–atmosphere exchanges of energy, water, and carbon at the regional scale. The question of how to scale up energy exchanges from point and small area sites to those of a GCM-grid square is crucial to future modelling efforts.

The final scientific results have not yet been published, but preliminary analyses suggest several findings with significant implications for studies of climate change. The energy and water exchange observations in the growing season reveal a system with surprisingly low evaporation. This seems to arise from two main sets of factors. First, the root zone for conifers is less than 0.4 m deep and is underlain by semi-impermeable materials. This causes infiltration to run off rather than be stored as soil moisture, so the water-holding capacity is low. Second, in response to high vapour pressure deficit (dry air) there is strong physiological control exerted by the trees, giving low conductances in the stem and stomata. Together with high absorptivity for solar radiation (coniferous albedos around 8 per cent), this water-conserving strategy leads to relatively large sensible heat fluxes (high Bowen ratios) and a deep mixing layer in the planetary boundary layer. The low transpiration rates also mean relatively low photosynthetic fixation of carbon, compared with more southerly temperate forests. Though it is too early to be sure, these results raise the possibilities that present GCMs grossly overestimate evaporation for this biome and that the boreal forest is not a major sink for atmospheric carbon (Sellers et al., 1995).

REFERENCES

Adams, R.S., Black, T.A., and Fleming, R.L. 1991. "Evapotranspiration and Surface Conductance in a High Elevation, Grass Covered Forest Clearcut." *Agricultural and Forest Meteorology* 56: 173–93.

Amiro, B.D. 1990. "Comparison of Turbulence Statistics within Three Boreal Forest Canopies." *Boundary-Layer Meteorology* 51: 99–121.

Amiro, B.D., and Johnston, F.L. 1991. "Some Turbulence Features within a Boreal Forest Canopy during Winter." In *Proceedings of the 20th Conference on Agricultural and Forest Meteorology*, American Meteorological Society, Boston, 135–8.

Amiro, B.D., and Wuschke, W.E. 1987. "Evapotranspiration from a Boreal Forest Drainage Basin Using an Energy Balance–Eddy Correlation Technique." *Boundary-Layer Meteorology* 38: 125–39.

Black, T.A., and Kelliher, F.M. 1989. "Processes Controlling Understory Evapotranspiration." *Philosophical Transactions of the Royal Society* (London) B 324: 207–31.

Black, T.A., Spittlehouse, D.L., Novak, M.D., and Price, D.T., eds. 1989. *Estimation of Areal Evapotranspiration*. International Association of Hydrological Science, Publication 177, Wallingford, England.

Borman, F.H. 1985. "Air Pollution and Forests: An Ecosystem Perspective." *Bioscience* 35: 434–41.

Bourque, C.P.A., Arp, P.A., and Dickison, R.B.B. 1989. "Destabilization of the Lower Atmosphere above a Forest: A Model." *Agricultural and Forest Meteorology* 47: 49–74.

Calder, I.R. 1990. *Evaporation in the Uplands*. New York: John Wiley and Sons Ltd.

Cofer, W.R., Levine, J.S., Winstead, E.L., and Stocks, B.J. 1990. "Gaseous Emissions from Canadian Boreal Forest Fires." *Atmospheric Environment* 24A: 1653–9.

Denmead, O.T. 1984. "Plant Physiological Methods for Studying Evapotranspiration: Problems of Telling the Forest from the Trees." *Agricultural and Water Management* 8: 167–89.

Desjardins, R.J., McPherson, J.L., Alvo, P., and Schuepp, P.H. 1985. "Measurements of Turbulent Heat and CO_2 Exchange over Forests from Aircraft." In B.A. Hutchison, and B.B. Hicks, eds., *The Forest-Atmosphere Interaction*, Dordrecht: D. Reidel, 645–58.

Flannigan, M., Lynham, T., and Ward, P. 1989. "An Extensive Blowdown Occurrence in Northwestern Ontario." In D.C. McIver, H. Auld, and R. Whitewood, eds., *Proceedings of the 10th Conference on Fire and Forest Meteorology*, Ottawa: Forestry Canada, 65–71.

Gao, W., Shaw, R.H., and Paw, U.K.T. 1989. "Observation of Organized Structure in Turbulent Flow within and above a Forest Canopy." *Boundary-Layer Meteorology* 47: 349–77.

Hall, J., and Addison, P. 1991. *Response to Air Pollution: ARNEWS Assesses the Health of Canada's Forests*. Natural Resources Canada, Canadian Forestry Canada Forestry Service, Information Report DPC-X-34, Ottawa.

Hare, F.K., and Ritchie, J.C. 1972. "The Boreal Bioclimates." *Geographical Review* 62: 333–65.

Harmon, M.E., Ferrell, W.K., and Franklin, J.F. 1990. "Effects on Carbon Storage of Conversion of Old-growth Forests to Young Forests." *Science* 247: 699–702.

Hetherington, E.D. 1987. "The Importance of Forests in the Hydrological Regime." In M.C. Healy, and R.R. Wallace, eds., *Canadian Bulletin of Fish and Aquatic Science* 215, Canadian Aquatic Resources, Department of Fisheries and Oceans, Ottawa, 179–211.

Hutchison, B.A., and Hicks, B.B., eds. 1985. *The Forest–Atmosphere Interaction*. Dordrecht: D. Reidel.

Iacobelli, A., and McCaughey, J.H. 1993. "Stomatal Conductance in a Northern Temperate Deciduous Forest: Temporal and Spatial Patterns." *Canadian Journal of Forest Research* 23: 245–52.

Jarvis, P.G., James, G.B., and Landsberg, J.J. 1976. "Coniferous Forest." In J.L. Monteith, ed., *Vegetation and the Atmosphere*, vol. 2, London: Academic Press, 171–240.

Jarvis, P.G., and McNaughton, K.G. 1986. "Stomatal Control of Transpiration: Scaling Up from Leaf to Region." *Advances in Ecological Research* 15: 1–49.

Kelliher, F.M., Black, T.A., and Price, D.T. 1986. "Estimating the Effects of Understory Removal from a Douglas-fir Forest Using a Two-layer Canopy Evapotranspiration Model." *Water Resources Research* 22: 1891–9.

Kurtz, W.A., and Apps, M.J. 1992. "Atmospheric Carbon and Pacific Northwest Forests." In G. Wall, ed., *Symposium on the Implications of Climate Change for Pacific Northwest Forest Management*, Departmental Geographical Publication Series, Occasional Paper No. 15, University of Waterloo, Waterloo, 69–80.

Larsen, J.A. 1980. *The Boreal Ecosystem*. New York: Academic Press.

– 1989. *The Northern Forest Border in Canada and Alaska: Biotic Communities and Ecological Relationships*. New York: Springer-Verlag.

Leclerc, M.Y., Beissner, K.C., Shaw, R.H., den Hartog, G., and Neumann, H.H. 1991. "The Influence of Buoyancy on Third-Order Turbulent Velocity Statistics within a Deciduous Forest." *Boundary-Layer Meteorology* 55: 109–23.

Lee, X., Black, T.A., Chen, J.M., and Sagar, R.M. 1991. "Turbulent Fluxes above and within a Douglas-Fir Stand." In *Proceedings of the 20th Conference on Agricultural and Forest Meteorology*, Boston: American Meteorological Society, 143–6.

McBean, G.A. 1968. "An Investigation of Turbulence within the Forest." *Journal of Applied Meteorology* 7: 410–16.

McCaughey, J.H. 1981. "Impact of Clearcutting of Coniferous Forest on the Surface Radiation Balance." *Journal of Applied Ecology* 18: 815–26.

– 1985. "Energy Balance Storage Terms in a Mature Mixed Forest at Petawawa, Ontario – a Case Study." *Boundary-Layer Meteorology* 31: 89–101.

– 1987. "The Albedo of a Mature Mixed Forest and a Clear-cut Site at Petawawa, Ontario." *Agricultural and Forest Meteorology* 40: 251–63.

– 1989. "Energy Exchange for a Forest Site and a Clear-cut Site at Chalk River, Ontario." *Canadian Geographer* 33: 299–311.

McCaughey, J.H., and Iacobelli, A. 1994. "Modelling Stomatal Conductance in a Northern Deciduous Forest, Chalk River, Ontario." *Canadian Journal of Forest Research* 24: 904–10.

McCaughey, J.H., and Saxton, W.L. 1988. "Energy Balance Storage Terms in a Mixed Forest." *Agricultural and Forest Meteorology* 44: 1–18.

McMinn, R.G., and Herring, L.J. 1989. "Site Preparation Ecology: Some Climatic Connections." In D.C. MacIver, R.B. Street, and A.N. Auclair, eds., *Climate Applications in Forest Renewal and Forest Production*, Proceedings of Forest Climate '86, Ottawa: Canadian Government Publishing Centre, 89–92.

Martin, H.C. 1971. "Average Winds above and within a Forest." *Journal of Applied Meteorology* 10: 1132–7.

Meidinger, D., and Pojar, J. 1991. *Ecosystems of British Columbia*. Special Report Series No. 6. Victoria: B.C. Ministry of Forests.

Price, D.T., and Black, T.A. 1989. "Estimation of Forest Transpiration and CO_2 Uptake Using the Penman-Monteith Equation and a Physiological Photosynthesis Model." In T.A. Black, D.L. Spittlehouse, M.D. Novak, and D.T. Price, eds., *Estimation of Areal Evapotranspiration*, International Association Hydrological Science, Publication 177, Wallingford, England, 213–27.

– 1990. "Effects of Short-term Variations in Weather on Diurnal Canopy CO_2 Flux and Evapotranspiration of a Juvenile Douglas-fir Stand." *Agricultural and Forest Meteorology* 50: 139–58.

– 1991. "Effects of Summertime Changes in Weather and Root-Zone Soil Water Storage on Canopy CO_2 Flux and Evapotranspiration of Two Juvenile Douglas-Fir Stands." *Agricultural and Forest Meteorology* 53: 303–23.

Raupach, M.R., and Finnigan, J.J. 1988. "'Single-Layer Models of Evaporation from Plant Canopies Are Incorrect But Useful, Whereas Multilayer Models are Correct But Useless': Discuss." *Australian Journal of Plant Physiology* 15: 705–16.

Robertson, A. 1987a. "The Centroid of Tree Crowns as an Indicator of Abiotic Processes in a Balsam Fir Wave Forest." *Canadian Journal of Forest Research* 17: 746–55.

– 1987b. "Use of Trees to Study Wind." *Arboretum Journal* 11: 127–43.

– 1991. "Some Effects of Wind on Northern Forests in a Changing Climate." *Commonwealth Forestry Review* 70: 47–55.

Rowe, J.S. 1972. *Forest Regions of Canada*. Canadian Forestry Service Publication No. 1300, Department of Environment, Ottawa.

Schmidt, R.A., and Gluns, D.R. 1991. "Snowfall Interception on Branches of Three Conifer Species." *Canadian Journal of Forest Research* 21: 1262–9.

Sellers, P., Hall, F., Margolis, H., Kelly, B., Baldocchi, D., den Hartog, G., Cihlar, J., Ryan, M., Goodison, B., Crill, P., Ransom, J., Lettenmaier, D., and Wickland, D. 1995. "The Boreal Ecosystem-Atmosphere Study (BOREAS): An Overview and Early Results from the 1994 Field Year." *Bulletin of the American Meteorological Society* 76: 1549–77.

Shaw, R.H. 1977. "Secondary Wind Speed Maxima inside Plant Canopies." *Journal of Applied Meteorology* 16: 514–21.

Shaw, R.H., Paw, U.K.T., and Gao, W. 1989. "Detection of Temperature Ramps and Flow Structures at a Deciduous Forest Site." *Agricultural and Forest Meteorology* 47: 123–38.

Shaw, R.H., Silversides, R.H., and Thurtell, G.W. 1974. "Some Observations of Turbulence and Turbulent Transport within and above Plant Canopies." *Boundary-Layer Meteorology* 5: 429–49.

Shuttleworth, W.J. 1989. "Micrometeorology of Temperate and Tropical Forest." *Philosophical Transactions of the Royal Society* (London) B 324: 299–334.

Shuttleworth, W.J., and Calder, I.R. 1979. "Has the Priestley-Taylor Equation Any Relevance to Forest Evaporation?" *Journal of Applied Meteorology* 18: 639–46.

Silversides, R.H., Taylor, S.W., and Hawkes, B.C. 1989. "Influence of Prescribed Burning on Seedling Microclimate and Its Potential Significance in Northern Interior British Columbia." In D.C. MacIver, R.B. Street, and A.N. Auclair, eds., *Climate Applications in Forest Renewal and Forest Production*, Proceedings of Forest Climate '86, Ottawa: Canadian Government Publishing Centre, 127–32.

Spittlehouse, D.L. 1985. "Determination of the Year-to-Year Variation in Growing Season Water Use of a Douglas-Fir Stand." In B.A. Hutchison and B.B. Hicks, eds., *The Forest–Atmosphere Interaction*, Dordrecht: D. Reidel, 235–54.

– 1989. "Estimating Evapotranspiration from Land Surfaces in B.C." In T.A. Black, D.L. Spittlehouse, M.D. Novak, and D.T. Price, eds., *Estimation of Areal Evapotranspiration*, International Assocociation Hydrological Science, Publication 177, Wallingford, England, 245–56.

Spittlehouse, D.L., and Black, T.A. 1979. "Determination of Forest Evapotranspiration Using Bowen Ratio and Eddy Correlation Measurements." *Journal of Applied Meteorology* 18: 647–53.

Spittlehouse, D.L., and Stathers, R.J. 1990. *Seedling Microclimate*. Land Management Report No. 65, Victoria: B.C. Ministry of Forests.

Szeicz, G., Petzold, D., and Wilson, R. 1979. "Wind in the Subarctic Forest." *Journal of Applied Meteorology* 18: 1268–74.

Toews, D.A.A., and Gluns, D.R. 1986. "Snow Accumulation and Ablation on Adjacent Forested and Clearcut Sites in Southeastern British Columbia." In *Proceedings 54th Annual Meeting of the Western Snow Conference*, Phoenix, Ariz., 101–11.

Turner, M.G., Gardner, R.H., Dale, V.H., and O'Neill, R.V. 1989. "Predicting the Spread of Disturbance across Heterogeneous Landscapes." *OIKOS* 55: 121–9.

CHAPTER TWELVE

Agricultural Surfaces

TERRY J. GILLESPIE

INTRODUCTION

Successful farming requires harmony with the climate. Sometimes this harmony has been achieved only after many seasons of trial and error in growers' fields. But on occasion the winding pathway to farming success can be shortened by scientific discovery of the biophysical principles that underlie interactions between crops and climate. This is the rationale behind the science of "agrometeorology," which seeks to understand crop performance in terms of exchanges of radiant energy, heat, water, and atmospheric gases between plants and the atmosphere.

This chapter first explores the physical principles that link energy and climate at the level of a leaf and a crop. The next section uses this knowledge to quantify the main climatic restrictions on agricultural production and therefore lay the basis for classifying the land's suitability for farming. The following section examines some aspects of surface climates that are distinct to agricultural environments, with examples from field crops, Prairie grain crops, and horticultural crops. The final section briefly addresses the complex interweaving of air pollution, greenhouse gases, climate change, and agriculture.

ENERGY BALANCES AND THE OHM'S LAW ANALOGY

The core principle of an energy balance emerges again and again in the exploration of surface climates. This is simply a statement that the contributions of energy "suppliers" to a surface must be balanced by the demands of energy "consumers."

For a single plant leaf during the day, the suppliers are the net fluxes of solar (about 0.3–3 μm) and longwave (about 5–50 μm) radiation from the sun, sky, and underlying surface. Net radiation (Q^*) is generally positive during the day and negative at night (Figure 12.1).

The consumers are heat, which convects into the surrounding air from the two leaf surfaces (Q_H); the latent heat of vaporization required to evaporate water from within the stomatal cavities (Q_E); and photosynthesis (Q_P). The latter two fluxes are crucial to agriculture, because their ratio defines the key concept of water use efficiency (WUE), or the amount of food produced per unit of water consumed by the crop. Despite its

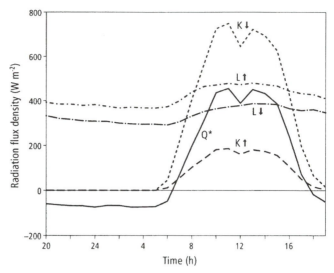

Figure 12.1
Diurnal variation of the radiation balance components for a maize crop in
Ontario on a sunny summer day. Incoming solar energy ($K\downarrow$) exceeds
reflected solar radiation ($K\uparrow$) during the day, but incoming longwave
energy ($L\downarrow$) is usually less than outgoing longwave radiation ($L\uparrow$) both day
and night. Therefore net radiation ($Q*$) is usually positive during the
day and negative at night.

importance, photosynthesis cannot be conveniently computed from the energy budget
because it consumes so little of the radiation, and its size is beyond the limits of error in
measuring the energy fluxes. Therefore a practical energy budget for a leaf is simply

$$Q* - Q_H - Q_E = 0. \tag{12.1}$$

But how does this expression of energy conservation lead to explanations and estimates of the surface climate variables themselves, such as temperature and humidity?
The link is provided by the Ohm's law analogy, which takes the simple and elegant
law governing current flow in a conductor and applies it to the "currents" of heat (Q_H)
or water vapour (Q_E) that are leaving the leaf during the day or returning to a dewy
leaf at night (Figure 12.2 and chapter 2). The "driving voltages" are the differences
in temperature (T_a) or vapour pressure (e) between the leaf and the surrounding air.
"Resistances" arise from the boundary layer of still air (r_b) that clings to the leaf's surface and from the guard cells of the stomata (r_s), which regulate flow of water vapour:

$$Q_H = C_a (T_l - T_a)/r_b \tag{12.2a}$$

and

$$Q_E = (C_a/\gamma) (e_l - e_a)/(r_b + r_s), \tag{12.2b}$$

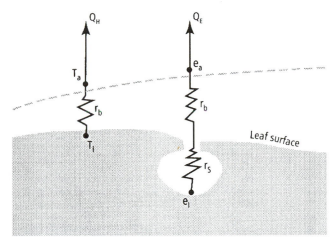

Figure 12.2
Sensible and latent heat fluxes from a leaf may be quantified using the Ohm's law analogy. Heat (Q_H) and water vapour (Q_E) flow through boundary layer (r_b) and stomatal resistances (r_s), driven by leaf–air temperature differences ($T_l - T_a$) and vapour pressure differences ($e_l - e_a$).

where T_l, C_a, and γ are leaf temperature, heat capacity of air, and the psychrometric constant, respectively.

Simultaneous solution of the energy budget and Ohm's law equations allows the course of a surface climate variable, such as leaf temperature, to be traced from knowledge of the radiant energy inputs (Q^*) and of the wind and water status of the plant (to compute r_b and r_s, respectively). Leaves exposed to the sun and sky are typically warmer than the air during the day and cooler at night (Figure 12.3). Water stress, which partially closes the stomates during the day, increases daytime leaf temperatures and enhances the temporal variability of leaf microclimate. Thus an insect or spore alighting on a leaf's surface may be warmed or cooled much more than the diurnal swing of ambient air temperature would imply.

When the energy budget (including the ground heat flux, Q_G) and Ohm's law concepts are expanded and combined to describe evaporation from the whole ensemble of leaves that form a crop (equation 2.18), the result is the combination model. This is an excellent diagnostic tool, which allows the behaviour of all the stomates acting in parallel to be encapsulated in a single variable, canopy resistance (r_c). In this fashion, we can trace the integrated response of the crop to surface energy flows (Q^*, Q_E, Q_H, and Q_G), humidity and temperature (vapour pressure deficit, or *vpd*), and the wind that affects the aerodynamic resistance (r_a). Examples from an Ontario alfalfa field are shown below. The same combination model has sometimes been used in a predictive, rather than diagnostic, fashion to estimate rates of crop evaporation from independent estimates of r_c. However, this procedure is problematic because the functional dependence of r_c on solar radiation and plant water status is very difficult to quantify.

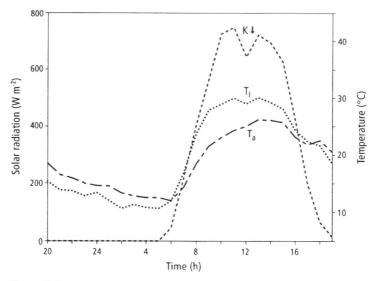

Figure 12.3
Solar radiation, air temperature, and leaf temperature in a maize crop in Ontario.
Leaves were warmer than the air during the day, so sensible and latent heat flowed
from crop to atmosphere. Heat flow reversed at night, when leaves were cooler than
the air.

AGROCLIMATIC CLASSIFICATIONS

The concept of canopy resistance emphasizes the close connection between crop productivity and crop water use. When stomates close in response to water stress, the incoming stream of CO_2 that drives photosynthesis is blocked. This connection leads to
use of a water budget (chapter 2) to help classify surface climate suitability for agricultural production. The underlying idea is simple. How well do available precipitation and soil water storage satisfy the water needs of the growing crop? The answer
determines the crop types and strategies that may be used. This concept helps us to
understand the present distribution of agriculture in Canada (Figure 12.4) and can
guide planning for new or expanded farming ventures.

Two Canada-wide agroclimatic classifications "The Climates of Canada for Agriculture" (Chapman and Brown, 1966) and "Agroclimatic Maps for Canada: Derived
Data" (Sly, 1977) use climatic moisture indices. Crops' water needs are estimated
from potential evapotranspiration (PE) – the amount of water that would be used by
soil evaporation and plant transpiration when there is no water stress. Soil moisture
storage for both these classifications is taken to be 100 mm at the beginning of the
growing season. Climatic moisture indices ($CMIs$) (Figure 12.5) are derived from:

$$CMI = [P/(P + DF + SM)] * 100,   (12.3)$$

where P is precipitation and DF is the deficit or amount by which precipitation and soil
moisture (SM) fail to meet crop water needs (PE). Thus CMI represents the percentage

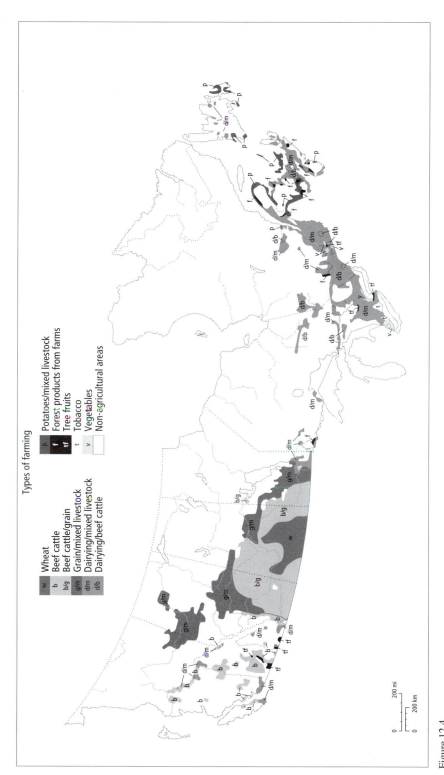

Types of farming

w	Wheat
b	Beef cattle
b/g	Beef cattle/grain
g/m	Grain/mixed livestock
d/m	Dairying/mixed livestock
d/b	Dairying/beef cattle

p	Potatoes/mixed livestock
f	Forest products from farms
tf	Tree fruits
t	Tobacco
v	Vegetables
	Non-agricultural areas

0   200 mi

0   200 km

Figure 12.4
Distribution of agricultural activities in Canada.

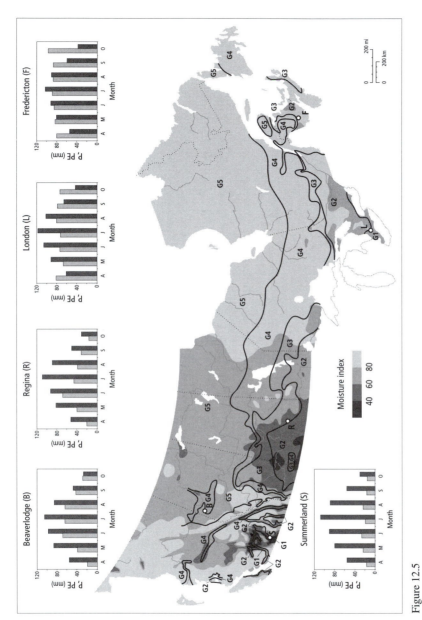

Figure 12.5

Climatic moisture index (*CMI*) and growing season classes (G1 to G5) for Canada. *CMI* represents the percentage contribution of rainfall to the water needs of an unstressed crop. Growing season classes are based on moisture, growing degree-day accumulation and frost-free periods. Station insets show average precipitation (lightly shaded bars) and unstressed water needs (dark bars).

contribution of rainfall to the total water needs of an unstressed crop. Clearly crop water use does not proceed at the potential rate for the whole season. However, treating all regions with the same assumptions about water use and soil moisture results in a good comparative rating of land areas for agriculture, based only on their moisture regime.

The *CMI* map (Figure 12.5) suggests four broad zones across Canada, separated according to whether precipitation satisfies more or less than 40, 60, or 80 per cent of seasonal water needs. The average monthly variation of precipitation and *PE* is also shown for selected stations from April to October. The driest regions, where *CMI* is less than 40 per cent, are found only near the southern Alberta-Saskatchewan border and in the southern BC interior. An example from the BC interior is Summerland (Figure 12.5), where the rain shadow of the Coast Mountain ranges, combined with abundant summer sunshine, allows *PE* far to outstrip available precipitation. These lands are best suited to sparse grazing, except where irrigation is practised. Regions where *CMI* lies between 40 and 60 per cent occupy a large part of southern Alberta and Saskatchewan, where wheat is commonly grown on better land and the remainder is used for grazing. Data from Regina, Sask. (Figure 12.5), show that the deficit of precipitation there is not quite so severe as at Summerland, but water-supply is still well below the needs of an unstressed crop from spring through autumn. In order to enhance the soil moisture portion of the water budget, wheat land is often planted only every second year so that the soil can "rest," collecting rain and snowmelt from a fallow year before producing a new crop of grain.

Although southern Ontario is the most diverse and productive agricultural area in Canada, it lies in a region where just 60 to 80 per cent of the growing season's water needs are met by precipitation. Beginning in May, in London, for example (Figure 12.5), *PE* normally begins to exceed rainfall, so periods of water shortage occur in most summers. These can seriously reduce yields when they happen near the time when plants are shifting from vegetative to reproductive growth and are most susceptible to stress. Much of the farmland in southern Manitoba and in the Peace River (for example, Beaverlodge, Alta, in Figure 12.5) region of north-central Alberta lies in the same *CMI* class, but agriculture there is significantly more restricted than in southern Ontario, because of cooler temperatures and a much shorter frost-free period.

The remainder of farmable land in Ontario, Quebec, and the Maritimes has more than 80 per cent of its water needs satisfied by precipitation. But this apparent blessing can also bring the curse of overly wet soils during planting and harvesting, especially at cooler and cloudier locations, and on clay soils. Data from Fredericton, NB (Figure 12.5), illustrate this moisture class. There precipitation exceeds normal evapotranspiration through April and May, *PE* is only slightly greater than rainfall in summer, and excess rainfall normally returns for September and October.

But climatic moisture indices alone cannot tell the full story of climatic suitability for farming, because a second major climatic factor also governs distribution of agriculture in Canada – namely, temperature. For example, London in southern Ontario and Beaverlodge in northern Alberta were seen above to fall into the same moisture class, but they clearly do not have the same potential for agriculture! Chapman and Brown (1966) have added sophistication to the moisture-index approach by also considering the frost-free period (average duration of minimum temperatures $\geq 0°C$) and growing degree–days (*GDDs*). This latter parameter assumes that temperatures below a threshold called the base temperature (T_{base}) do not contribute to plant growth and development, so the *GDD* value for each day is calculated from the daily mean temperature ($\bar{T}$) as

$$GDD = \bar{T} - T_{base}.$$ (12.4)

Each day's *GDD* value is added to the next to form a seasonal heat total. The T_{base} chosen depends on the crop, but Chapman and Brown chose 42°F (about 5°C), which is typical for perennial plants and results in seven heat zones for Canada. These zones range from the warmest in southern Ontario and small areas of interior British Columbia, through middle values in the Maritimes and southern Prairies, to the coolest regions, found in Newfoundland, higher elevations in southern British Columbia, and the northern portions of all mainland provinces.

Some of the temperature categories of Chapman and Brown have been combined before being superimposed on the moisture patterns shown in Figure 12.5. The growing season isolines delineate the following classes of land (with *GDD* converted to 5°C base from the original 42°F base):

G1: more than 1,950 *GDDs*;
G2: 1,450 to 1,950 *GDDs*;
G3: 1,230 to 1,450 *GDDs*, except in Alberta and Saskatchewan, where more than 90 frost-free (f.f.) days served as a category boundary;
G4: 1,000 to 1,230 *GDDs*, or 75–90 f.f. days in Alberta and Saskatchewan;
G5: less than 1,000 *GDDs*, or less than 75 f.f. days in Alberta and Saskatchewan.

Now Beaverlodge, for example (category G4), is clearly separated from London (category G1), and the vast area of Quebec and the Maritimes that lies in the wettest moisture regime is suitably stratified according to temperature limitations on farming. In British Columbia, growing season isolines cannot conveniently be drawn at the scale of Figure 12.5 to delineate the thin ribbons of agricultural land in the lower mainland and interior valleys, but eastern Vancouver Island and the Fraser Delta would fall in G2, along with northern portions of the Okanagan Valley, while the southern Fraser and Okanagan valleys would fit in G1.

The above, broad discussion of surface climate suitability for farming leaves out much important detail. For readers requiring more information, "The Climates of Canada for Agriculture" Chapman and Brown (1966) remains a highly recommended document, which weaves the elements of temperature and moisture in Canadian agricultural regions into forty classes, with maps and commentary provided.

SURFACE CLIMATES ASSOCIATED WITH SELECTED SITES AND CROPS

This section considers some selected examples of Canadian surface climates and their interactions with agricultural surfaces. The discussions fall into three main categories:

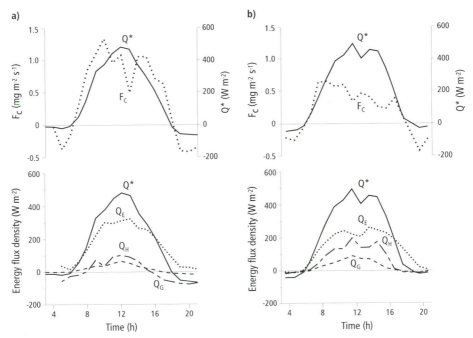

Figure 12.6
Energy fluxes for (a) unstressed (1 Sept. 1988) and (b) water-stressed (8 Aug. 1988) alfalfa. Upper graphs show photosynthesis (flux of CO_2, F_C) and net radiation; lower graphs show use of net radiation (Q^*) by fluxes of sensible (Q_H), latent (Q_E), and ground (Q_G) heat.
Source: Barr (1991).

field and forage crops in zones with mostly adequate moisture supply (*CMI* > 0.6), dryland crops on the Prairies, and horticultural crops at some special climate sites.

Field and Forage Crops

Figure 12.6 (left-hand graphs) shows use of net radiation, Q^*, received by an alfalfa crop that was adequately supplied with water on a sunny day. These data were obtained by measuring a vertical gradient of temperature and water vapour over the crop so that the Bowen ratio could be computed (equation 2.15), and then Q_H and Q_E, are found using the energy balance–Bowen ratio method. By measuring a vertical CO_2 gradient above the crop as well, the flux of CO_2 (F_C; downward during the day because of photosynthesis) was found in a similar fashion. The partitioning of energy among evaporation (Q_E), heating the overlying air (Q_H), and heating the soil (Q_G) would be similar for many other field crops, such as corn and beans, provided that they were not under water stress. Unstressed agricultural surfaces in Canada are usually strong consumers of water, as seen by the large fraction of Q^* consumed by Q_E on the left-hand side of Figure 12.6.

The Q_H flux was negative (downward) before sunrise, when a temperature inversion delivered energy to the cool crop surface from the slightly warmer overlying air. The crop became warmer than the air at about 0800 h, and the daily input of heat from crop to atmosphere began (Q_H became positive). Evaporation increased rapidly with strengthening net radiation and claimed a vast share of the available radiant energy throughout the day, giving an average Bowen ratio of 0.2. Q_G was often the smallest of the three heat fluxes. Q_H turned downward again (a temperature inversion formed) at about 1600 h, thereby augmenting the energy supply to the crop, so that evaporation actually exceeded net radiation for the last two hours before sunset. On this day photosynthesis was not impeded by water stress, so F_C tracked net radiation quite well and the plants were able to use available solar irradiance for good growth.

The graphs on the right-hand side of Figure 12.6 tell a different story for the same alfalfa crop a few weeks earlier, before rain had arrived to end a period of dry weather. This crop was stressed because of lack of water. Evaporation and photosynthesis were choked off by partly closed stomates, as plants struggled to avoid desiccation. With the Bowen ratio measurement of Q_E as input, equation (2.18) can be a diagnostic tool to find canopy resistance. Afternoon resistances were about 400 s m^{-1} for this day, compared to 140 s m^{-1} in the unstressed case of 1 September. Q_E could not consume its usual large share of $Q*$ on the stressed day, so more energy went into heating soil and air, giving a daily Bowen ratio of 0.48. Photosynthesis (F_C) could not track net radiation, and crop growth suffered. Such days are frequent during the growing season in areas with $CMI < 0.6$, leaving these regions less favourable for rain-fed agriculture in Canada.

Standard climate data in Canada are generally obtained from sensors located about 1.5 m above a grass surface or from a 10-m tower, in the case of wind. Although these data give a reasonable first approximation to actual crop climates, inaccuracies may be significant when one is considering pest activity or plant performance near critical temperatures such as 0°C. The largest differences are found for wind, which decreases rapidly as the crop top is approached from above and is almost extinguished at about two-thirds of crop height. These wind differences (Figure 12.7) are important in studies of such matters as leaf temperatures, evaporation of intercepted rain, and dispersal of disease spores. Crop air temperatures are typically 3 to 5 Celsius degrees warmer by day and 1 to 3 degrees cooler by night than the standard screen-level temperatures at 1.5 m. Major exceptions are young plants poking through dry mulches, where the diurnal temperature range is extreme. In this case, daytime temperatures can be more than 25 Celsius degrees warmer than a nearby weather station, and marked nighttime cooling can lead to cold damage. Carbon dioxide concentrations also fluctuate more in a plant canopy than in the free air, because of daytime sinks and nighttime sources for CO_2 in plant leaves. In summary, the microclimate in a plant canopy is more harshly variable than is revealed by standard weather station data. Readers wishing more details can find canopy microclimates well described by Oke (1987).

Despite differences between weather station and crop climates, agroclimatic studies must use standard data because they range over sufficiently large temporal and spatial scales. Thus the corn heat unit system, based on standard air temperatures, has been developed to rate agricultural land according to its ability to produce forage or grain corn (Brown, 1975). This more sophisticated version of the GDD concept (equation 12.4) takes account of growth retardation caused by higher-than-optimal temperatures. It has been used primarily in southern Manitoba, Ontario, Quebec, and the Maritimes to guide farmers in their choice of varieties of corn seed best suited to their zone.

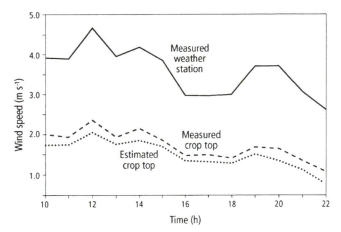

Figure 12.7
Winds at the top of a corn crop and from a standard 10-m tower nearby at a weather station. Large differences in wind speed between standard observations and crop-level can be well estimated from micrometeorological theory.
Source: Brown and Gillespie (1991).

Climatic data have also guided agricultural planning in other, innovative ways. A method has been developed for estimating forage yield potential in Canada using standard air temperature and precipitation in a soil moisture budgeting procedure, similar to that described above in connection with the *CMI* classification for Canada (Bootsma and Boisvert, 1991). Standard station data have also been used to estimate the effect of weather on the quality of cut forage as it lies drying in the field before baling (Smith and Brown, 1994). In addition, guidance concerning field tractability and subsequent choice of planting and harvesting machinery has been developed from standard data. The number of days in the spring or autumn when soil is likely to be dry enough for heavy equipment to work in the field may be judged (for example, Baier et al., 1978). Finally, the critical autumn period when alfalfa should not be harvested so that plants may store sufficient root reserves to survive the winter can also be determined from data on air temperature (Bootsma and Suzuki, 1985). In all cases of analysis with climatic data, it is possible to give various levels of probability of occurrence, which allows growers to manage their operations according to the risks they wish to take.

Examination of data on radiation and temperature (Bailey, 1981) helps to explain an interesting anomaly – existence of a viable farming region at the surprising location of about 56°N in the Peace River region of Alberta and British Columbia. Isolines of net radiation curve northwestward in Canada (see Figure 1.7), so values at the BC/Alberta/NWT border (60°N) are as large as those near the tip of James Bay (52°N). This results in southward depression of the boreal forest region in eastern Canada and an extension of the Great Plains in the northwestern Prairies. Bailey (1981) has shown that agricultural and forest lands in the Peace River region are well separated by the 14°C isotherm of mean growing season temperature and that much of the region's farmland meets the ninety-day frost-free criterion often used to demark agricultural areas further east on the Prairies (Chapman and Brown, 1966). Early snowcover, and low frequency of snowmelt during winter, help prevent winter kill and low soil temperatures in spring. Barley, wheat, oats, and forages can be grown with success, though wet soils at harvest are a frequent problem.

Agriculture on the Canadian Prairies

It is clear from *CMI*s (Figure 12.5) that major differences between agricultural practices on the Canadian Prairies and those in eastern Canada must be linked to moisture. Generally, less than 60 per cent of atmospheric demand for water is met by precipitation on the Prairies (*CMI* < 0.6). Spring wheat, summerfallow, oats, and barley are the dominant agricultural land uses, with the former two dominating in the driest and warmest regions, while the latter two generally increase toward the north (Dumanski and Kirkwood, 1988).

Water contents that are less than 50 per cent of the soil's water holding capacity occur with a probability of at least 80 per cent at the critical time when grain heads are appearing in most areas. Therefore some water stress is experienced in most years, and the grain and oilseed crops chosen must be tolerant of this stress. The practice of leaving fields uncropped every second year (summerfallow) can reduce the probability of stress by 10 to 20 per cent on clay soils but hardly helps with the struggle for water on sandy soils (de Jong and Bootsma, 1988).

The chinook is a special climatic event that affects the agriculture of southern Alberta. When the synoptic-scale weather pattern produces a pronounced westerly flow at high levels over the Rockies, down-slope surface winds and strong adiabatic warming may occur in the lee of the mountains, particularly in winter. The area around Lethbridge shows the highest frequency of winter days warmer than 5°C on the Prairies (Dzikowski and Heywood, 1990), and there are more than 500 h per year when winds exceed 50 km h^{-1}. There the chinook truly lives up to its Indian meaning of "snow eater," but moisture losses resulting from sublimation of snow or evaporation of meltwater put extra strain on the water budget of crops in the following summer.

Figure 12.8 shows data obtained over a typical dryland field of spring wheat in July 1989. Comparison with Figure 12.6 shows that the Prairie crop is much more jealously guarding its supply of soil water than is the alfalfa field of rain-blessed Ontario. Evaporation was clearly the winning consumer of net radiation, even when the crop was under some water stress in the east, but on the Prairies the largest share of $Q*$ was taken to heat the overlying air (Q_H). On average, in Figure 12.8, only 31 per cent of net radiation was used for Prairie evapotranspiration, while more than 80 per cent of radiant energy was consumed by water vapour flux from the unstressed Ontario alfalfa. The canopy resistance averaged 260 s m^{-1} for the wheat, as compared to 140 s m^{-1} for the alfalfa, indicating that the wheat was holding its stomates considerably closed in order to conserve moisture. Of course, this miserly metering of water also restricts the incoming flow of CO_2, so productivity suffers.

At some Prairie locations, reservoirs permit irrigation. The abundant sunshine of clear Prairie skies, when combined with adequate water, provides high yield potential. But a dilemma arises, because the cost in evaporative consumption can be huge when arid air from surrounding dry land sweeps over a wet, irrigated oasis. The crop may transpire so vigorously that it cools below air temperature; the sensible heat

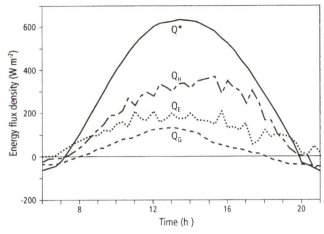

Figure 12.8
Energy fluxes for a Prairie wheat crop. Evaporation (Q_E) here consumes much less net radiation ($Q*$) than is the case for a well-watered crop. *Source*: Data from S. McGinn, Agriculture Canada.

flux then travels downward (becomes an energy source rather than a sink) and fuels evaporation to a value that exceeds net radiation. In addition, if the irrigation is applied with sprinklers, nearly 50 per cent of the water may evaporate in flight or blow away before reaching target plants and soil. A careful study of centre pivot irrigation (500-m boom and delivery rate about 3,800 l min^{-1}) during a growing season at Birsay, Sask. (51°N, 107°W), showed that water reaching the target was usually lost by evaporation and transpiration within thirty-six hours, with loss rates reaching as high as 14 mm per day (data from N. Livingston, pers. comm., 1992). In moister regions, typical daily losses might be near 5 mm.

Horticultural Crops

Horticultural crops are grown across an incredible range of climates in Canada, from the moist Maritimes to the parched BC interior and the cool Peace River region. On a small scale, ingenious growers can produce vegetables or fruits despite the weather by making use of protected south-facing locations, mulches, tunnels of clear plastic, greenhouses, and irrigation. However, commercial horticultural production on larger scales is often found where special combinations of soil and climate exist.

A major constraint to fruit production in Canada is often the depth of midwinter cold. Damage to dormant buds of peach and grape occurs with temperatures of −24°C or colder, while well-hardened pear buds may survive to −30°C and apples can tolerate −29°C to −35°C. Blackburn (1984) has examined apple tree losses in Canada from winter injury using minimum temperature criteria that slide lower as trees harden into January (−29°C) but rise to −18°C in April as trees awaken for the new season's production. He also considered the damaging consequences of rapid temperature changes (> 1°C per hour) in winter that sometimes accompany major arctic outbursts. The frequency with which Blackburn's criteria are met pinpoints the favoured

Figure 12.9
Diverse agricultural landscape of the Annapolis Valley of Nova Scotia.

locations for fruit production in Canada. The probability of apple injury in winter is less than 10 per cent in the southern BC interior, the Annapolis Valley of Nova Scotia (Figure 12.9), and the Niagara Peninsula of Ontario. It rises to nearly 40 per cent in other areas of southern Ontario and almost 100 per cent, according to climate statistics, in southern Quebec and the Saint John Valley of New Brunswick.

Southern areas of British Columbia, Ontario, and Quebec, and the Annapolis and Saint John valleys, all grow some fruit crops, despite Blackburn's finding. Their suitability rests on special local or meso-scale climatic conditions. The Atlantic Ocean and Bay of Fundy flank the Annapolis Valley, while the Great Lakes protect southwestern Ontario. They provide slow spring warming and gentle autumn cooling, which delays spring blossoming until most of the frost danger has passed, allows slow hardening of trees in autumn, and buffers arctic airmasses in winter. The Rockies usually prevent frigid winter air masses from entering the BC interior.

In addition to these larger-scale considerations, the tenderest fruits are grown on locally sloping land such as the walls of the Niagara Escarpment or valley sides in British Columbia. Growers there are taking advantage of the air drainage that occurs on freezing nights, as parcels of air near the slope are chilled by contact with the cooling ground, become more dense, and slide down the landscape like oozing molasses,

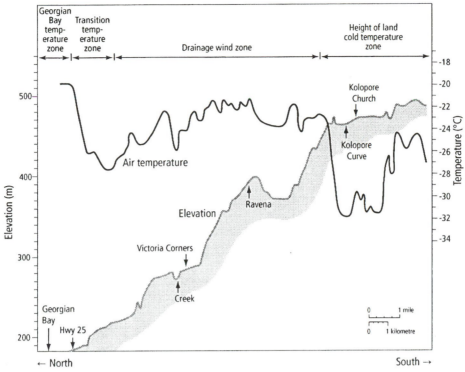

Figure 12.10
A temperature transect down the slopes of the Beaver Valley, Ont., on a clear winter night. Temperatures are warmer on the steeper slopes than at the top or bottom of the valley.
Source: Data from L.L. Wark, Atmospheric Environment Service.

via katabatic flow. This motion stirs the lower atmosphere and prevents formation of a deep temperature inversion. The valley sides are kept considerably warmer than surrounding flat lands, or valley bottoms, where the cold air ponds (Figure 12.10). Such local phenomena have been discovered by experienced growers but may not show up in the statistics derived from standard climate stations located away from water or on flat land.

Choice of a location for crop establishment is a long-term decision that may be aided by analysis of multi-year climatic data. Schemes to make short-term decisions based on daily weather data also exist and are best-developed for horticultural crops. These crops have high values, and therefore the economic opportunity exits for active management of factors such as water and pests. Translation of weather information into estimates of evaporation and soil moisture is an important management tool where irrigation must be practised, as in the BC interior and the sandy, potato and tobacco soils of southern Ontario. North American ginseng (*Panax quinquefolium L.*) is so highly valued that growers can afford to provide the costly shaded microclimate needed for production. Its roots are sought throughout Asia for their medicinal and herbal properties. A wood lath roof or polypropylene shade canopy must be erected and constantly maintained over the entire production area, and soil temperatures must be carefully managed (Proctor and Bailey, 1987). This specialty agricultural production is expanding rapidly in southwestern Ontario, in the semi-arid regions of British Columbia, and throughout other agricultural regions in Canada.

Pest Management

Development of insect pests can be monitored using weather data and the same degree-day concept that aids in climatic classification of crops. Each insect has a biological clock that runs according to the accumulation of heat above a particular base temperature (as opposed to mammals' life clocks, which run according to time). Therefore arrival of a crop-damaging life stage can be predicted by modifying the *GDD* equation (equation 12.4) to incorporate the target insect's T_{base}. At the appropriate time, some corrective action can be taken, such as applying a biological control agent or pesticide or harvesting before significant damage occurs.

Outbreaks of crop diseases can also be anticipated from weather data. Usually data on both temperature and moisture are needed, because the speed with which disease spores can penetrate to a leaf's interior depends on the presence of liquid water and adequate warmth when they land on a leaf surface. The correct signal from weather observations can allow a grower to swing into combat with accurate timing that minimizes costs and environmental threat. The World Meteorological Organization has compiled a good review on meteorological aspects of crop protection (WMO, 1988).

DISTINCTIVE FEATURES OF AGRICULTURAL ENVIRONMENTS

Agricultural surface climates are often distinct because they are purposefully selected or modified for farming. This feature is well illustrated by four examples: soil-surface treatments, frost risk and protection, windbreaks, and irrigation.

Surface Treatments

Surface treatments range from change of tillage practices to application of mulches. Their mode of action can be qualitatively explained with the concept of resistance from the Ohm's Law analogy. First, a surface treatment often provides extra resistance to evaporation of water from soil. This resistance may be the fluffy, air-filled layer of dry soil that develops soon after surface tillage, the tortuous pathway through the air spaces in a layer of straw (Figure 12.11), or the solid physical barrier provided by a plastic film (often black polyethylene). Mulches therefore conserve some soil moisture.

Subsequent reduction in evaporation allows more of the radiation absorbed by the mulch to be used by air (Q_H) and ground (Q_G) heating. The high thermal resistance of still air trapped within or under the mulch layer ensures that only a small portion of the energy is carried away by Q_G; most of the energy flows into Q_H. In this fashion, the thermal regime of the soil may be significantly manipulated. Generally, mulched soils have a moderated microclimate because of lower maximum and higher minimum temperatures than their bare, undisturbed counterparts. There are

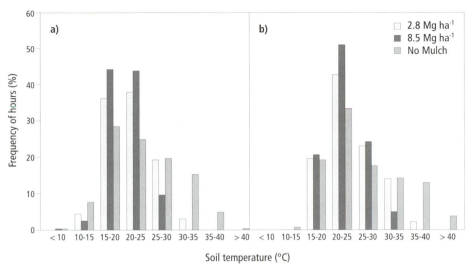

Figure 12.11
Frequency of various classes of soil temperatures in (a) June and (b) July at a depth of 50 mm under soybeans growing in bare soil or with two densities of straw mulch. Frequency of extreme temperatures is reduced under mulch.
Source: Wagner-Riddle (1992).

fewer occurrences of the coldest and warmest categories of temperature under soybeans planted through a medium (early-kill) or heavy (late-kill) rye mulch, compared to beans planted conventionally on bare soil (Figure 12.12).

An exception is found in application of clear plastic as a mulch. In that case radiant energy is absorbed right at the soil's surface, since the plastic is transparent to solar radiation. The thermal resistance of the mulch is now in the Q_H pathway, rather than in the Q_G pathway as described above, so heat flux to the atmosphere is choked off. This condition allows soils under clear plastic to warm rapidly in spring. Clear plastic mulch might become more widespread in spring, except that weeds too love the warm, sunny microclimate under the mulch.

Cold Protection

Sub-zero temperatures are a widespread hazard to crop production in Canada. Late-spring or early-autumn frosts on buds, flowers, fruits, or grains are an important aspect of cold damage. In addition, frigid airmasses that push temperatures below −25°C in the grip of winter may also damage autumn-planted annuals, perennial crops, or buds and branches of tender tree fruits and grapes. Maps of last spring and first autumn frosts are widely available. Brown and Blackburn (1987) present an excellent review of the effects of freezing temperatures on agriculture. They remind us that the "average" time of last frost is near the 50 per cent probability date, so that spring frost will occur later than this in fully five of every ten years! This level of risk may be too high for many farming operations, where the odds of replanting must be reduced to less than 50: 50. Experience with frost statistics suggests the following helpful planning rules. The probability of spring frost is reduced to about one year in four by waiting

Figure 12.12
A surface mulch created by growing rye as a winter cover crop to prevent wind and
water erosion, and then mowing the rye in spring. The machine is planting soybean seeds
through the mulch.

one week after the "average" frost date, and to one year in ten by waiting two weeks.
The same rules work in autumn, when applied before the average first frost date.

Brown and Blackburn (1987) reached five conclusions concerning cold damage
to Canadian crops. First, spring grain losses from late frosts occur about once every
three to four years in Saskatchewan and the Peace River region; less frequently in
Alberta, Manitoba, and northern Ontario and Quebec; and seldom elsewhere in Canada. Second, autumn freezes frequently affect corn and soybean yields only where
short-season varieties are grown at the northern fringes of their range. Third, winter
grains and perennial forages are damaged more often by smothering ice sheets than
by low winter temperatures. Fourth, major declines in production of tender fruit,
caused by severe winter temperatures or frost at spring blossom time, occur once or
twice every five to ten years (depending on crop and location), even in the few
favoured meso-climatic sites, such as the Niagara peninsula and southern British
Columbia. And fifth, loss of vegetables to freezing seldom occurs because of lengthy
experience in timing and locating these crops and use of frost protection where eco-
nomically feasible.

In the battle against cold damage, several "passive methods" have already been
suggested. The site should be chosen to match the crop to the frost-free season. Sloping land kept free of hedgerows or buildings allows cold air to ooze away downhill,
as illustrated by the planting of grapes and peaches on the flanks of the Niagara
Escarpment in southern Ontario. Mulches or recent tillage generally accentuates the
danger of frost damage to young plants by impeding nighttime flow of life-saving
heat from storage in the soil.

Of course there are also "active" methods of cold protection. Every home gardener
knows about covering tomatoes on a chilly autumn night or burying strawberry plants

Figure 12.13
Rows of trees planted perpendicular to the prevailing wind act as windbreaks between farm fields.

in hay for winter. On calm, frosty nights, cold air often pools below warmer air to form a temperature inversion. For very high-value crops such as commercial orchards or vineyards, heaters, helicopters, or a surplus aircraft engine mounted on a tower have been used to stir up this inversion and bring warmer air down from aloft, but operating costs are high! Constant, uniform sprinkling has protected some low crops and tree fruits because latent heat released as the water freezes keeps the plant parts very close to 0°C. Also, some special situations, such as a cranberry bog, may allow the crop to be temporarily submerged in water by flooding. A more detailed discussion of the methods and mechanisms of cold protection can be found in Oke (1987).

Windbreaks

Windbreaks are practised in some agricultural operations, usually to prevent soil erosion when the land is not covered by crops. Natural windbreaks (rather than artificial barriers such as snowfences) are the normal method. They can range from narrow strips of winter grains (such as rye or barley), planted between beds of vegetables on fluffy organic soils, to rows of evergreen trees planted perpendicular to the prevailing winds in large fields of erosion-prone sandy soils (Figure 12.13).

The size of the wind-protected area depends on the height and porosity of the barrier elements. Significant wind reduction ends at a downwind distance eight to fifteen times the height of the barrier. Perhaps surprisingly, partly porous barriers provide protection over the longest downwind distance, although solid barriers offer the greatest wind reductions in the immediate lee of the shelter. These general principles were elucidated by Naegeli as early as 1943 (see, for example, Oke, 1987). True porosity to the wind (aerodynamic porosity) is difficult to judge in the field, but the number of visual gaps in the shelter (optical porosity) may be used as a practical guide.

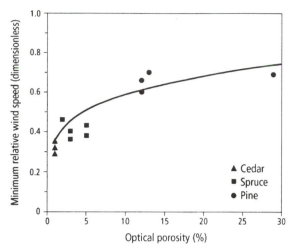

Figure 12.14
The relationship between minimum relative wind speed and optical porosity of the bottom half of several windbreaks in southern Ontario. Relative wind speed compares the slowest wind observed behind the windbreak to the wind in the open field ahead of the barrier.
Source: Loeffler, Gordon, and Gillespie (1992).

Figure 12.14 shows the relationship between the weakest leeward winds (expressed as a fraction of the open-field winds ahead of the barrier) and several tree species used as windbreaks on sandy land in southern Ontario. The most important factor governing protection offered to surrounding crops by a line of trees is the porosity of the lower half of the trees – considerable wind jets through the trunk space when the porosity exceeds about 10 per cent.

There is some observational evidence that windbreaks can slightly enhance yields over their area of influence, except right next to the barrier, where heavy shading from the sun reduces photosynthesis (Figure 12.15). Attributing exact cause to the yield increases is difficult because there are competing influences at work. For example, equation (12.2b) suggests that evaporation should be reduced (and water conserved) because reduced leeward winds increase resistance to water vapour flow. Conversely, plants are likely to be warmer in the sheltered region, and this may increase evaporation (enhancing drought) because the plant-to-air difference in vapour pressure is increased. Fortunately, the balance of competing factors often results in a small productivity gain, to complement the soil conservation afforded by windbreaks.

Irrigation

A fourth major manipulation of local agricultural climates is provided by irrigation. The intention is to permit agriculture in perennially dry regions or to carry crops through dry periods that occur as part of natural precipitation fluctuations in moister

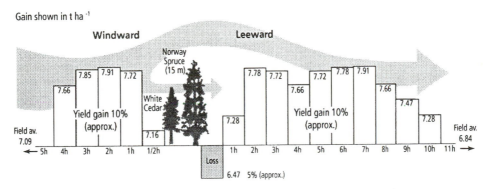

Gain shown in t ha^{-1}

Figure 12.15
Corn yield may be enhanced by windbreak protection. Small gains in yields are sometimes observed within five tree heights' distance upwind, and ten tree heights' downwind from the windbreak.
Source: Baldwin (1989).

regions. The practicality of irrigation therefore depends on an economic balance among frequency of dry periods, cost of irrigation equipment, and productivity gains possible with greater soil moisture. In the very driest regions, such as the BC interior (Figure 12.16), the benefits of irrigation are clear. But when the *CMI* lies between 0.6 and 0.8, deciding if irrigation is worthwhile is a formidable task, because it requires a comprehensive crop growth simulation model that can be run with many years of climatic input and various possible irrigation scenarios. Because of their complexity, such studies have been done for only a few sites in Canada. Based on the climatic data from Harrow in southern Ontario, for example, irrigation would most frequently (about eighteen years out of sixty in this study) increase yields by about 20 per cent. More such studies are clearly needed to permit rational decisions on use of precious irrigation water, especially where drier growing seasons may begin to prevail as a result of climate change.

SPECIAL ASPECTS OF SURFACE CLIMATES
FOR AGRICULTURE

Air Quality and Agriculture

Ground-level ozone and acid precipitation are Canada's major air-quality problems. Acidification is less serious for agriculture, because most farm soils in areas with high risk of acid rain or snow are alkaline and can buffer the acid input. Fortunately, acid concentrations in intercepted rainwater are seldom high enough to damage crop leaves directly. There are only a few locations where sulphur dioxide or other industrial emissions such as fluorine have localized effects.

The Lower Fraser Valley of British Columbia, the Windsor–Quebec corridor, and parts of southern New Brunswick and western Nova Scotia have become chronic sites

Figure 12.16
Irrigation permits intensive fruit farming in the Okanagan Valley of British Columbia,
which has a very low climatic-moisture index.

of summer oxidant smog. More than half of all Canadians are routinely exposed
to summer ozone levels that are known to have adverse effects on health and crops
(CCME, 1990). The required ingredients in the ground-level oxidant soup are gaseous
oxides of nitrogen (NO_x) and volatile organic compounds (VOCs). These must be
brewed in the presence of solar irradiance and temperatures exceeding 18°C to produce ozone (O_3) concentrations exceeding the maximum acceptable one-hour quality
objective of 82 ppb. In a complex set of photochemical reactions, nitrogen dioxide
(NO_2) donates an atom of oxygen, which pairs up with O_2 to form O_3. (This undesir-
able process is different from the desirable natural production of ozone in the stratosphere.) Ozone enters plants through open stomates, its powerful oxidant capabilities
destroy plant tissue, and the photosynthetic capability of the crop is impaired. It has
been estimated that ozone injury to crops costs about $70 million each year in south-
ern Ontario and $10 million in British Columbia.

What are the sources of the ground-level brew? Oxides of nitrogen result from
all kinds of combustion, since high temperature allows the two most plentiful atmospheric gases, oxygen and nitrogen, to combine. Therefore NO_x is strongly anthropogenic in origin. VOCs are also produced by human sources such as burning of gasoline,
evaporation of organic chemicals, and refineries. However, there are also significant
natural sources of VOCs, particularly forests.

The Canadian Council of Ministers of the Environment (CCME) has recognized the
seriousness of the ozone problem and has established a plan of attack (CCME, 1990).
Research is under way to provide the information needed to set final NO_x and VOC
emission targets, which must be met by the year 2005, so as to drop ground-level
ozone concentration by 15 to 35 per cent. This will achieve a 40 to 100 per cent reduction in exposure of crops to unacceptable O_3 levels by 2005.

Agriculture and Climate Change

This chapter concludes with a glimpse into the future. What is the potential influence of climate change on agriculture? First, increased concentration of atmospheric carbon dioxide (CO_2) may directly affect crop performance, since CO_2 is the fuel that drives the photosynthetic engine. Also, there would be very strong additional effects on agricultural production if the present temperature and precipitation regimes in Canada are altered significantly and swiftly.

One of the few certainties in climate change is the increasing global concentration of CO_2. Accelerating combustion of fossil fuels has increased the average CO_2 concentration about 25 per cent in the past 100 years (Houghton, Jenkins, and Ephraums, 1990). Greenhouse horticulturalists have purposely used elevated CO_2 levels to enhance the productivity of crops such as roses for some time, so it is quite logical to ask if growth gains might be achieved in the out-of-doors as the whole atmosphere is pumped up with carbon dioxide. Studies in controlled environments have already shown that crop types vary in their response to elevated CO_2 and, in particular, that certain crops show a significant increase in water use efficiency (*WUE*) with carbon dioxide enrichment. For example, in one indoor study, each unit of water used increased the amount of dry matter produced by maize by 35 per cent (King and Greer, 1986), when CO_2 concentrations were doubled (as is projected for the atmosphere before the middle of the next century). This process may be explained roughly as follows. Higher CO_2 tends to close stomates slightly, thus restricting water loss. The same stomatal restriction attempts to reduce the incoming flow of carbon dioxide, but the higher CO_2 concentration outside the leaf provides a compensating enhancement of the "driving force" (equation 2.14). So the carbon dioxide flux changes less than the water flux, and the *WUE* ratio increases.

There are two reasons to be cautious about how much of the enhanced *WUE* effect observed indoors will be realized in crop fields. First, Jarvis and McNaughton (1986) have pointed out that outdoor plants are relatively less responsive to stomatal control of water loss, and more responsive to control of transpiration by solar energy, than are their indoor counterparts. They would say that outdoor plants are less well "coupled" to the atmosphere, so the energy term in equation (2.21) dominates over the term containing air humidity and canopy resistance. Thus the authors "do not expect the current global rise in CO_2 concentrations to lead to significant reductions in transpiration at the regional scale," and they "would not expect to find any increase in water use efficiency on this scale." Second, reduced transpiration on a large scale would lower air humidity, which would have the effect of returning transpiration towards its original value. Such feedbacks are generally not functioning in controlled-environment experiments.

Certainty becomes even more elusive when trying to predict the possible effects of higher concentrations of CO_2 and other greenhouse gases on temperature and precipitation in farming regions of Canada. In addition to CO_2, two other greenhouse gases directly linked to agriculture are methane (CH_4) and nitrous oxide (N_2O). These are released from the digestive tracts of cattle and/or the microbiological processes responsible for nutrient and carbon cycling in the soil. Although agriculture has been blamed as a significant producer of these gases, we are only now beginning to have the technology capable of effectively comparing such fluxes from various managed and natural surfaces. There is strong hope that these new measurement capabilities will show us how to manipulate agricultural practices to reduce emissions of greenhouse gases.

We know that concentrations of several greenhouse gases are definitely increasing in the atmosphere, but it is still difficult to detect a clear signal that climate change is occurring, and impossible to be sure than any detected changes are anthropogenic. Bootsma (1994) has studied five of Agriculture Canada's research stations, each with about one hundred years of weather data: Agassiz, BC; Indian Head, Sask.; Brandon, Man.; Ottawa, Ont.; and Charlottetown, PEI. He searched their records for statistically significant time trends in a number of climate variables important to agriculture. He did find that the last frosts of spring have become significantly earlier with time, at all stations except Charlottetown's. The frost-free period and number of growing degree-days increased with time only at the western stations.

Changes in precipitation are much harder to identify, because there is so much variability from year to year and the signal contains a lot of "noise." A suitable analogy might be to throw a stone in the Pacific Ocean off Vancouver Island and hope to detect the ripple in Tokyo harbour. Bootsma did find weak evidence of a precipitation decrease at Agassiz and an increase at Charlottetown. No trend across the mid-continental stations was identified. There was a highly significant decrease in water stress potential at Agassiz and Ottawa, primarily due to a decrease in potential evapotranspiration. But are these changes the result of increased greenhouse gases in the atmosphere, or of regional urbanization (possible at all sites studied except Indian Head), or of slow, natural pulsations in the climate?

The only tools available for large-scale, long-time climate prediction are the GCMs, which are very complex, numerical simulations of the atmosphere that are run on the world's most powerful computers. Despite their massive complexity, such models still do not adequately mimic such fundamental processes as cloud dynamics and atmosphere–ocean coupling, so there is disagreement among GCMs constructed at different research centres. The Intergovernmental Panel on Climate Change (IPCC) of the United Nations (Houghton, Jenkins, and Ephraums, 1990) tried to pull some consensus from the several available simulations. For example, it states that doubling CO_2 in the various models predicts that for central North America "the warming varies from 2 to 4°C in winter and 2 to 3°C in summer. Precipitation increases range from 0 to 15 % in winter whereas there are decreases of 5 to 10 % in summer. Soil moisture decreases in summer by 15 to 20 %."

However, when several of the models that contributed to the IPCC consensus are run individually for Canada, predicted effects on crop yields from climate warming caused by doubling of CO_2 vary markedly. Depending on the model chosen, changes in crop revenues across the Prairie provinces varied by ±1–7 per cent in one study, and a more pessimistic −16 per cent for Saskatchewan alone, in another. All studies acknowledge serious inability to assess the likelihood that warmer temperatures would be accompanied by higher variability in precipitation, thus accentuating the frequency of severe drought. It is expected that adverse effects of climate change could be offset to some degree by shifts in crop types and farming methods.

A detailed study of the effect of climate change scenarios on food production opportunities in Ontario (Brklacich and Smit, 1992) has suggested that a longer growing season with more total growing season rainfall will result in an overall average increase of 24 per cent in net farm revenue. But there will be significant areas of sandy soils, where increased evaporation will force major adjustments to combat drought. Growers in Ontario will need to cope with greater variability in farm income – perhaps doubled profit in the best years, but at least 25 per cent more losses in the poor years.

To close, we can review Environment Canada's synthesis of "Climate Warming and Canada's Comparative Position in Agriculture" (Smit, 1989). Consideration of the likely effects of climate change on other farming regions of the world produces several conclusions. First, Canada's position in the production and trade of wheat and grain corn may improve. Second, export opportunities for Canadian wheat to Europe, Africa, and Asia could expand considerably. Third, new markets for Canadian grain corn could open in Central America. Fourth, Canada may be able to reduce imports of grain corn from the US midwest. Fifth, conclusions are uncertain regarding changes in Canada's competitive position for other crops such as barley, oats, soybeans, and rice. Sixth, effects are likely to vary among climatic zones in Canada, so policies will need to recognize local sensitivities of economics and societies to environmental change.

The central message is that climate change may present opportunities to Canadian agriculture, provided that it is prepared to change farming locations and practices in ways that minimize the pain in disadvantaged areas and maximize the gain in places where more favourable farming climates emerge.

ACKNOWLEDGMENTS

The author of this chapter gratefully acknowledges suggestions and contributions from the following colleagues: W.G. Bailey, Simon Fraser University; A. Barr, National Hydrology Research Centre; A. Bootsma, Agriculture Canada; D.M. Brown, University of Guelph; P. Dzikowski, Alberta Agriculture; N. Livingston, University of Victoria; and S. McGinn, Agriculture Canada.

REFERENCES

Baier, W., Dyer, J., Hayhoe, H.N., and Bootsma, A. 1978. *Spring Field Workdays in the Atlantic Provinces.* Atlantic Commission on Agrometeorology, Publication ACA 1978–1.

Bailey, W.G. 1981. "The Climate Resources for Agriculture in Northwestern Canada." *Agriculture and Forestry Bulletin* 4: 11–17.

Baldwin, C.S. 1989. "Windbreaks on the Farm." Ontario Ministry of Agriculture and Food, Publication No. 527, Toronto.

Barr, A. 1991. "Local and Regional Estimates of Sensible and Latent Heat Flux Densities from a Patch-work Surface." PhD thesis, University of Guelph.

Blackburn, W.J. 1984. *Apple Tree Losses in Canada Due to Winter Injury.* Agriculture Canada R & E Report 84–3.

Bootsma, A. 1994. "Long term (100 yr) Climatic Trends for Agriculture at Selected Locations in Canada." *Climate Change* 26: 65–88.

Bootsma, A., and Boisvert, J.B. 1991. *Modelling Methodology for Estimating Forage Yield Potential in Canada.* Agriculture Canada Technical Bulletin 1991–6E.

Bootsma, A., and Suzuki, M. 1985. "Critical Autumn Harvest Period for Alfalfa in the Atlantic Region Based on Growing Degree Days." *Canadian Journal of Plant Science* 65: 573–80.

Brklacich, M., and Smit, B. 1992. "Implications of Changes in Climatic Averages and Variability on Food Production Opportunities in Ontario, Canada." *Climatic Change* 20: 1–21.

Brown, D.M. 1975. "Heat Units for Corn in Southern Ontario." *Ontario Ministry of Agriculture and Food.* Fact Sheet 75–077.

Brown, D.M., and Blackburn, W.J. 1987. "Impacts of Freezing Temperatures on Crop Production in Canada." *Canadian Journal of Plant Science* 67: 1167–80.

Brown, R.D., and Gillespie, T.J. 1991. "Estimating Crop Top Microclimate from Weather Station Data." *Atmosphere-Ocean* 29, 110–32.

Canadian Council of Ministers of the Environment (CCME). 1990. *Management Plan for Nitrogen Oxides and Volatile Organic Compounds.* CCME-EPC/TRE-32E.

Chapman, L.J., and Brown, D.M. 1966. *The Climates of Canada for Agriculture.* Canada Land Inventory Report No. 3. Ottawa: Queen's Printer.

de Jong, R., and Bootsma, A. 1988. "Estimated Long Term Soil Moisture Variability on the Canadian Prairies." *Canadian Journal of Soil Science* 68: 307–21.

Dumanski, J., and Kirkwood, V. 1988. *Crop Production Risks in the Canadian Prairie Region in Relation to Climate and Land Resources.* Agriculture Canada Technical Bulletin 1988–5E.

Dzikowski, P., and Heywood, R.T. 1990. *Agroclimatic Atlas of Alberta.* Edmonton: Alberta Agriculture.

Houghton, J.T., Jenkins, G.J., and Ephraums, J.J., eds. 1990. *Climate Change: The IPCC Assessment.* Cambridge: Cambridge University Press.

Jarvis, P.G., and McNaughton, K.G. 1986. "Stomatal Control of Transpiration: Scaling up from Leaf to Region." *Advances in Ecological Research* 15: 1–49.

King, K.M., and Greer, D.H. 1986. "Effects of Carbon Dioxide Enrichment and Soil Water on Maize." *Agronomy Journal* 78: 515–21.

Loeffler, A.E., Gordon, A.M., and Gillespie, T.J. 1992. "Optical Porosity and Windspeed Reduction by Coniferous Windbreaks in Southern Ontario." *Agroforestry Systems* 17: 119–33.

Oke, T.R. 1987. *Boundary Layer Climates.* 2nd edn. London: Routledge.

Proctor, J.T.A., and Bailey, W.G. 1987. "Ginseng: Industry, Botany and Culture." *Horticultural Reviews* 9: 187–236.

Sly, W.K. 1977. *Agroclimatic Atlas – Canada.* Agrometeorology Research and Service Section, Research Branch, Agriculture Canada, Ottawa.

Smit, B. 1989. *Climate Warming and Canada's Comparative Position in Agriculture.* Climate Change Digest 89–01, Environment Canada, Toronto.

Smith, D.M., and Brown, D.M. 1994. "Rainfall-Induced Leaching and Leaf Losses from Drying Forage." *Agronomy Journal* 86: 503–10.

Wagner-Riddle, C. 1992. "The Effect of Rye Mulch on Soybean Yield: A Field and Modelling Study." PhD thesis, University of Guelph.

World Meteorological Organization (WMO). 1988. *Agrometeorological Aspects of Operational Crop Production.* World Meteorological Organization Technical Note No. 192.

CHAPTER THIRTEEN

Urban Environments

TIMOTHY R. OKE

INTRODUCTION

The study of the physical climates of cities is young. A summary of Canadian work in 1965 started with the statement: "Until fairly recently, meteorologists have turned their backs on cities" (Thomas, 1971). Even today coverage is patchy in terms of both subject and geographical location. Therefore, in order to give a balanced treatment, this chapter draws on non-Canadian examples and general principles. It also relies heavily on work from Vancouver, BC.

Almost 80 per cent of Canadians live in urban areas, and part of the surface climate they experience is something they have inadvertently created for themselves. Every lot cleared, road paved, stream diverted, or house built radically modifies the surface characteristics of a small area. Thus the radiative, roughness, thermal, and moisture properties are changed, and a new microclimate results. Of course the overall climate of a given city is set and modulated by the macro- and regional climate, but each apparently minor change is an incremental contribution to creation of a new climate at a larger scale (the building, block, neighbourhood, land-use zone, city, and city-region).

The resulting climate of each city is unique and, in the minds of Canadians, a key element of its character. The overall role of climate in livability is crudely expressed by the severity index (Figure 13.1), often called the "misery index" because it describes the uncomfortable, depressing, confining, and hazardous aspects imparted by the climate of a place. Many more subtle nuances of climate within and between towns and cities are the result of local topography and the detailed character of the built environment itself. Most cities are located on interesting, often complex sites, such as a coast or shoreline, the confluence of river valleys, a hill, or a sheltered basin. Each carries with it a set of local or topoclimates, including local winds, frost pockets, warm slopes, and zones of greater cloud and precipitation. The climate is also set by the character of the built environment, including street pattern, building density, building materials, amount of vegetation left, need for space heating and irrigation, and degree of industrialization. This chapter concentrates on the effect of these physical controls and explains the way in which the climate of Canadian cities is altered by urban development.

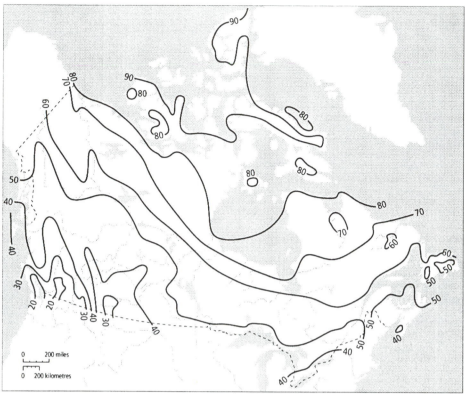

Figure 13.1
Distribution of the "misery" index across Canada: habitability as determined by the macro- (not urban) climate on a scale from 0 (pleasant) to 100 (severe). The ranked list for the major cities is: Victoria 13, Vancouver 18, Calgary 34, Toronto 35, Edmonton 37, Ottawa 43, Montreal 44, Regina 47, Saint John 48, Winnipeg 51, Quebec City 52, and St John's 56.
Source: Phillips (1990).

Urban systems possess characteristics that distinguish them from the other surfaces studied in this book. The surface of a city is complex. Every facet of every building has a unique miroclimate, depending on its colour, material, and exposure to radiation, wind, and precipitation. Add to these myriad vegetation types, roads, natural and paved ground, and water bodies, and even the definition of "the surface" becomes moot. Probably the two most unusual features of urban systems are the abruptness of surface change and the pervasive role of human activity. Unlike most natural systems, the city has sharp breaks and straight-line order. Sharp building edges shed vigorous vortices and create abrupt shadows. Street canyons channel flow in preferred directions. These spatial discontinuities occur at all scales (for example, an irrigated lawn adjacent to a paved driveway, a large parking lot next to a grove of trees, a golf course abutting an industrial park, a city in an agricultural valley). This fact leads to a complex distribution of sources and sinks in both the horizontal and the vertical; thus advection is the norm. The release of heat and water vapour associated with combustion of fuel, or industrial cooling, and irrigation of urban vegetation actively supplement the energy and water balances.

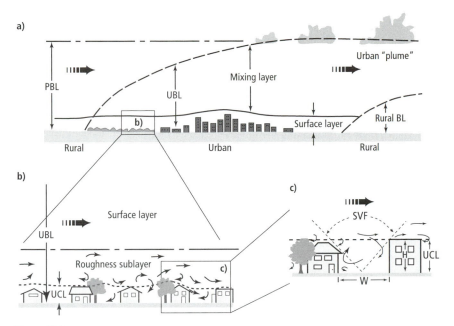

Figure 13.2
Idealized vertical structure of the urban atmosphere over (a) an urban region at the scale of the whole city (mesoscale), (b) a land-use zone (local scale), and (c) a street canyon (microscale).
Source: Modified after Oke (1984).

The sheer mass of the buildings and trees creates an urban "canopy" analogous to that of a forest, but less continuous. Within this urban canopy layer (UCL) (Figure 13.2b, c) is a mosaic of individual microclimates, whose character is determined by the properties of nearby surfaces. These in turn contribute to an hierarchy of climates, organized by the geometric arrangement of the surfaces into buildings, street canyons, city blocks, and so on. The heterogeneity of the UCL is blended above roof level by the action of turbulent mixing (Figure 13.2b). Overlying this transitional "roughness layer," the city's influence extends vertically to form the urban boundary layer (UBL) (Figure 13.2a) – a concept articulated in 1964 by Summers (a Canadian who developed the first numerical model of the urban heat island). During the day, most of this layer is well mixed by turbulence over the rough, warm city. At night a shallower, mixed layer often remains, or its lower portion may be weakly or intermittently stable. The lowest 10 per cent of the UBL, just above the roughness layer, is the turbulent surface layer. There the spatial chaos of the UCL is absent, and one can apply constant-flux layer theory and evaluate spatially averaged fluxes at the scale of land-use zones.

AIRFLOW

When one walks around the base of a large building on a windy day, it becomes obvious that urbanization affects the wind climate in the UCL. It is sometimes more sheltered, and sometimes more gusty and windy, than an open site. While everything appears chaotic, with rapidly changing speed, gustiness, and direction as one moves about, there is in fact a pattern. The flow around an isolated, low building fluctuates,

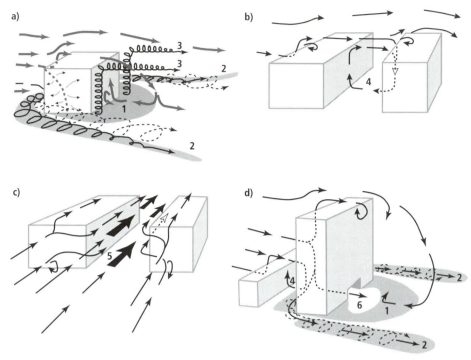

Figure 13.3
Idealized flow in the vicinity of (a) an isolated low building; (b) a street with flow perpendicular to its long axis; (c) a street with flow parallel to its long axis; and (d) a building much taller than those around. Numbered flow features: 1: lee cavity, 2: 'horseshoe' vortex system, 3: elevated corner vortices, 4: canyon vortex circulation, 5: constriction jet, 6: through-flow jet.

but the general features are as shown in Figure 13.3a, with a horseshoe-shaped wake of vortex activity surrounding a sheltered cavity in its lee, and elevated corner vortices trailing downwind. If the buildings are arranged in rows along streets, and the wind is approximately normal to the street, wakes tend to interact, and an in-street vortex circulation is set up (Figure 13.3b). If the flow is parallel to the rows, the flow is constricted and a jet is channelled along the street (Figure 13.3c). The two patterns meet at street intersections, where winds can be very erratic – as at the famous junction of Portage and Main in the windy Prairie city of Winnipeg, Man.

Unusually tall buildings produce severe pedestrian-level problems because their windward face deflects the high momentum flow aloft down to ground level, where it can escape only around the base (Figure 13.3d). Corner jets of the 'horseshoe' vortex system are then so strong that they make walking difficult, even hazardous; pressures created can make doors inoperable, waste paper swirl around, and windchill almost unbearable. Suction on the leeward face has been known to pluck out window glass. The problem is severe enough to warrant by-laws in many Canadian cities requiring new, large structures to be subjected to wind-tunnel analysis before construction is permitted (Figure 13.4). With light winds, the wind climate in the deep canyons of the city core can create an opposite problem. Shelter may decouple the UCL air from that above roof-level and stifle dispersion of noxious pollutants from heavy traffic.

Figure 13.4
A wind-tunnel model of downtown Vancouver used to assess the wind climate around
a proposed building; the model is on a turntable so that the impact of flow from different
directions can be simulated.

If we step back from this level of detail and view the roughness array presented
by a whole residential subdivision, we might argue that, except for the absolute scale
involved, the surface is no more heterogeneous than, say, a field of cabbages. Indeed,
from the limited data available, it appears that the structure of turbulence and the
form of the vertical profile of mean horizontal wind speed, in the surface layer over a
city (above the roughness layer), behave essentially in accord with micrometeorological expectations over natural surfaces. The roughness length (z_o) for a typical Canadian suburban subdivision is about 0.5 m, and values up to about 1.5 m are possible
for a dense, inner-city residential district. However, given the difficulty of measuring
z_o over a city, it is common to use the power law to compute wind speed at some
height z, such that $\bar{u}_z = \bar{u}_{\text{ref}}(z/z_{\text{ref}})^a$, where $\bar{u}_{\text{ref}}$ is wind speed at a reference station.
The exponent a depends on roughness and stability, with typical values of 0.16 for
open rural land, 0.20–0.25 for suburban terrain, and up to 0.35 for a high-rise, downtown district (Figure 13.5).

From yet another perspective, the city as a whole affects overall regional airflow.
These UBL effects are best classified according to the strength of the approach flow.
When winds are strong, the influence of the increased roughness and the barrier
represented by the mass of the city dominate. As air crosses the city, increased drag
slows the lowest layers (Figure 13.5). This deceleration causes mass convergence and
a compensating uplift throughout the UBL. Deceleration also weakens the Coriolis
deflecting force, so the trajectory of the air turns to the left (compared to its rural path)
after entering the city. On return to the smoother, downstream, rural area, the process
is reversed, and subsidence is accompanied by a turn to the right. Therefore rural
flow in the lee of the city resumes its upstream direction but is offset to the left (see
Figure 5.18e in Oke, 1987). Unfortunately, except for some near-surface observations

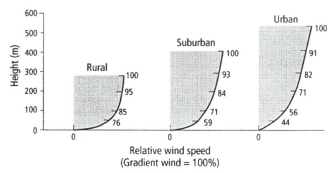

Figure 13.5
Schematic representation of changes in wind profile as air moves from
rural to suburban to urban roughness, under neutral stability (windy,
cloudy). Numbers are percentages of the gradient wind at top of boundary
layer, the height of which also depends on roughness. Notice how the
increase in roughness deepens the layer from which momentum is
extracted and how the speed at any given height decreases as air moves
into the city.
Source: Modified from Davenport (1965).

of reduction in wind speed, there are no examples of Canadian research on these features, despite their potential relevance to both plume trajectories and cloud development.

When winds are weak, thermal effects often dominate. The characteristic warmth of
the city (see the heat island section, below) generates an atmospheric pressure field
akin an oceanic island in daytime. The potential for sea breeze–like, centripetal flow
from country to city therefore exists. In Winnipeg, this circulation has been linked to
nocturnal infestations of mosquitoes, which breed in country areas and drift into
the city in the evening (Thorsteinson, 1988). But it is hard to detect in cities such as
Halifax, Toronto, or Vancouver, where urban breezes compete with other local winds
associated with proximity to a coast, lakeshore, or valley slopes. Intraurban thermal
breezes are also important. Cool breezes emanate from large urban parks within the
heat island, and cold air drains down into ravines and valleys, such as those in Edmonton, Saskatoon, Toronto (see Figure 15.1), and Brandon, Man. (Suckling, 1981), and
down the slopes of Mount Royal in Montreal and the escarpment in Hamilton.

RADIATION BALANCE

Urban atmospheres are polluted and therefore radiatively active. They scatter, absorb,
and reflect solar radiation, and absorb and emit longwave radiation. Therefore radiative fluxes may vary with height in the UBL, but here, for reasons of brevity, we
consider fluxes only at the surface. Since the urban–atmosphere interface is highly
convoluted and the heat balance involves a volume, all fluxes are initially referred to
an active surface plane at the top of the UCL.

The attenuation of solar radiation due to urban pollution has been inferred from
measurements at urban-rural station pairs in several Canadian cities. Typical average

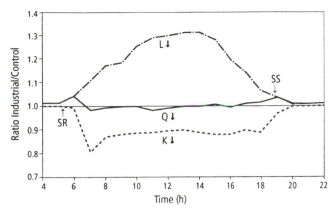

Figure 13.6
Ratio of incoming radiant fluxes (solar $K\downarrow$, longwave $L\downarrow$, and total $Q\downarrow$) observed at an industrial site in Hamilton to those at a rural control site. SR: sunrise SS: sunset.
Source: Rouse et al. (1973).

values are 12 per cent for Hamilton (Figure 13.6), 9 to 12 per cent for Windsor (Brazel and Osborne, 1976), 9 per cent for Montreal (East, 1968), 5 to 7 per cent for Toronto (Emslie, 1964; Yamashita, 1973), and 1 to 2 per cent for Vancouver (Hay, 1983). Greater depletion is observed on individual days. As these values suggest, the size of the effect is related to the type and degree of industrial development. Historical records also indicate that the attenuation is becoming less with time because of introduction of pollution controls and a general shift from particulate to gaseous emissions. Observations elsewhere show that urban air pollution increases the proportion of radiation received as diffuse radiation and is particularly effective at removing ultraviolet wavelengths. The latter effect helps reduce rays harmful to skin but it also is the energy used in photodissociation of oxygen, leading to ozone pollution.

Cities are relatively good absorbers of solar radiation (Oke, 1988). Lewis and Carlson (1989) map the surface albedo of Montreal and show the downtown and dense commercial areas to have values in the range of 0.11 to 0.14, while the rest of the city lies below 0.17. The industrial area of Hamilton has a mean albedo of about 0.12 (Rouse and Bello, 1979), but the value increases with solar zenith angle, rather like smoother natural surfaces. No Canadian measurements of albedo have been made during periods with snowcover. On the basis of data from other cities, urban–rural albedo differences are increased by snowcover, because of soiling by pollution, snow removal, and the increased radiative role of the sides of buildings when the sun is close to the horizon.

The first observations to confirm the expected increase in incoming longwave radiation over a city were conducted in Montreal (Oke and Fuggle, 1972). At night the increase was between 6 and 40 W m^{-2}. This excess was attributed to the heat island rather than to the greenhouse effect of the pollutants – meaning that it is an effect rather than a cause of the heat island. Subsequent work in other cities confirms the increase and shows that it can be larger in the daytime. Remarkably large effects are reported for the industrial atmospheres of Hamilton and Windsor (Rouse, Noad, and McCutcheon, 1973; Brazel and Osborne, 1976). In Hamilton, the peak excess averages 31 per cent (about 70 W m^{-2}) (Figure 13.6), and individual peaks of almost 70 per cent (> 140 W m^{-2}) occur. The authors suggest that this is due to emission from pollutants warmed by absorption of solar radiation and note that total incoming radiation ($Q\downarrow = K\downarrow + L\downarrow$) is almost the same at the two sites (Figure 13.6).

This offsetting between urban effects on radiation seems to be a common feature, as seen in the summary diagram (Figure 13.7). The reduction in $K\downarrow$ seems to be largely offset by a lower urban albedo; the larger $L\downarrow$ is approximately matched by greater emission from the surface heat island. Therefore, despite large effects on individual radiative streams, differences of $Q*$ between city and country are usually rather small. Further, there is relatively little intra-urban variability of $Q*$ (Schmid et al., 1991). By day, both positive and negative differences of urban-rural $Q*$ have been reported, but at night the city almost always seems to experience a greater radiative loss (see Figure 13.11 for the case of Vancouver).

The foregoing applies to the urban system viewed from above. Within the UCL, the radiative environment is spatially and temporally complex because of the geometrical orientation, and radiative interaction between buildings and the wide range of surface radiative properties involved. Nevertheless, distinctive patterns can be seen around an individual building or in street canyons, and, since radiation travels in straight lines, these otherwise-complex radiative patterns can be modelled if the dimensions, albedo, emissivity, and temperature of the surface are specified (for example, Verseghy, 1987). In general, the UCL radiation climate is rich in diffuse radiation caused by multiple reflection between facets. The longwave exchange is dominated by the ability of a point on the surface to "view" the much colder sky, rather than other urban surfaces, which are of approximately similar temperature. The smaller this "sky view factor" (SVF) (Figure 13.2c) of a point, the less it loses radiation; for example, the floor of a deep canyon is unable to cool at the same rate as a roof or an open parking lot.

ENERGY BALANCE

Canadians are among the largest consumers of energy to heat and cool homes and offices, drive vehicles, and run industrial activities. These uses are concentrated in the city, and the heat released to the atmosphere as a by-product can be an important term (Q_F) in the urban surface's energy balance:

$$Q* + Q_F = Q_H + Q_E + Q_S, \tag{13.1}$$

where Q_S is net heat storage in the urban volume (from roof level to the depth where the surface-generated temperature wave disappears, expressed as an equivalent flux density through a plane; see Oke (1988). The largest release of Q_F for a whole city in Canada is for Montreal. In the winter of 1961, an average of 153 W m^{-2} was released; at certain times and places, the value was much greater. Since the corresponding net radiation input was only 13 W m^{-2}, the heat balance was largely attributable to human activities. In summer, the value of Q_F was 57 W m^{-2}, and of $Q*$, 92 W m^{-2}.

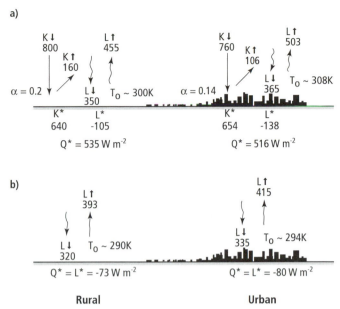

Figure 13.7
Schematic representation of radiation balance components at rural and urban locations on a cloudless summer day: (a) midday; (b) night.
Source: Daytime values from Oke (1988).

In contrast, in Vancouver, where heating and cooling demand and industrial activity are much less, the winter value of $Q*$ is 6 and of Q_F about 23 W m^{-2}. In summer $Q*$ is 107 and Q_F about 15 W m^{-2}.

The daily cycle of human activities sets the hourly variation of Q_F (Figure 13.8). The effect of heat release on the urban climate is not a matter simply of the amount of energy used by people. It is also related to the density of sources and site of release. For example, in the arctic town of Inuvik, NWT, per-capita use is very high, but because the settlement is spread out, the energy flux density in winter is only about 46 W m^{-2}. In the steel works of Hamilton, however, the heat flux density is huge (about 370 to 560 W m^{-2}), because of the concentrated sources. The impact on the surrounding thermal climate in the two cases needs careful assessment. During "arctic night" at Inuvik, with no solar input, any low-level heat source is significant. At Hamilton the releases from tall chimneys may be important to the UBL heat plume but less likely to affect the ULC climate.

Heat Storage

The mass of the building fabric of a city is a potentially huge reservoir for heat storage. It is sometimes described as an inertial "fly-wheel," which takes up heat during the day and releases it slowly later at night. Until recently this role has been hard to verify because it is impossible to measure directly an areally averaged value of Q_S. Now, with the use of eddy-correlation measurements of both Q_H and Q_E, a radiometer

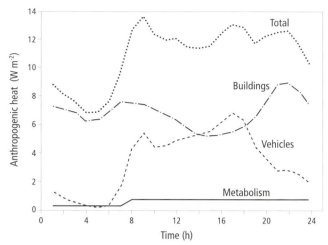

Figure 13.8
Daily variation of anthropogenic heat flux density (Q_F) in a residential
suburb of Vancouver on 22 January 1987. The total is the sum of
heat released from buildings, vehicles, and people.
Source: Grimmond (1992).

for Q^*, and calculated values of Q_F, this term can be found as a residual in equation (13.1). Results from a residential area of Vancouver (Figures 13.9 and 13.10 and Table 13.1) show that Q_S sequesters about 18 to 20 per cent of Q^* by day, and its release at night is approximately equal to Q^*. This is in the middle of the range of other suburban sites in North America (Grimmond and Oke, 1997). Storage at more urbanized sites (industrial, central city) is greater. At a light industrial site in Vancouver daytime Q_S/Q^* was 0.46, and at the old city centre of Mexico City, as high as 0.57 (Grimmond and Oke, unpubl.).

Sensible and Latent Heat

The warm and rough city creates an ideal source of atmospheric turbulence. But the spatial heterogeneity of the surface and the dichotomous availability of water led many to doubt the existence of consistent partitioning of energy between sensible and latent heat. Nevertheless, observations show that while variability of turbulent fluxes is much greater than for net radiation (Schmid et al., 1991) partitioning of Q_H and Q_E at suburban sites is consistent from hour to hour (Figures 3.9, 3.10). Summertime balances from suburban Vancouver, and several U.S. cities, indicate that evapotranspiration remains a major energy sink, despite the removal of vegetation that accompanies urban development. At the residential site in Vancouver, depending on the dryness of the season, the Bowen ratio (β) ranges from about 0.5 (wet) to about 2.0 (dry) (Figure 13.10). Two notable features deserve attention. First, Q_H tends to remain positive well into the evening hours (Figure 13.9). This may be associated with the rather large release of sensible heat from storage at the same time. Second, there are often large day-to-day variations of β, which do not seem to be tied to precipita-

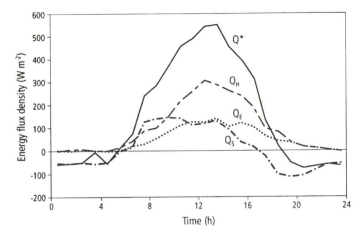

Figure 13.9
Ensemble hourly average energy balance for a suburban site in Vancouver,
based on data from seven summer days.
Sources: Roth (1991), Roth and Oke (1994).

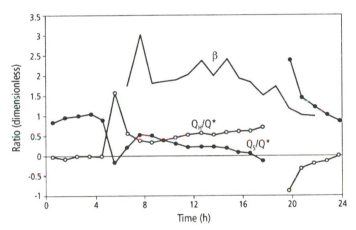

Figure 13.10
Observed values of the flux density ratios: Q_S/Q^* Q_H/Q^*, and $\beta = Q_H/Q_E$
for a residential suburb of Vancouver. Turbulent terms measured by eddy
correlation. Data for same period as Figure 3.9. Absent data indicate ratio very
large or erratic.
Source: Roth and Oke (1994).

tion or expected patterns of drying. This behaviour may be the result of one or
both of two processes – provision of external water for garden sprinkling (discussed
below) or the efficient coupling of the rough and warm urban surface to a deep layer
of air, including the mixed layer of the UBL, and perhaps above, through convective
cells and cumulus cloud formation. In Vancouver, surface evaporation is sometimes
driven by dry downdrafts rather than by moist updrafts (Roth and Oke, 1995). This
reminds us that controls on surface climate include more than just the surface layer,
particularly in cities where turbulent coupling is large (typical values of McNaughton

Table 13.1
Flux ratios for daytime (times with $Q* > 0$) at the Sunset residential site in Vancouver, BC, for three summers with different surface water availability

Year	$Q_H/Q*$	$Q_E/Q*$	$Q_S/Q*$	$\beta = Q_H/Q_E$	Q_H/Q_S
1983	0.44	0.34	0.22	1.3	2.0
1989	0.54	0.27	0.19	2.0	2.8
1992	0.61	0.21	0.18	2.9	3.4

Sources: Cleugh and Oke (1986); Roth and Oke (1994); and Grimmond and Oke (unpubl.).

Note: 1983: normal rainfall plus garden irrigation; 1989: end of the summer with mild drought but garden irrigation; 1992: drought and a ban on garden irrigation. All quantities are dimensionless.

and Jarvis's coupling factor Ω are 0.20 to 0.25; Roth and Oke, 1995). Whether these characteristics hold for all Canadian cities in summer is not known. The winter case with snow remains totally unexplored.

Given the difficulty of measurement over cities, it is encouraging to note the relatively stable nature of simple parameterizations for Q_H and Q_S in terms of $Q*$ (and therefore probably $K\downarrow$) alone (Figure 13.10), since these radiation quantities are relatively easy to measure or calculate.

Some gauge of the impact of urban development on the heat balance is given by Figure 13.11. It summarizes differences between hourly fluxes measured at the suburban residential site and a grass-covered ("rural") control site in Greater Vancouver. The positive daytime $Q*$ difference is due to the lower albedo of the city (solar attenuation is small), and the negative nocturnal difference is the result of extra long-wave emission from the surface heat island. At night the additional radiative drain is supplied entirely from the extra heat stored in the urban fabric during the daytime. At night turbulent exchange is very weak in both environments, but by day reduction in evaporation (latent heat) in the city is the dominant feature.

Some idea of the role of surface water availability in energy sharing is provided by Table 13.1. It shows flux ratios for three years at the same site in Vancouver; as water becomes more restricted, because of drought or restrictions on irrigation, Q_E declines and Q_H grows.

In summary (Figure 13.12), it is clear that urbanization acts to channel more heat into sensible forms, both conductive (Q_S) and turbulent (Q_H), which must lead to a warmer and drier urban atmosphere than its surrounding countryside in summer.

EVAPORATION, WATER BALANCE, AND HUMIDITY

Water availability is a powerful control on urban surface climate. Thermal images of cities show remarkable differences of surface temperature resulting from contrasts between vegetated (especially if irrigated) and built surfaces, such as roofs and paved areas. The energy balances of such UCL surface facets can differ dramatically, as illustrated by an almost-dry street canyon and an irrigated lawn (Figure 13.13). The

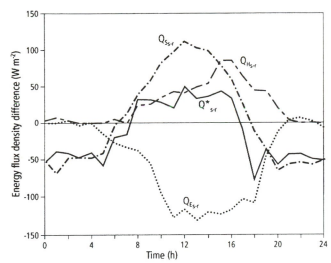

Figure 13.11
Differences of ensemble average energy fluxes between a suburban and a rural site in Vancouver for thirty summer days. Values are positive if suburban > rural.
Source: Cleugh and Oke (1986).

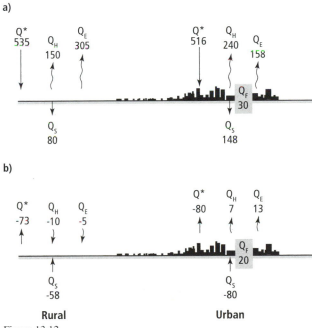

Figure 13.12
Schematic representation of energy balance components at rural and urban locations on a cloudless summer day: (a) midday; (b) night.
Sources: Midday values from Oke (1988); night values from Oke (unpubl.).

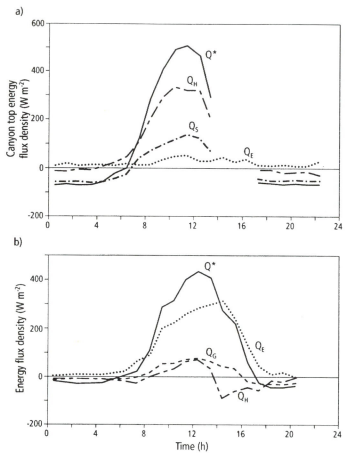

Figure 13.13
Diurnal energy balances of (a) an urban canyon and (b) an irrigated suburban lawn. Notice the different sizes of Q_E in the two cases. In the afternoon some of the energy for evaporation is drawn from the air as it wafts from warm, dry built surfaces across the cooler grass.
Source: Oke (1988).

former channels its energy almost completely into heating the surrounding air and ground; the latter, mostly into vaporizing water. Differences of more than 40 Celsius degrees in T_o between such surfaces are common in summer, which suggests that use of vegetation and water can be a powerful design tool. The effects are both local and at the meso-scale. For example, each tree helps cool a building lot, and the assemblage of all trees into an urban forest modifies the entire UBL (Oke, 1989). Though a city tends to evaporate less than its surroundings (Figures 13.11 and 13.12), its loss of water is far from negligible (Figure 13.9). In Vancouver, evaporation constitutes 38 per cent of losses in the annual water balance and 81 per cent in summer (Grimmond and Oke, 1986). In the latter case, the main source of water is irrigation. Indeed, there is a remarkable correspondence between evaporation and external water use

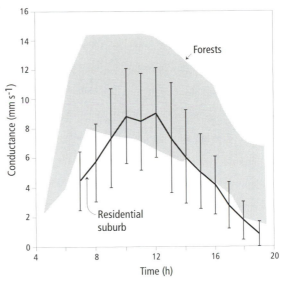

Figure 13.14
Diurnal variation of surface conductance (g_c) for water vapour for a residential suburban site in Vancouver. Shaded area is observed range of g_c for forests presented by Shuttleworth (1989). Bars are standard deviation of measured value.
Source: Grimmond and Oke (1991).

by householders. Intuition suggests that the surface conductance (g_c) of a city will be much less than vegetation surfaces, because the surface is partially sealed by construction materials and the loss of plant cover. In fact, g_c for residential areas of Vancouver is only slightly less than that of most forests (Figure 13.14). It is quite possible that some cities evaporate more water than their surroundings. For example, in the Prairies or the Okanagan Valley of British Columbia, an irrigated city may act like an oasis if its rural surroundings are relatively dry.

Grimmond and Oke (1991) developed the first evaporation model for cities, based on an approach used in forest climatology. The model tracks the water balance of six types of urban surface to assess their water status and calculates evapotranspiration via the Penman-Monteith approach. It incorporates the role of lawn sprinkling and uses a parameterized surface conductance (g_c). It has been validated in Vancouver. In order to generalize the model for Canadian cities, it needs to incorporate snowcover.

The altered fluxes of water to and from the atmosphere of cities create humidity effects. Observations in Edmonton provide one of the best illustrations of this fact. The daily graph shows a classic humidity cycle for a rural site at a continental location in summer (Figure 13.15a). After sunrise, evaporation pumps vapour into a fairly stable lower atmosphere and humidity rises sharply. As the mixed layer grows, the rate of upward mixing exceeds input from below, and humidity concentrations decrease until evening. Evaporation into the newly developing nocturnal stable layer again causes an increase in humidity, but later removal by dewfall onto the cooler surface depletes moisture content again. The UCL curve follows the trend of the rural case for much of the daytime, but concentrations are lower due to the smaller evaporation rates and enhanced

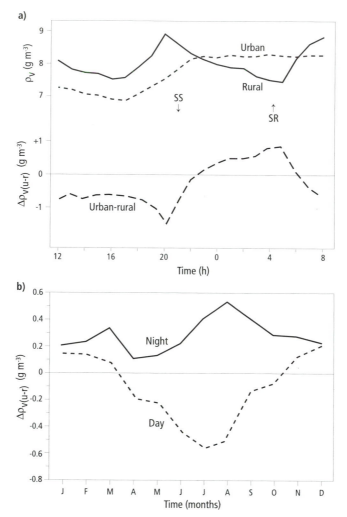

Figure 13.15
Urban effects on humidity in Edmonton. (a) Hourly variation of absolute humidity
(ρ_v) at urban and rural stations and their difference ($\Delta\rho_{v(u-r)}$), on cloudless summer
nights; (b) annual variation of urban–rural absolute humidity differences.
Source: Constructed from data in Hage (1975).

mixing. At night a more obvious urban–rural difference appears – humidity does not
drop as rapidly in the city, because of continued input of moisture by weak evaporation
and combustion releases. Further, it is thought that dewfall is lighter and much less frequent in the city, thus depleting less of the atmospheric moisture in the lowest air layers.
The net effect is that the city is drier by day and more moist at night (Figure 13.15a).

The seasonal pattern in Edmonton is thought to be characteristic of cold-climate
cities (Figure 13.15b). The results show the city to be more moist at all times of day
in winter. This is attributed to virtual cessation of evaporation in the snow-covered
countryside and a large release of vapour associated with combustion of natural gas

for space heating in the city. This forced input of water vapour into a cold atmosphere (often below −25°C) produces the unpleasant ice fog that bedevils many Canadian communities in the depths of winter (Hage, 1972b).

HEAT ISLAND

Surface Heat Island

The altered energy balance produces a warmer city on most occasions. This heat island effect is present at the surface, in the UCL and in the UBL. The surface heat island is easily observed in infrared images (see Roth, Oke, and Emery, 1989). It is best displayed on cloudless days, and a muted version occurs at night. The genesis of these distributions is closely linked to the surface energy balances outlined above. This fact is exploited in Carlson's surface energy balance model, which is coupled with satellite temperatures to produce maps of energy exchange components and surface properties, such as moisture availability and thermal admittance (Lewis and Carlson, 1989). The method is potentially invaluable, although there are questions concerning what "surface" temperature is seen from a vertically viewing platform (Roth, Oke, and Emery, 1989; Voogt and Oke, 1997).

UCL and UBL Heat Islands

The UCL heat island, based on air temperatures from standard climate stations or automobile surveys, is the most clear-cut urban climate effect. Its existence has been reported in cities and towns from every region of Canada. The spatial distribution shows a marked correspondence with patterns of land use and building density. These thermal features include relatively sharp gradients at the rural–urban border, warmer and cooler patches in commercial-industrial and urban-park areas, and a maximum in or near the downtown core (Figure 13.16). In the absence of synoptic weather changes, there is a daily cycle, with the peak heat island occurring about three to four hours after sunset and a minimum in the early afternoon (Figure 13.17). It is therefore principally a nocturnal phenomenon.

The magnitude of the heat island effect is strongly influenced by weather controls and is inversely related to wind speed and cloud cover (type and amount). Wind speed is probably a surrogate for turbulence and advection effects which smear thermal differences, and cloudiness stands for the role of longwave exchange, which is a control on radiative cooling potential. The seasonal variation in synoptic weather – especially the mix of wind and cloud – is a significant control on the seasonal variation of the UCL heat island, together with the state of the rural surroundings and the magnitude of Q_F. The rural surroundings are important because the heat island is usually approximated by differences in the air temperature at urban and rural sites ($\Delta T_{u\text{-}r}$). Therefore the heat island depends not only on the properties of the urban area but also on rural conditions. Rural sites are usually more open to the influences of wind and cloud: they are less sheltered, and their horizon is not as obstructed; there are major changes of surface cover because of the cycle of plant growth; and snowcover is more continuous and lasts longer. Fresh snowcover, fully developed crops, and dry surface soils all

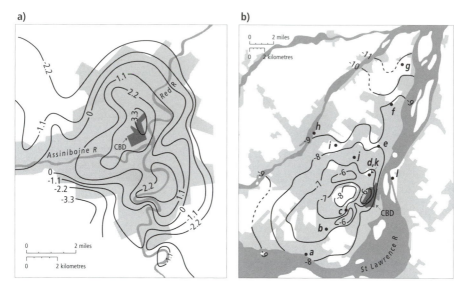

Figure 13.16
Urban canopy layer (UCL) heat island distributions of (a) Winnipeg; and (b) Montreal, with little cloud or wind near sunrise. Numbered points on (b) are sites of helicopter temperature soundings in Figure 13.19.
Sources: (Winnipeg) Einarsson and Lowe (1955); (Montreal) Oke and East (1971).

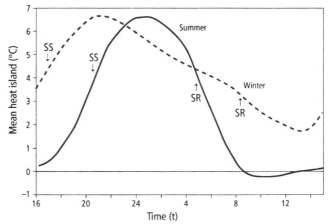

Figure 13.17
Temporal variation of heat island intensity for Edmonton in summer (May–July) and winter (December–February). Data are 5-h running means of temperature differences between the city airport and rural airport stations (ΔT_{u-r}) for nights when these differences were large. SR: sunrise; SS: sunset.
Source: Constructed from Hage (1972a).

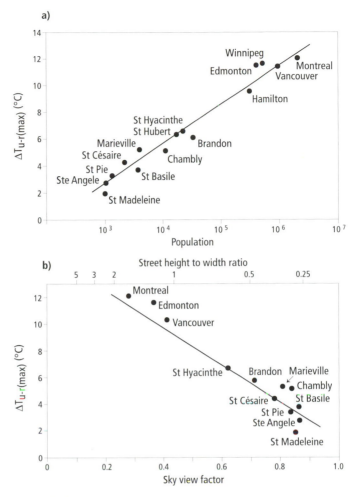

Figure 13.18
Maximum heat island intensity ($\Delta T_{\text{u-r max}}$) in Canadian cities and its relationship to (a) population and (b) geometry of street canyons in city core (H/W and SVF – see text for explanation). Data from Oke (1973) use only summer observations to minimize effects of Q_F. Regression lines are from the original studies and are based on larger North American data sets.
Sources: Modified from Oke (1973; 1981).

lead to lower surface thermal admittance and therefore favour nocturnal cooling and hence large ΔT_{u-r}. Heavy clay soils, bare rock, wet soils, and water bodies dampen the heat island potential (Oke et al., 1991). The relatively slow decline of the winter heat island in Edmonton (Figure 13.17) may be the result of the enhanced role of Q_F caused by space heating in this cold-winter city (Hage, 1972a). A similar explanation is proposed by Nkemdirim, Truch, and Leggat (1977) for a strange mid-morning peak in Calgary's UCL heat island.

As might be expected, the magnitude of the heat island positively correlates with city size (Figure 13.18). Estimates of $\Delta T_{u\text{-}r}$ show good relations with the logarithm of the city's population and the geometry of its downtown street canyons – both the ratio of street height to width (H/W) and sky view factor (SVF), for a point in the middle of the canyon's floor (see Figure 13.2c). Population is no more than a convenient surrogate for size; canyon geometry is more physically meaningful in terms of wind shelter and radiative screening controls. For the largest Canadian cities, with the deepest canyons, maximum heat islands, observed by car surveys on calm, cloudless nights, are therefore as large as 12 Celsius degrees. On an annual mean basis, which includes all the dampening effects of wind and cloud, a city of about one million inhabitants may have a heat island of about 1 to 2 Celsius degrees. The island's growth over time follows the pattern of population growth (Landsberg, 1981).

A complete explanation for UCL heat island causation ties it not only to differences in urban and rural surface energy balances, as in Figure 13.11, but also to differences between the convergence of sensible heat in the UCL and the equivalent rural surface boundary layer. Such differences are governed by complex patterns of turbulent diffusion and advection. In the simplest case, with no wind at night, the most powerful urban controls are decreased net longwave loss from canyons due to horizon screening (small SVF) and release of heat stored from building materials (Q_S) (Oke et al., 1991). This is the case for summer. It is possible that release of anthropogenic heat (Q_F) dominates the thermal climate of some Canadian cities in winter.

Summers's concept of a developing UBL (Figure 13.2a) resulting from greater urban roughness and warmth has proved correct (for a review see Oke, 1995). It is especially apt for the case that he postulated – a stable airmass traversing the city has its lowest layers modified to become an adiabatic mixing layer, its depth growing with distance from the upwind urban–rural border. This essentially describes the picture of the UBL heat island over Montreal (Figure 13.19). The along-wind cross-section and profiles (Figures 13.19a, c, respectively) show how the upwind rural inversion is eroded as it crosses the city. The inversion base occurs at a height of about 300 to 400 m over the downtown high-rise buildings. The across-wind cross-section shows that the heat island is shallower on either side, where fetch over the city is less (Figure 13.19b). On closer scrutiny the mixing layer is not uniformly adiabatic; the lowest portion is slightly unstable, and much of the rest is slightly stable – but this is only detail. Most significant, the city, unlike the countryside, maintains a nocturnal mixing layer, which entirely changes the environment for pollutant dispersion – fumigation is the norm. By day a mixing layer is present over urban and rural surfaces, but that over the city is deeper as a result of the extra convergence caused by its greater roughness and larger turbulent sensible heat flux.

The extra warmth of the UBL at night is the result of two influences. First, air warmed by the UCL is entrained up into the layer above roof level, because of both its inherent instability and the mechanical turbulence generated by flow over the buildings. Second, the more turbulent UBL also causes warm air to be entrained downward from the overlying inversion. In fact, if the approach flow is stable, roughness by itself can generate a warm layer (a heat island). By day these processes may be joined by warming of the polluted UBL through absorption of solar radiation.

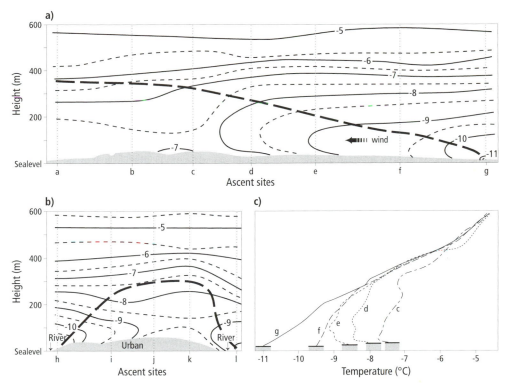

Figure 13.19
Urban boundary layer (UBL) heat island structure over Montreal at the same time as the UCL pattern in Figure 13.16b. (a) Thermally modified layer (heavy dashed line) grows in the along-wind direction. (b) Layer is domed over the city in the cross-wind direction. (c) Vertical profiles progressing into the city at the ascent sites labelled in (a) and in Figure 13.16b. Isotherms are potential temperature (°C).
Source: Oke and East (1971).

URBAN CLIMATE AND GLOBAL CLIMATE CHANGE

These inadvertent modifications to urban climate may be thought of as microcosms of the changes that human activities may beget at the global scale. The density of the urban influence is very much greater than anything to be expected for the globe, but, as Oke (1997) notes, there are several useful analogues and lessons to be drawn from asking the following questions about effects of humans on climate at the two scales.

Are urban changes responsible for observed global changes? The answer is indirectly yes, in large measure because cities are the source of most of the greenhouse gases and much of the aerosols that may lead to global warming. The indirect radiative and cloud physics implications of these pollutants may be at the root of the human role in global climate change. The answer is no, however, in terms of any suggestion that the heat islands of cities are directly warming the global atmosphere to any significant extent. Urban areas cover only about 0.25 per cent of the earth's surface; their extra heat added to that of the global atmospheric heat content is minuscule. This leads to a significant side issue. The measured global temperature record (Figure 14.1) is based on historical observations from weather stations, most of which have been sited in or near settlements and airports. It is possible that the global temperature record has been "contaminated" by overrepresentation of urban-scale thermal effects. Accordingly, the records of some urban stations have been culled from, or corrected in, the most-used global data sets. The consensus of global climatologists is that urban effects in the global temperature signal are now small; not all urban climatologists are convinced.

Oke (1997) also asks whether cities are useful "laboratories" in which to study the processes, effects, and social influences of human activities. Clearly, the answer is positive. The processes and effects are those responsible for much of the human role in global climate change, except that they are more concentrated or strongly felt. Changes to the surface energy and water balances, the radiative effects of aerosols and gases, and the alterations to cloud microphysics attributable to human activities can all be studied in cities or their downwind urban "plumes" (Figure 13.2a). It is these urban, in some cases megalopolitan, plumes that diffuse to become the polluted global atmosphere.

The effects are clear and large. Large cities have experienced warming of 1 to 2 Celsius degrees in their annual mean air temperature during the last century or so. Therefore urban populations and administrations have experienced the magnitude and speed of thermal change projected by GCMs for the future global climate system. That means that they have also seen any accompanying biometeorological and societal effects and should be able to use that information to plan appropriate responses to any future change. Whether cities recognize and are prepared act on these facts at present is doubtful. Toronto and Vancouver have adopted plans to reduce greenhouse gas emissions that also have the desirable side benefit of improving local air quality. Whether they meet their stated targets for reduction will be watched with interest; unfortunately the national record of cutting consumption of energy and thereby reducing greenhouse gas emissions is disappointing.

CONCLUSION

The study of urban climates has therefore come a long way since 1965. Unlike then, today there are observations of urban albedo, incoming and outgoing longwave radiation, radiation budget, evaporation, heat storage, energy balance, water balance, and urban mixing layer, and there are numerical models capable of simulating several aspects of urban climate. In 1969 I wrote a review of knowledge which included a schematic diagram of urban and rural energy balances. It had to be entirely conceptual; the fluxes had no magnitude, and for some even their direction was in doubt. Now it is possible to put real numbers into the equivalent figure (Figure 13.12). Clearly our understanding has come a long way, but it is still only rudimentary by comparison with that for most other surfaces in this book. Having "turned their backs" on cities until about 1965, meteorologists and climatologists have at least glanced over their shoulder, but much remains to be discovered about the atmospheres of Canadian cities and their interactions with global climates.

REFERENCES

Brazel, A.J., and Osborne, R. 1976. "Observations of Atmospheric Thermal Radiation at Windsor, Ontario, Canada." *Archiv für Meteorologie, Geophysik und Bioklimatologie*, B 24: 189–200.

Cleugh, H.A., and Oke, T.R. 1986. "Suburban–Rural Energy Balance Comparisons in Summer for Vancouver, B.C." *Boundary Layer Meteorology* 36: 351–69.

Davenport, A.G. 1965. "The Relationship Of Wind Structure to Wind Loading." In *Proceedings Conference Wind Effects on Buildings and Structures*, National Physics Laboratory, London: HMSO, 53–102.

East, C. 1968. "Comparaison du rayonnement solaire en ville et à la campagne." *Cahiers Géographie, Québec* 25: 81–9.

Einarsson, E., and Lowe, A.B. 1955. *A Study of Horizontal Temperature Variations in the Winnipeg Area on Nights Favouring Radiational Cooling*. Meteorology Division, Department of Transport, CIR-2647, TEC-214, Toronto.

Emslie, J.H. 1964. *The Reduction of Solar Radiation by Atmospheric Pollution at Toronto, Canada*. Meteorology Branch, Deptartment of Transport, CIR-4094, TEC-535, Toronto.

Grimmond, C.S.B. 1992. "The Surburban Energy Balance: Methodological Considerations and Results from a Mid-latitude West Coast City under Winter and Spring Conditions." *International Journal of Climatology* 12: 481–97.

Grimmond, C.S.B., and Oke, T.R. 1986. "Urban Water Balance: 2. Results from a Suburb of Vancouver, British Columbia." *Water Resources Research* 22: 1397–1403.

– 1991. "An Evapotranspiration-Interception Model for Urban Areas." *Water Resources Research* 27: 1739–55.

Hage, K.D. 1972a. "Nocturnal Temperatures in Edmonton, Alberta." *Journal of Applied Meteorology* 11: 123–9.

– 1972b. "Urban Growth Effects on Low-Temperature Fog in Edmonton." *Boundary-Layer Meteorology* 2: 334–47.

– 1975. "Urban–Rural Humidity Differences." *Journal of Applied Meteorology* 14: 1277–83.

Hay, J.E. 1983. "An Assessment of the Mesoscale Variability of Solar Radiation at the Earth's Surface." *Solar Energy* 32: 425–34.

Landsberg, H.E. 1981. *The Urban Climate*. New York: Academic Press.

Lewis, J.E., and Carlson, T.N. 1989. "Spatial Variations in Regional Surface Energy Exchange Patterns for Montréal, Québec." *Canadian Geographer* 33: 194–203.

Nkemdirim, L.C., Truch, L.C., and Leggat, K. 1975. *Calgary's Urban Heat Island 1975 – Surface Features*. Weather Research Monograph 1, University of Calgary, Calgary.

Oke, T.R. 1973. "City Size and the Urban Heat Island." *Atmospheric Environment* 7: 769–79.

– 1981. "Canyon Geometry and the Urban Heat Island: Comparison of Scale Model and Field Observations." *International Journal of Climatology* 1: 237–54.

– 1984. "Methods in Urban Climatology." In A. Ohmura, H. Wanner, and W. Kirchhofer, eds., *Applied Climatology*, Züricher Geographische Schriften 14: 19–29.

– 1987. *Boundary Layer Climates*. 2nd edn. London: Routledge.

– 1988. "The Urban Energy Balance." *Progress in Physical Geography* 12: 471–508.
– 1989. "The Micrometeorology of the Urban Forest." *Philosophical Transactions Royal Society of London* B324: 335–49.
– 1995. "The Heat Island of the Urban Boundary Layer." In J.E. Cermak, A.G. Davenport, E.J. Plate, and D.X. Viegas, eds., *Wind Climate in Cities*, Dordrecht: Kluwer Academic Publishers, 81–107.
– 1997. "Urban Climates and Global Change." In A. Perry and R. Thompson, eds., *Applied Climatology: Principles and Practices*, London: Routledge, 273–87.
Oke, T.R., and East, C. 1971. "The Urban Boundary Layer in Montréal." *Boundary-Layer Meteorology* 1: 411–37.
Oke, T.R., and Fuggle, R.F. 1972. "Comparison of Urban/Rural Counter and Net Radiation at Night." *Boundary-Layer Meteorology* 2: 290–308.
Oke, T.R., Johnson, G.T., Steyn, D.G., and Watson, D. 1991. "Simulation of Surface Heat Islands under 'Ideal' Conditions at Night. Part 2: Diagnosis of Causation." *Boundary-Layer Meteorology* 56: 339–58.
Phillips, D. 1990. *Climates of Canada*. Environment Canada, Ottawa: Canadian Government Publishing Centre, Supply and Services Canada.
Roth, M. 1991. "Turbulent Tansfer Characteristics over a Suburban Surface." PhD thesis, University of British Columbia, Vancouver.
Roth, M., and Oke, T.R. 1994. "Comparison of Modelled and 'Measured' Heat Storage in Suburban Terrain." *Contributions to Atmospheric Physics* 67: 149–56.
– 1995. "Relative Efficiencies of Turbulent Transfer of Heat, Mass and Momentum over a Patchy Urban Surface." *Journal of the Atmospheric Sciences* 52: 1863–74.
Roth, M., Oke, T.R., and Emery, W. 1989. "Satellite-Derived Urban Heat Islands from Three Coastal Cities and the Utilization of Such Data in Urban Climatology." *International Journal of Remote Sensing* 10: 1699–1720.
Rouse, W.R., and Bello, R.L. 1979. "Short-wave Radiation Balance in an Urban Aerosol Layer." *Atmosphere-Ocean* 17: 157–68.
Rouse, W.R., Noad, D., and McCutcheon, J. 1973. "Radiation, Temperature and Atmospheric Emissivities in a Polluted Urban Atmosphere at Hamilton, Ontario." *Journal of Applied Meteorology* 12: 798–807.
Schmid, H.-P., Cleugh, H.A., Grimmond, C.S.B., and Oke, T.R. 1991. "Spatial Variability of Energy Fluxes in Suburban Terrain." *Boundary-Layer Meteorology* 54: 249–76.
Shuttleworth, W.J. 1989. "Micrometeorology of Temperate and Tropical Forest." *Philosophical Transactions of the Royal Society* (London) B 324: 299–334.
Suckling, P.W. 1981. "Nocturnal Heat Island Observations for Small Urban Centres." *Geographical Perspectives* 48: 35–40.
Summers, P. 1964. "An Urban Ventilation Model Applied to Montreal." PhD thesis, McGill University, Montreal.
Thomas, M.K. 1971. *A Survey of the Urban Effect on the Climates Of Canadian Cities*. Climate Division, Atmospheric Environment Service, CDS No. 11–71, Downsview, Ont.
Thorsteinson, A.J. 1988. "Urban Airflow Dynamics and Mosquito Infestations." *Bulletin Society of Vector Ecology* 13: 97–101.
Verseghy, D. 1987. "On the Measurement and Modelling of Radiative Exchange for Building Surfaces." PhD thesis, University of Toronto.

Voogt, J.A., and Oke, T.R. 1997. "Complete Urban Surface Temperatures." *Journal of Applied Meteorology* 36 (in press).

Yamashita, S. 1973. "Air Pollution Study from Measurements of Solar Radiation." *Archiv für Meteorologie, Geophysik und Bioklimatologie* B 21: 243–53.

CHAPTER FOURTEEN

Climatic Change

L.D. DANNY HARVEY

INTRODUCTION

During the last two hundred years human activity has raised atmospheric concentration of carbon dioxide (CO_2) by over 25 per cent, from perhaps 280 ppmv at the beginning of the industrial revolution to 360 ppmv in 1996. Further, methane concentration has more than doubled, from 0.7 ppmv to over 1.7 ppmv. Current concentrations of these gases already greatly exceed any level that has existed during the last 160,000 years. Recent rates of change appear greater than those that have occurred naturally, as demonstrated by the record of air bubbles trapped in antarctic ice (Chappellaz et al., 1990). Carbon dioxide and methane are both greenhouse gases, and the increase in their concentration and that of other greenhouse gases – principally tropospheric ozone, chlorofluorocarbons (CFCs), and nitrous oxide – has already committed the earth to an average warming of 1 to 2 Celsius degrees, based on present understanding. Under business-as-usual scenarios of global energy use, a globally averaged temperature increase of 3 to 6 Celsius degrees may occur by the end of the next century. For aggressive scenarios of reduced emission of greenhouse gases (leading to stabilization of atmospheric CO_2 at 450 ppmv), globally averaged warming of 1.2 to 2.4 degrees is still possible (Harvey, 1989b; 1990; Kattenberg et al., 1996).

Such projected, globally averaged warming will be accompanied by significant regional variations in warming and shifts in precipitation belts. Climate models generally predict greater warming than the global average at middle to high latitudes, including Canada, and a greater warming in winter than in summer at middle to high latitudes. They anticipate more winter precipitation at mid-latitudes but differ widely on changes in summer precipitation. Significant disagreements occur with regard to predicted change in summer soil moisture.

This chapter outlines the basis for confidence in projections of significant warming as a result of human greenhouse gas emissions – warming that will swamp the effect of natural variability over the coming decades – while suggesting that preparation of reliable projections of regional climate change will require at least one to two decades of further research. It presents a hierarchy of increasingly complex surface-process parameterizations, used in general circulation models (GCMs). It also discusses the difficulties in applying point-process descriptions to GCM grid boxes. In the absence

of reliable regional projections of climatic change, it is impossible to make quantitative predictions of how a given surface type in a given region of Canada will be
affected. This chapter is therefore limited to discussion of the magnitude and character of possible climatic change, of the effect of such changes on different surface
types in Canada, and of atmosphere–surface feedback affecting climatic change.

Scientific Basis of Confidence

Climatic response to any perturbation, including increases in greenhouse gases, can
be broken into two parts: the magnitude of the heating perturbation and the radiative
damping parameter.

The appropriate heating perturbation, governing both the surface and the tropospheric responses, is the globally averaged change in net radiation at the tropopause
before adjustment of temperature or any other climatic variable has occurred. The
change in absorbed longwave radiation due to increased CO_2 depends in part on the
temperature, water vapour content, and cloudiness of the unperturbed atmosphere,
as these parameters determine the magnitude of the absolute fluxes available for absorption. In the case of a hypothetical CO_2 doubling (used as a common reference for
comparison purposes in climate-modelling research), the upward longwave radiative
flux at the tropopause decreases by about 2.9 to 3.2 W m^{-2}, while the downward
flux from the stratosphere increases by about 1.4 W m^{-2}, giving a total surface–
troposphere heating perturbation of about 4.3 to 4.6 W m^{-2} in one GCM (Schmitt and
Randall, 1991). It is this perturbation, not the increase in downward longwave radiation at the earth's surface, that governs the temperature response of the surface.

Indeed, the globally averaged increase in downward radiation at the earth's surface
resulting from CO_2 doubling is rather small – about 1 W m^{-2}. It is close to zero in
tropical regions, as a result of absorption of extra downward radiation by water
vapour, which is concentrated in the lowest kilometre of the atmosphere (Kiehl and
Ramanathan, 1982). The reason why the surface temperature response is driven by the
net surface–troposphere heating perturbation, rather than by that of the surface alone,
is that the surface and troposphere are tightly coupled through convective exchanges
of sensible and latent heat flux. Furthermore, the increase in atmospheric temperature
in response to the initial surface–troposphere heating perturbation leads to an increase
in downward longwave radiation at the surface. This is an order of magnitude larger,
in the global average, than the perturbation in downward longwave radiation due to
doubling of CO_2 (Cess and Potter, 1984; Gutowski, Gutzler, and Wang, 1991).

The radiative damping parameter is equal to the increase in net energy emitted to
space per degree of temperature increase. The more easily the earth can radiate excess
heat to space by increasing its temperature – that is, the larger the radiative damping
parameter – the smaller the temperature increase must be to offset a given trapping of
heat by greenhouse gases. For a blackbody at the earth's effective radiating temperature, the parameter has a value of 3.8 W m^{-2} °C^{-1}. Thus, for CO_2 doubling, the planet
must warm by 1.2 Celsius degrees (4.4 W m^{-2}/3.8 W m^{-2} °C^{-1}) to restore balance.

Climate models and a wide variety of observational evidence indicate that, as
the climate warms, the water vapour content of the atmosphere will increase (Harvey,
1996). Since water vapour itself is a greenhouse gas, this increase makes it harder for

the earth to radiate excess energy to space, thereby necessitating a larger temperature increase if radiative balance is to be restored. This is a positive feedback, and climate models are in close agreement in indicating that the water vapour feedback cuts the radiative damping parameter roughly in half. This would almost double the global mean climatic response from 1.2 to about 2.4 Celsius degrees for a doubling of CO_2.

Changes in clouds can also act as a powerful feedback on climatic change. However, clouds have two competing effects on climate: cooling it through reflection of solar radiation and heating it through absorption of longwave radiation. Because cloud amounts, heights, locations, and microphysical properties change with climate, both the heating and cooling effects of clouds will change. The net effect is therefore difficult to predict and will depend on the difference between potentially large but opposite effects. Not surprisingly, climate models disagree over the net effect of cloud feedbacks. In some models, the net effect is close to zero, but in most, climatic sensitivity to heating perturbation increases. Because of uncertainties in net cloud feedback, climate models generally predict a globally averaged warming of 1.5 to 4.5 Celsius degrees for doubling of CO_2, with most results falling between 2 and 4 degrees.

To constrain empirically the potential effect of cloud feedbacks, one can turn to evidence from past climates. During the peak of the last ice age, about 18,000 years ago, global mean temperature was about 4 to 5 Celsius degrees colder than today. Greenhouse gas concentrations were also lower, dust content of the atmosphere was higher, and large areas of continents and oceans were covered by ice and snow. Altogether, these conditions resulted in a globally averaged radiative deficit of 6 to 10 W m^{-2} (Harvey, 1989a). This implies a radiative damping parameter of 1.5 to 2 W m^{-2} °C and, by extrapolation, a climatic sensitivity of 2 to 3 degrees for a doubling of CO_2.

About 100 million years ago, during the Cretaceous epoch, the concentration of atmospheric CO_2 is estimated to have been six to seven times higher than today's, based on multiple constraints (Berger and Spitzy, 1988; Berner, 1990), although there is large uncertainty in this estimate. The earth's climate was about 4 to 10 Celsius degrees warmer on average than today (Hoffert and Covey, 1992), with much greater differences at high latitudes. The difference in ocean fraction and continental positions alone would have caused average climate to be somewhat warmer than today, but solar luminosity was about 0.5 per cent less than it is today. Higher concentration of CO_2 is a believable explanation of warmth during the Cretaceous. Indeed, throughout the three billion years of the earth's geological history, warm climates are consistently associated with periods of inferred higher concentration of CO_2 (Young, 1991). Concentrations of CO_2 and temperatures during the Cretaceous also imply a likely climatic response of 2 to 3 degrees for doubling of CO_2, based on interpolation rather than extrapolation (Hoffert and Covey, 1992).

In summary, two independent empirical estimates of climatic sensitivity to heating perturbations – based on a climate colder and on a climate warmer than the present – indicate a likely global average climatic response of 2 to 3 Celsius degrees for a doubling of CO_2. This is within the lower half of the temperature range of 2–4 degrees predicted by most climate models, and brackets the value expected assuming that net cloud feedback is close to zero, or, equivalently, based on the water vapour feedback acting alone.

Transient Climatic Change

A sudden doubling of CO_2 has been used as a standardized perturbation for comparing climate models, and the focus has been on the statistical steady-state or equilibrium climatic response to this perturbation. However, actual greenhouse gas increases are gradual. The ultimate heating perturbation due to anthropogenic emissions of greenhouse gases could eventually far exceed that associated with a CO_2 doubling, and climatic response will lag behind the perturbation as a result of the large heat capacity of the oceans. The time-varying climatic response to greenhouse gas increases – the transient response – has been reviewed by Harvey (1989c).

Increases in greenhouse gases since the industrial revolution have already caused a globally averaged surface–troposphere heating of 2 W m^{-2}. Given a steady-state climatic response to CO_2 of 2 to 4 Celsius degrees and a lag of one to three decades due to oceanic influences, one would expect the globally averaged climate to have warmed already by 0.7 to 1.0 degrees. Figure 14.1 shows the variation in globally averaged surface air temperature over the past 120 years, after accounting for changing observational practices, which can produce spurious trends (urbanization effects are thought to be less than 0.1 degrees). It can be seen that the warming so far, of about 0.5 degrees, is less than one might expect. However, the decrease in stratospheric ozone during the past one to two decades has probably had a cooling effect, which largely cancels the heating effect of CFCs (Lacis, Wuebbles, and Logan, 1990). CFCs are responsible for about 12 per cent of the increased greenhouse heating since the industrial revolution (Shine et al., 1990). Emissions of sulphur dioxide (SO_2) associated with use of fossil fuels also appear to be partially masking the effect of greenhouse gas increases (Kaufman, Fraser, and Mahoney, 1991; Charlson et al., 1992). This masking will decrease if and when stratospheric ozone recovers and if SO_2 emissions decrease as part of a strategy to control acid rain. Thus the climate will tend to "catch up" to the concentrations of greenhouse gases already in the atmosphere.

Natural Causes of Climatic Change

In addition to a number of anthropogenic causes of climatic change, global mean climate and year-to-year temperatures can change as a result of the El Niño (the periodic occurrence of unusually warm surface waters in the equatorial Pacific Ocean), volcanic eruptions, solar variability, and changes in heat flow between the ocean surface layer and deep ocean. Changes in ocean surface currents can also cause large regional changes in climate without necessarily affecting global mean temperature.

El Niño can cause fluctuations in globally averaged air temperature of 0.1 to 0.2 Celsius degrees, with much larger shifts in some regions (Ropelewski and Halpert, 1991). Major SO_2-emitting volcanic eruptions can cause a hemispherically averaged cooling of up to 0.4 Celsius degree one year after the eruption in the hemisphere in which the eruption occurs (Mass and Portman, 1989; Hansen et al., 1992). Direct observations of the sun indicate that variations in its energy output can cause changes in the average amount of energy absorbed by the earth of possibly 1 W m^{-2} (Hoffert, Frei, and Narayanan, 1988; Baliunas and Jastrow, 1990; Kelly and Wigley, 1990; Reid, 1991). Modelling studies suggest that ocean fluctuations can cause temporary heating

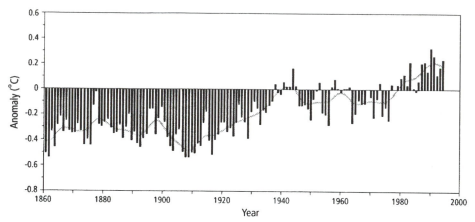

Figure 14.1
Variation in globally averaged surface air temperature anomalies, 1861–1994.
Source: Nicholls et al. (1996).

variations of the same size (Harvey, 1992). In both cases, the possible heating fluctuations are very small compared to the extra heat trapping (5 to 10 W m^{-2}) that could occur by the end of the next century if little is done about emissions of greenhouse gases.

Observed Temperature Trends in Canada

Figure 14.2 shows the change in average temperature over Canada during the past century (as determined by Environment Canada) for regions where statistically significant trends have occurred The long-term trends cannot be unequivocally attributed to the build-up of greenhouse gases during the same period. As noted above, the heating perturbation caused by build-up of greenhouse gases has not yet risen clearly above the size of heating or cooling perturbations resulting from natural causes of climatic change. Other anthropogenic factors (stratospheric ozone depletion, pollution by oxides of sulphur, and so on) have a cooling effect on climate. Nevertheless, Figure 14.2 clearly indicates that climate is already changing in parts of Canada, although with strong year-to-year variability superimposed on long-term trends.

Summary of Evidence

There is a considerable body of empirical, theoretical, and palaeoclimatic evidence to indicate that ongoing increases of greenhouse gas concentrations will cause warming of the earth's climate during the next few decades. Observed warming during the past century is less than would be expected based on climate models and greenhouse gas increases observed so far. There is increasing evidence that this apparent discrepancy is a result of partial masking of the increased greenhouse gases by short-term anthropogenic and natural effects on climate. El Niño and volcanic eruptions are two major causes of year-to-year variation in global and regional temperatures that will be superimposed on longer-term trends. Other perturbations of the surface–troposphere energy balance and solar and oceanic variability will or may already be exceeded by the heat-

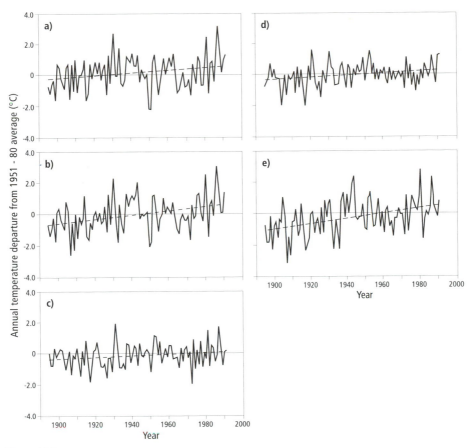

Figure 14.2
Change in surface air temperature over Canada, 1895–1991, for regions showing statistically significant trends: (a) Prairies; (b) northwestern forest (east of Ontario); (c) northeastern forest; (d) Great Lakes–St Lawrence Lowlands; (e) Mackenzie District.
Source: Gullett and Skinner (1992).

ing perturbation due to greenhouse gas increases. Hence natural causes of climatic change will most likely be overshadowed by the effect of more greenhouse gases, leading to long-term warming of the climate.

GENERAL CIRCULATION MODELS (GCMs) AND MODELLING GROUPS

The impact of increases of greenhouses gases is estimated using computer-based, mathematical models of climate. The most sophisticated models are three-dimensional GCMs, coupled to oceanic models of varying sophistication. These models divide the atmosphere into several thousand grid boxes with typical horizontal dimensions ranging from 200×200 km to 500×500 km and two to twenty layers in the vertical. The models simulate day-to-day variation of such variables as wind, temperature, cloudiness, precipitation, soil moisture, and ice and snow extent.

Table 14.1
Selected general circulation–modelling (GCM) groups that have simulated effects of an increase in CO_2

AES	Canadian Climate Centre, Atmospheric Environment Service, Downsview, Ont., Canada
GFDL	Geophysical Fluid Dynamics Laboratory, Princeton, NJ, USA
GISS	Goddard Institute for Space Physics, Greenbelt, Md., USA
HAM	Max Planck Institute, Hamburg, Germany
NCAR	National Center for Atmospheric Research, Boulder, Col., USA
OSU	Oregon State University, Corvallis, Ore., USA
UKMO	Meteorological Office, Bracknell, United Kingdom

Table 14.1 lists the major GCM groups that have performed simulations with increased CO_2. A more extensive listing of modelling groups, along with information about model characteristics, can be found in Gates et al. (1996).

Scenarios of Climatic Change over Canada

GCMs are in general agreement in simulating a greater equilibrium temperature warming at high latitudes than in the global average, and greater warming in winter than in summer at high latitudes. The polar amplification is in part a result of positive feedback at high latitudes from melting of ice and snow, but a contributing factor to greater winter warming is the thinning of sea ice (Harvey, 1988). GCMs simulate intensification of the hydrologic cycle by 9 to 15 per cent for CO_2 doubling (Grotch, 1991) and generally project an increase in winter precipitation in mid- to high latitudes. Models disagree as to the sign and magnitude of summer precipitation changes in mid-latitudes (30 to 60° latitude). The mid-latitude cyclonic precipitation belt is projected to shift poleward as climate warms. The magnitude of the shift and the change in its intensity depend in part on the amount of polar amplification of warming and thus, to some extent, on the initial amounts of ice and snow (Rind, 1988).

There are reasons for confidence in projections of significant warming over the next century and of the broad latitudinal and seasonal patterns of climate change. However, quantitative model projections of temperature and precipitation changes in specific regions, such as Canada, are highly uncertain. Figures 14.3 and 14.4 show the winter and summer surface-air temperature increase over Canada for a climate in equilibrium with doubled CO_2, as computed by four climate models. Figure 14.5 shows calculated changes in summer soil moisture. Significant disagreements among models are evident. Projected winter warming ranges from as little as 2 Celsius degrees over western and central Canada, according to the NCAR model, to warming in excess of 6 degrees over most of the country according to the AES, GFDL, and GISS models. For summer, the NCAR model projects a warming near 2 degrees over most of the country, while GFDL projects warming greater than 6 degrees over most of southern Canada. AES and GISS fall between these two extremes. All the models, except NCAR, project a decrease in average summer soil moisture for doubling of CO_2 (Figure 14.5).

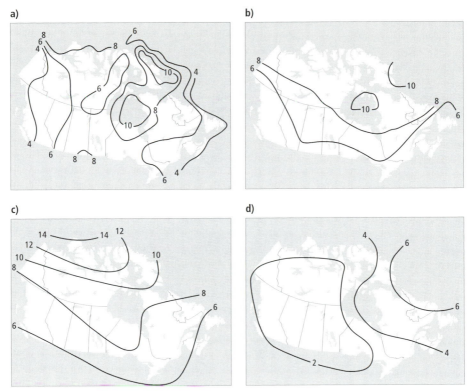

Figure 14.3
Change in winter surface air temperature (Celsius degrees), in Canada for a doubling of atmospheric CO_2 as computed by four GCMs: (a) Canadian Climate Centre; (b) GISS; (c) GFOL; (d) NCAR.
Source: Environment Canada (1992).

Conditions associated with doubling of CO_2 could occur by the middle of the next century. None of the model results shown in Figures 14.3 through 14.5 should be regarded as predictions. Rather, they serve to suggest the likely magnitude of future climatic change. Furthermore, under realistic, gradual change in climate and greenhouse gas concentrations, different regions will warm at different rates. This situation is likely to result in patterns of temperature change that are different from those obtained by the equilibrium simulations shown in Figures 14.3 through 14.5. This in turn could produce completely different precipitation and soil moisture changes during the transition to warmer climates than are expected for an equilibrium climate.

A number of diagnostic studies have shown that differences in climate-model simulations of regional climates for present-day conditions are a large factor in causing disagreements in projections of regional changes when CO_2 is increased (Mitchell, Wilson, and Cunnington, 1987; Meehl and Washington, 1988). Accurate simulation of the present-day distribution of precipitation, cloudiness, soil moisture, and ice and snow amounts is therefore a necessary, but not sufficient, condition for reliable simulation of the regional changes that will accompany global-scale climatic change.

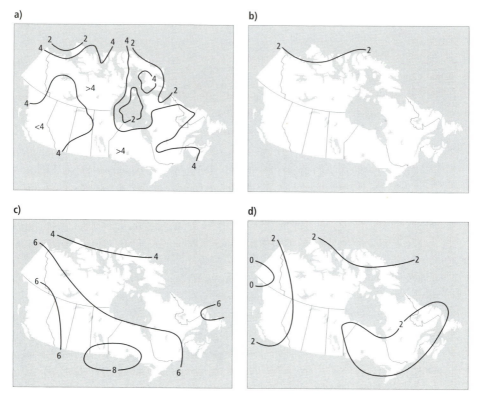

Figure 14.4
Change in summer surface air temperature (Celsius degrees), in Canada for a doubling of atmospheric CO_2 as computed by four GCMs: (a)–(d) as in Figure 14.3.
Source: Environment Canada (1992).

Many important processes, such as cloud development, precipitation, and land surface hydrological exchanges, occur at scales finer than the grid resolution of climate models, and therefore cannot be explicitly represented. Rather these processes must be parameterized which involves representing the effect of subgrid-scale processes in terms of the grid-resolved variables and hence arbitrary adjustment or "tuning" of input parameters to obtain reasonable results. However, even if one were to obtain an accurate simulation of present climatic variables, this could be the result of a fortuitous cancellation of errors in parameterization. Such cancellation might not occur as soon as the grid-resolved climate begins to change, thereby leading to incorrect projections of regional and global climatic change.

MODELLING LAND SURFACE PROCESSES IN GLOBAL MODELS

We begin with a description of point-process models that have been applied to GCM grids, which range in size from 8° latitude × 10° longitude (GISS model) to 2.5° latitude by 3.75° longitude (UKMO).

a) Canadian Climate Centre

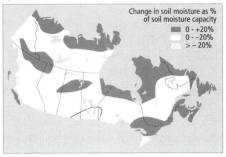

b) Goddard Institute for Space Studies

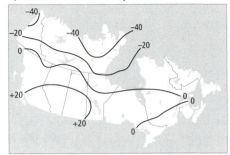

c) Geophysical Fluid Dynamics Laboratory

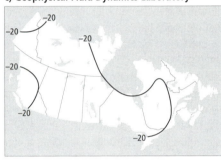

d) National Centre for Atmospheric Research

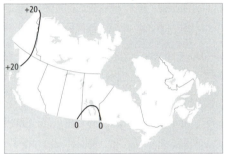

e) Oregon State University

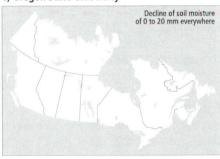

f) United Kingdom Meteorological Office

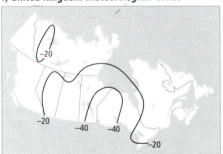

Figure 14.5
Change in summer soil moisture (mm) in Canada for a doubling of atmospheric CO_2, as computed by six GCMs: (a)–(d) as in Figure 14.3; (e) OSU; (f) UKMO.
Source: Environment Canada (1992).

Zero Heat Capacity and Bucket Models

The simplest method for computing land surface temperature T_o, used in the GFDL and NCAR models, is to treat the land surface as having no heat capacity and to use bulk aerodynamic parameterizations of turbulent fluxes of sensible and latent heat. That is, for given atmospheric conditions, one solves for T_o such that:

$$f(T_o) = K^* + L\downarrow - L\uparrow - Q_E - Q_H = 0. \tag{14.1}$$

The function $f(T_o)$ is non-linear, and equation (14.1) must be solved iteratively.

A variation of the above approach, used in the OSU model, is to assign a finite heat capacity to the land surface and solve:

$$z_s C_s \, (dT/dt) = K* + L\downarrow - L\uparrow - Q_H - Q_E, \tag{14.2}$$

where z_s is the thickness of a soil slab that is assumed to undergo isothermal temperature change. The product $z_s C_s$ is referred to as "thermal inertia," since the larger its value, the more slowly temperature responds to a given energy imbalance. In equation (14.1), land surface temperature is determined diagnostically by assuming that it adjusts instantaneously, so as to give zero on the right-hand side of the surface energy balance. In equation (14.2), it is determined prognostically by extrapolating the rate of temperature change over a short period, adding the extrapolated change to the previous temperature, and then continuing in this manner. More sophisticated land surface parameterizations, presented below, use a mix of diagnostic and prognostic variables.

The AES, GFDL, NCAR, OSU, and UKMO GCMs use the "bucket method" for calculation of soil moisture, first introduced by Manabe (1969). Precipitation P fills a bucket below each grid cell, with runoff R occurring if and when the bucket is filled. Changes in soil moisture SM are given by:

$$\partial SM/\partial t = P - E - R \qquad (0 \le SM \le SM_{max}), \tag{14.3}$$

where E is evaporation rate (computed from the bulk aerodynamic formula) and SM_{max} is usually set to a geographically uniform value of 0.15 m.

Two Layer Force-Restore Models

For models driven by diurnally and seasonally varying solar radiation, a two-layer soil model is required, with the upper layer having a thickness corresponding to heating and cooling on a diurnal time-scale and the lower layer having a thickness corresponding to seasonal heating and cooling. The governing equations are

$$z_{s1} C_s \, (dT_1/dt) = f(T_1) - a(T_1 - T_2) \tag{14.4}$$

and

$$z_{s2} C_s \, (dT_2/dt) = a(T_1 - T_2), \tag{14.5}$$

where subscript 1 refers to the upper layer, subscript 2 refers to the lower layer, a is a coupling coefficient, and $f(T)$ is the same function as appears in equation (14.1).

A similar approach can be applied to soil moisture (Deardorff, 1977). Both evaporation and surface temperature respond quite differently to short, intense rainfall events using the bucket model and the force-restore method (Hunt, 1985). In the bucket model, evaporation varies with total soil moisture, which changes more slowly than that of the upper layer used in the force-restore method.

Modelling Snowmelt

Snowmelt is modelled very crudely in most GCMs. If snow is present and the computed surface temperature is above 0°C, the temperature is reset to 0°C and the excess energy computed and used to melt snow. Usually no distinction is made between ground and snow temperatures or between ground and snowpack heat capacities. The processes of melting and refreezing as the snowpack ripens, which can lead to sudden rather than gradual release of snowmelt once the pack collapses, are not represented. In most models, snowmelt is assumed to infiltrate the soil unless the soil is saturated. In practice, relatively little infiltration will occur if the snowmelt is released suddenly or the soil is frozen.

Mitchell and Warrilow (1987) compared the effect on summer soil moisture of a CO_2 doubling using two versions of the UKMO GCM. In one, all snowmelt infiltrates the soil unless the soil is saturated. In the second, all snowmelt is assumed to produce runoff if either of the middle two soil layers are below freezing. In the first case, mid-latitude soil conditions for present levels of CO_2 are simulated to be saturated in spring. The increase in winter precipitation under doubled CO_2 therefore does not increase initial amounts of soil water prior to the period of enhanced evaporation in late spring and summer. Late summer soil moisture therefore decreases when CO_2 is increased. In the second case, soils are not simulated to be saturated in spring for present-day CO_2. Thus the winter precipitation increase under doubled CO_2 leads to greater soil moisture in spring, which largely offsets the increased spring-summer evaporation. This results in only a minor decrease in soil moisture by late summer. Clearly, these two parameterizations represent extreme cases, but they demonstrate the importance of snowmelt modelling through its control on the unperturbed climate.

Current snowmelt algorithms assume a horizontal surface. In mountainous regions, significant differences occur in the timing of snowmelt, runoff contributions, and evaporative moisture fluxes into the atmosphere from north- and south-facing slopes. Incorporation of slope aspect variability would spread out in time the grid average snowmelt, runoff, and evaporative fluxes, with possible subsequent feedbacks on precipitation (depending on the relative importance of regional and local moisture sources).

CLASS – A Canadian Land Surface Scheme

Verseghy (1991) and Verseghy, McFarlane, and Lazare (1992) have developed a land surface parameterization called CLASS (Canadian Land Surface Scheme), which addresses some of the problems in modelling snowmelt and infiltration identified above. CLASS contains three soil layers and an optional fourth layer to represent snow, when present. Soil heat capacities, thermal and hydraulic conductivities, and maximum infiltration rates are a function of soil moisture content, and separate heat capacities and thermal conductivities are specified for the snowpack. The mean temperature of each of the soil layers and of the snowpack is determined prognostically, based on heat flows between the layers or to the atmosphere, but a separate temperature T_o at the surface–air interface is determined diagnostically by requiring the heat flux into (or out of) the surface to balance the net atmosphere–surface heat exchange (this is equivalent to solving equation 14.1 for T_o, with a term added representing conductive heat flux into the soil).

Melting of the snowpack occurs when the solution for T_o is above freezing. The meltwater (and any rainfall) are assumed to percolate into the snowpack, and if the mean snowpack temperature is $< 0°C$, refreezing and latent heat release occur. In this way ripening of the snowpack takes place, with runoff or soil infiltration happening only when the snowpack's temperature reaches $0°C$. Snowmelt can also occur at the base of the snowpack if the temperature of the upper soil layer exceeds $0°C$. This can happen in the presence of snow because, unlike previous GCM snow parameterizations, partial penetration of solar radiation through the snowpack and absorption by the underlying soil surface are modelled. Snowmelt or rainfall in excess of the moisture-dependent, maximum permissible infiltration rate is assumed to form ponds on the soil surface, rather than immediately producing runoff, as in other GCMs.

Comparison of the CLASS model with the standard surface scheme used in the AES GCM indicates that CLASS simulates a much lower latent heat flux and a much greater sensible heat flux, except during the first week after a rainfall event, when the two schemes give comparable fluxes. T_o is permitted to drop below freezing in CLASS, even while the top soil layer contains liquid water. The ability of T_o to drop below freezing while liquid water freezes reduces near-surface atmospheric stability and reduces heat loss from the soil to the atmosphere, thereby prolonging the autumn period of soil water freezing. Fresh snow on unfrozen ground also persists longer in CLASS, since the snow's surface temperature can drop well below freezing while the top soil layer is unfrozen, thereby reducing sublimation losses. The greatest impact on runoff arises from the assumption of ponding when infiltration capacities are reached and underlines the need for field validation of this parameterization.

In CLASS, maximum infiltration capacity depends on both liquid-moisture content and frozen-water content. Thus partitioning of snowmelt into runoff and infiltration, based on the soil's thermal state, is also addressed.

More Complex Models

One can envisage more complex land-surface models incorporating several canopy layers and more detailed treatment of radiative and evaporative fluxes. But is it necessary to use more complex models than have already been coupled to GCMs, given the inaccuracies in the GCM forcing of the surface submodel, uncertainties in the submodel parameters, and the relative insensitivity of the coupled GCM-surface model to some of the surface model parameters? There appears to be a good case for further improving the treatment of snowmelt and its partitioning into infiltration and runoff. However, Xue et al. (1991) compared the original SIB (Simple Biosphere) model, which requires forty-four input parameters at each grid point, with a simplified version requiring twenty-one input parameters and found little difference in simulated surface–air fluxes of heat, moisture, and momentum.

Scaling from Point- to Grid-Size Variables

The foregoing discussion outlined a hierarchy of point process models that have been used in global-scale climate models. However, application of such models to GCM grid boxes can result in significant errors because of the non-linear model response coupled with subgrid-scale variability of precipitation intensity, soil and vegetation properties, and the soil's initial water content. Model non-linearity can be significant for some ranges of a model parameter, but not for others. Hence it can be important at some times but not at others, or in some regions but not in others (Band et al., 1991).

Precipitation in GCMs occurs either as convective or as large-scale precipitation – the former whenever moist convective instability occurs, and the latter, when large-scale ascent leads to supersaturation of a model layer. The usual practice is to distribute the precipitation during a given time step uniformly within the grid box, resulting in a drizzle that rarely exceeds the soil's infiltration capacity. Allowance for distribution of rainfall intensities within the grid cell, with intense convective precipitation over a small fraction of the cell, produces greater runoff than does uniform drizzle. Generation of surface runoff in nature involves two distinct processes – first, overland flow ("Hortonian" runoff), when the precipitation rate exceeds the infiltration rate; and second, precipitation on saturated regions, such that relatively small areas often contribute most or all of the surface runoff, leading to the concept of partial contributing areas (Dunne and Leopold, 1978).

A number of authors (Warrilow, Sangster, and Slingo, 1986; Entekhabi and Eagleson, 1989; Pitman, Henderson-Sellers, and Yang, 1990) have assumed a rainfall intensity P having a distribution of the form

$$P = \left(\frac{\varkappa}{\bar{P}}\right) e^{-\varkappa P/\bar{P}} \tag{14.6}$$

where $\varkappa$ is the fraction of the grid cell over which precipitation occurs and P is the precipitation rate averaged over this region ($\bar{P}/\varkappa$ is the precipitation rate averaged over the entire grid cell). $\varkappa$ depends on the typical scale of precipitation events and hence on the model resolution. For synoptic-scale precipitation $\varkappa = 1$, while for convective rainfall $\varkappa$ has been set at 0.3 to 0.66.

Entekhabi and Eagleson (1989) considered a spatial distribution of soil moisture content along with the above precipitation distribution in computing runoff but assumed uniform soil hydraulic properties. They used a gamma function to represent the probability distribution of soil moisture content, which allows soils in some parts of the grid to be saturated even when the grid's average moisture content is below saturation. By combining the precipitation and soil moisture distributions and an expression for the soil's hydraulic conductivity, they derived an analytical expression for runoff during a given period, which involves contributions from saturated subregions and from Hortonian overland flow.

Incorporation of subgrid-scale precipitation variation can dramatically change the simulated surface climatology, as shown by Pitman, Henderson-Sellers, and Yang (1990). Much more could be done to incorporate subgrid-scale variability than described above. As in the further refinement of point-process models, it is not clear to what extent this is justified. The necessary data may simply not be available or reliable, and in any case large errors occur in the simulated grid-averaged precipitation. Nesting of regional, meso-scale atmospheric models, with a typical grid resolution of 60 km, can substantially improve simulated precipitation fields in regions of mountainous relief (Giorgi, Brodeur, and Bates, 1994), thereby justifying more accurate land surface parameterizations than are used at present in GCMs.

Subgrid-scale variability is also important in computing grid-average evaporation from grid-average soil moisture content (Wetzel and Chang, 1987) and in determining the vertical momentum flux (Beljaars and Holtslag, 1991). In the former case, evaporation varies non-linearly with soil moisture content, such that use of grid-mean soil moisture overestimates grid-mean E at intermediate values of soil moisture content. In the case of the momentum roughness height, the effective values at a scale of 10 km can be an order of magnitude larger than the local value, because form drag on dispersed obstacles covers only a small fraction of the grid cell. However, the effective roughness height for heat (and water vapour) tends to decrease with increasing scale, since heat transfer is not concentrated on obstacles. Beljaars and Holtslag (1991) present data on the change in these roughness heights, in going from point to 10-km scales for one particular terrain type. Variation in roughness heights with scale, in particular when increasing to the scale of the GCM grid cell (several hundred km), depends on the degree of landscape heterogeneity at different scales.

Plant Physiological Effects

As discussed in preceding chapters, there is increasing evidence that the direct physiological effect on plants of a higher atmospheric concentration of CO_2 is to cause partial constriction of plant stomata and reduce water loss, all else being equal (Bazzaz, 1990). This effect is particularly strong in C_4 plants (which include many grasses) and could moderate the tendency for transpiration rates to increase as climate warms, with potentially significant effects on soil moisture levels and runoff. Furthermore, reduction in plant transpiration would tend to increase leaf surface temperatures, especially for broad-leafed plants. However, the stimulatory effect on plant photosynthesis of higher CO_2 concentrations could increase the leaf area index (Allen, Gichuki, and Rosenzweig, 1991). Both effects would tend to drive transpiration rates back up enough to nullify the effect of stomatal constriction. The partial control of humidity within a forest canopy, by forest transpiration itself, could also significantly weaken the reduction in transpiration rates from stomatal constriction (Jarvis and McNaughton, 1986). These feedbacks can be accounted for, at least crudely, in schemes in which canopy foliage temperature and canopy and air vapour pressure are explicitly computed, as both variables in this case would respond to changes in stomatal conductance. Note, however, that Jarvis and McNaughton's Ω factor may account for feedback between foliage transpiration and canopy vapour pressure following a change in stomatal resistance, but, like the Penman-Monteith equation, it does not include temperature feedback.

EFFECTS OF CLIMATIC CHANGE ON CANADIAN SURFACES

The direct effects of anticipated global warming include changes in the frequency, amount, and seasonal distribution of precipitation; a tendency for increased evaporation and drying of continental interiors; a rise of sea level; changes in oceanic currents and fisheries; thawing of permafrost; and melting of snow and sea ice. Each of these would profoundly affect Canadian surface climates. Canada has the longest shoreline of any country in the world, contains permafrost under half its land area, and has vast interior lakes. Canada also is highly dependent on climatically sensitive activities such as agriculture, forestry, and marine fisheries.

Forests and Ecoclimatic Regions

Figure 14.6 compares present-day distribution of ecoclimatic provinces in Canada with the distribution that would be in balance with the equilibrium climate for CO_2 doubling, as projected by the GISS GCM. Such a climate could be realized as early as 2050. Canadian forests will be profoundly affected by climatic change. Much of the present boreal forest in western Canada would have a climate that supports only grassland or transitional forest-grassland ecosystems, according to the scenario shown in Figure 14.6. The southern limit of the boreal forest would be displaced by as much as 1,100 km, and the northern limit by as much as 800 km. The cool, temperate forest in eastern Canada would be displaced as much as 500 km northward.

The ecoclimatic distributions given in Figure 14.6 do not, however, consider two key non-climatic limitations: soils and the time required for migration of forest limits. If conditions associated with doubled CO_2 occur by 2050, rates of forest migration of 100 to 200 km per decade would be required. This is far in excess of natural rates of migration, which are on the order of 20 to 50 km per century (Davis, 1989). As a result, dieback of trees along their southern margins in response to heat and/or moisture stress and replacement with "weedy" trees that are good at migrating and colonizing are likely (Solomon, 1986). Fire frequency and severity are likely to increase as a result of increased drought and a greater supply of dead wood. On the positive side, removal of maladapted forests through fires will allow faster replacement by species better suited to the changing climate, although there could be several centuries of ecological imbalance.

Forests at the present treeline (chapter 9), where climatic conditions are harshest, might not respond to climatic warming by advancing northward. Rather, climatic warming might initially induce an increase in tree density and possibly a shift from asexual to sexual modes of reproduction.

Agricultural Surfaces

Climatic warming can both enhance and hinder Canadian agriculture. Longer and warmer growing seasons will tend to increase crop yields in some cases, but moisture shortages or warming beyond ideal temperatures will tend to reduce yields. The

a)

Ecoclimatic provinces
1 Arctic
2 Subarctic
3 Boreal
   3a Maritime
   3b Moist continental
   3c Dry continental
4 Cool Temperate
5 Moderate Temperate
6 Transitional Grassland
7 Grassland
8 Semi-desert
9 Unclassified, Cordilleras
   not shown

b)

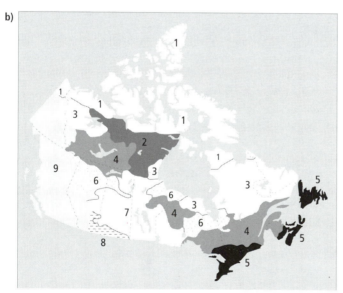

Figure 14.6
(a) Present-day ecoclimatic provinces in Canada; (b) ecoclimatic provinces projected to be in equilibrium with the equilibrium climatic response for CO_2 doubling, as computed by the GISS model.
Source: Environment Canada (1992).

tendency for evaporation rates to increase with warmer temperatures will cause soils to become drier if precipitation increases are not large enough. Impact studies suggest that the direct effect of climatic change associated with a CO_2 doubling will be small when averaged over an entire province and over several years. Much larger changes are likely in smaller regions within provinces and during individual drought years (Smit, 1987; Arthur, 1988). The effects would probably become more negative for a variety of reasons if climate continues to warm beyond the level associated with doubling of CO_2 (Harvey, 1996).

Coastal Zone Surfaces

Sea level is expected to rise by one-third of a metre to one metre by the end of the next century (Warrick et al., 1996). A one-metre rise in sea level would lead to significant losses of coastal wetlands. These wetlands play important ecological roles and occur in a number of large estuaries along the Pacific coast, along the western edge of Hudson Bay and the islands of the Foxe Basin in the Arctic, in the Mackenzie Delta, and in a number of Atlantic estuaries and embayments.

Lake Surfaces

Changes in lake levels will depend on regional changes in precipitation and evaporation and their effect on runoff. Though precipitation is expected to increase in many areas, so too will evaporation rates. The tendency for plant evapotranspiration rates to decrease as a direct physiological response to higher atmospheric CO_2 could counteract the tendency for warmer temperatures to increase evapotranspiration. This could leave more water for groundwater recharge of stream flow and eventual input to lakes.

It is likely that water supplies will decrease in many parts of Canada as the climate warms. Within the Great Lakes basin, net water-supply is expected to decrease by about 15 per cent for CO_2 doubling (Environment Canada, 1988) and mean water levels by 0.3 to 0.8 m (Sanderson, 1987).

Winter warming in particular will reduce the intensity of late-autumn and winter overturning in lakes in the southernmost part of Canada. If and when a doubled CO_2 climate is reached, surface waters in Lake Ontario might not cool to 4°C in most winters (Boyce et al., 1993). Water below the thermocline, currently near 4°C year-round (chapter 6), would become progressively warmer. An autumn/winter overturn would still occur whenever the surface water became denser than the underlying water, but at progressively warmer temperatures as the subsurface waters warmed.

Permafrost Surfaces

Permafrost occurs under about half the Canadian land surface, and much of it contains substantial amounts of ground ice. Climatic warming is projected to be largest at high latitudes and will result in thawing of significant areas of permafrost. Melting of the associated ground ice will lead to unstable terrain and significant erosion in some areas (Woo, Lewcowitz, and Rouse, 1992). Development of a deeper seasonally thawed zone will allow percolation. This, combined with earlier melting of snow banks, may lead to disappearance of many small wetlands (Woo, 1992). However, new wetland areas are likely to be generated by melting permafrost, which will enhance habitat for some species but will not qualitatively replace coastal wetlands lost to rising sea level or Prairie wetlands lost to summer drying.

Urban Surfaces

Urban environments tend to be warmer than surrounding regions – the urban "heat island" (chapter 13). Efforts to reduce the magnitude of the summer urban heat at ground level, through strategic planting of shade trees, choice of light-coloured building surfaces, and attention to urban form and design, could offset part of future warming in urban centres as regional climates warm.

ROLE OF SURFACE CLIMATE IN GLOBAL CLIMATIC FEEDBACK PROCESSES

Changes in surface climate could exert an important feedback on global and regional climatic change through the effect of surface temperature and moisture on biogeochemical cycles, particularly those involving carbon dioxide and methane, and through the effect of continental runoff on oceanic circulation.

Biogeochemical Cycles

Canadian soils and peatlands are estimated to contain 100–150 billion tonnes of carbon, representing about 10 per cent of the world's total estimated soil and peatland carbon (Environment Canada, 1992). Much of this carbon is currently frozen in permafrost. With climatic warming and thawing of permafrost, some of this carbon could be released to the atmosphere through oxidation. Since this would add to the original increase of atmospheric CO_2, it represents a positive feedback. The extent to which stored carbon would be released is likely to depend largely on concurrent changes in soil moisture. If soils remain saturated, relatively little carbon might be released. If soils become unsaturated, the potential exists for large releases (Harvey, 1996).

Canadian wetlands are also believed to represent significant sources of methane (CH_4). The methane flux from wetlands is thought to depend strongly on both temperature and water level (Nisbet, 1989; Gorham, 1991) and could therefore change significantly as climate warms, depending on the sign and magnitude of concurrent changes in wetland extent. Releases of both methane and CO_2 in response to climatic warming would exert positive feedback on global climate, tending to amplify the initial warming. However, the conditions that would maximize CH_4 emissions (high water table) would tend to minimize emissions of CO_2, so that strong positive feedbacks on climate warming involving both substances are not likely to occur simultaneously.

Role of Continental Runoff and Sea Ice

The density of seawater depends on temperature and salinity through the equation of state. The equation of state for seawater is non-linear and is such that salinity differences are more important than temperature differences in determining density at cold temperatures and hence at high latitudes (Friedrich and Levitus, 1972). Differences in vertical density determine the vertical thermal-diffusion coefficient and occurrence of convective overturning, while horizontal density differences drive thermohaline overturning. Associated with these mixing processes are significant horizontal and vertical heat fluxes, which profoundly influence regional climates. Changes in the oceanic salinity field, resulting in part from changes in continental hydrology, could transform oceanic heat transports and thereby significantly alter regional climates.

The surface waters of the Arctic Ocean today are cold and fresh but are underlain by intermediate-depth water of Atlantic origin, which is warm and saline. Mixing between cold surface water and warm intermediate water is inhibited by a strong vertical salinity gradient, or halocline, which is maintained by freshwater runoff, brine rejection by ice freezing over arctic continental shelfs, and sliding of the saline brine down the continental slopes and subsequent mixing with Atlantic intermediate water (Aagaard, Coachman, and Carmack, 1981; Bjork, 1989). A breakdown of the arctic halocline as a result of decreased runoff or, more likely, as a result of decreased ice formation over continental shelves could result in a significant upward heat flux to Arctic surface waters. This would further amplify high-latitude surface warming.

The surface waters of the North Atlantic Ocean, in contrast, are relatively saline, which permits convective overturning and large upward heat fluxes without the surface water cooling to the freezing point. An increase in continental runoff to the North Atlantic could reduce convective overturning and the associated upward heat flux, thereby inhibiting warming in the North Atlantic. This behaviour has been observed in a CO_2-increase experiment with the NCAR coupled atmosphere–ocean GCM (Washington and Meehl, 1989).

CONCLUDING COMMENTS

Global-scale warming as a result of increasing concentrations of greenhouse gases is a near-certainty. The regional distribution of temperature and precipitation changes that will accompany global warming is highly uncertain. Compounding the uncertainties in regional-scale forcing of surface climate are further uncertainties and errors arising from use of highly simplified, and generally inadequate, parameterizations of surface processes in GCMs. Not only are these parameterizations inadequate as point-process models, they fail to account for subgrid-scale variability in precipitation, soil, and vegetation properties and for soil moisture content. However, improved point-process models for incorporation into GCMs have recently been developed. Sensitivity tests indicate substantial improvement in the simulation of key variables such as surface temperature, evaporation, soil moisture, and runoff. Methods to incorporate some elements of subgrid-scale variability have also been developed and tested. None of the improved parameterizations has yet been used in the effects of experiments on greenhouse gas increase.

The simulated changes in surface climate and soil moisture depend in part on the accuracy of the simulation of present climate, prior to increasing greenhouse gas concentrations. Some of the disagreements concerning regional climatic changes accompanying global warming between the current generation of GCMs can be traced to differences in their simulation of the present climate. Accurate simulation of future changes in surface climate therefore ultimately depends not only on improved parameterizations of surface processes but on improvement in the simulation of all the key processes that influence climate – for example, precipitation, cloud physics processes, and ocean circulation. Each of these in turn can be strongly influenced by surface processes: precipitation, through the effect of evaporation on moisture supply; cloud processes, through the effect of surface temperature and partitioning of latent/sensible heat fluxes on atmospheric stability and through the effect of surface roughness on airmass flux convergence; and ocean circulation, through the impact of continental runoff on ocean salinity. More accurate simulation of regional and global climate, and of individual processes, will ultimately require parallel improvements in the parameterizations of all the processes that interactively determine climate.

REFERENCES

Aagaard, K., Coachman, L.K., and Carmack, E. 1981. "On the Halocline of the Arctic Ocean." *Deep Sea Research* 28A: 529–45.

Allen, R.G., Gichuki, F.N., and Rosenzweig, C. 1991. "CO_2-induced Climatic Changes and Irrigation Water Requirements." *Journal of Water Resources Planning and Management* 117: 157–78.

Arthur, L. 1988. *Implications of Climatic Change for Agriculture in the Prairie Provinces.* Climate Change Digest 88–01. Atmospheric Environment Service, Downsview, Ont.

Baliunas, S., and Jastrow, R. 1990. "Evidence for Long Term Brightness Changes of Solar-type Stars." *Nature* 348: 520–3.

Band, L.E., Peterson, D.L., Running, S.W., Coughlan, J., Lammers, R., Dungan, J., and Nemani, R. 1991. "Forest Ecosystem Processes at the Watershed Scale: Basis for Distributed Simulation." *Ecological Modelling* 56: 171–96.

Bazzaz, F.A. 1990. "The Response of Natural Ecosystems to the Rising Global CO_2 Levels." *Annals of the Review of Ecological Systems* 21: 167–96.

Beljaars, A.C.M., and Holtslag, A.A.M. 1991. "Flux Parameterization over Land Surfaces for Atmospheric Models." *Journal of Applied Meteorology* 30: 327–41.

Berger, W.H., and Spitzy, A. 1988. "History of Atmospheric CO_2: Constraints from the Deep Sea Record." *Paleoceanography* 3: 401–11.

Berner, R.A. 1990. "Atmospheric Carbon Dioxide Levels over Phanerozoic Time." *Science* 249: 1382–6.

Bjork, G. 1989. "A One-Dimensional Time-Dependent Model for the Vertical Stratification of the Upper Arctic Ocean." *Journal of Physical Oceanography* 19: 52–67.

Boyce, F.M., Hamblin, P.F., Harvey, L.D.D., Schertzer, W.M., and McCrimmon, C.R. 1993. "Response of the Thermal Structure of Lake Ontario to Deep Cooling Water Withdrawals and to Global Warming." *Journal of Great Lakes Research* 19: 603–16.

Cess, R.D., and Potter, G.L. 1984. "A Commentary on the Recent CO_2-Climate Controversy." *Climatic Change* 6: 365–76.

Chappellaz, J., Barnola, J.M., Raynaud, D., Korotkevich, Y.S., and Lorius, C. 1990. "Ice Core Record of Atmospheric Methane over the Past 160,000 Years." *Nature* 345: 127–31.

Charlson, R.J., Schwartz, S.E., Hales, J.M., Cess, R.D., Coakley, J.A., Hansen, J.E., and Hofmann, D.J. 1992. "Climate Forcing by Anthropogenic Aerosols." *Science* 255: 152–63.

Davis, M. 1989. "Lags in Vegetation Response to Greenhouse Warming." *Climatic Change* 15: 75–82.

Deardorff, J.W. 1977. "A Parameterization of Ground-Surface Moisture Content for Use in Atmospheric Prediction Models." *Journal of Applied Meteorology* 16: 1182–5.

Dunne, T., and Leopold, L.B. 1978. *Water in Environmental Planning.* New York: W.H. Freeman.

Entekhabi, D., and Eagleson, P.S. 1989. "Land Surface Hydrology Parameterization for Atmospheric General Circulation Models Including Subgrid Scale Spatial Variability." *Journal of Climate* 2: 816–31.

Environment Canada. 1988. *CO_2 Induced Climate Change in Ontario: Interdependencies and Resource Strategies.* Climate Change Digest 88–09. Atmospheric Environment Service, Downsview, Ont.

– 1992. "Climatic Change." In *State of the Environment Report 1991*, Ottawa: State of the Environment Reporting, chap. 22.

Friedrich, H., and Levitus, S. 1972. "An Approximation to the Equation of State for Sea Water Suitable for Numerical Ocean Models." *Journal of Physical Oceanography* 2: 514–17.

Gates, W.L., Henderson-Sellers, A., Boer, G.J., Folland, C.K., Kitho, A., McAvaney, B.J., Semazzi, F., Smith, N., Weaver, A.J., and Zeng, Q.-C. 1996. "Climate Models – Evaluation." In J.T. Houghton, L.G. Meira Filho, B.A. Callander, N. Harris, A. Kattenberg, and K. Maskell, eds., *Climate Change 1995: The Science of Climate Change*, Cambridge: Cambridge University Press, 229–84.

Giorgi, F., Brodeur, C.S., and Bates, G.T. 1994. "Regional Climate Change Scenarios over the United States Produced with a Nested Regional Climate Model." *Journal of Climate* 7: 375–99.

Gorham, E. 1991. "Northern Peatlands: Role in the Carbon Cycle and Probable Responses to Climatic Warming." *Ecological Applications* 1: 182–95.

Grotch, S. 1991. "A Statistical Intercomparison of Temperature and Precipitation Predicted by Four General Circulation Models with Historical Data." In M.E. Schlesinger, ed., *Greenhouse-Gas-Induced Climatic Change: A Critical Appraisal of Simulations and Observations*, Amsterdam: Elsevier, 3–16.

Gullett, D.W., and Skinner, W.R. 1992. *The State of Canada's Climate: Temperature Change in Canada 1895–1991.* SOE Report 92–2, Atmospheric Environment Service, Downsview, Ont.

Gutowski, W.J., Gutzler, D.S., and Wang, W.-C. 1991. "Surface Energy Balances of Three General Circulation Models: Implications for Simulating Regional Climate Change." *Journal of Climate* 4: 121–34.

Hansen, J., Lacis, A., Ruedy, R., and Sato, M. 1992. "Potential Climate Impact of Mount Pinatubo Eruption." *Geophysical Research Letters* 15: 323–6.

Harvey, L.D.D. 1988. "On the Role of High Latitude Ice and Snow." *Climatic Change* 13: 191–224.

– 1989a. "An Energy Balance Climate Model Study of Radiative Forcing and Temperature Response at 18 Ka BP." *Journal of Geophysical Research* 94: 12873–84.

– 1989b. "Managing Atmospheric CO_2." *Climatic Change* 15: 343–81.

– 1989c. "Transient Climatic Response to an Increase of Greenhouse Gases." *Climatic Change* 15: 15–30.

– 1990. "Managing Atmospheric CO_2: Policy Implications." *Energy* 15: 91–104.

– 1992. "A Two-Dimensional Ocean Model for Long Term Climatic Simulations: Stability and Coupling to Atmospheric and Sea Ice Models." *Journal of Geophysical Research* 97: 9435–53.

– 1996. "Development of a Risk-Hedging CO_2 Emission Policy, Part I: Risks of Unrestrained Emissions." *Climatic Change* 34: 1–40.

Hoffert, M.I., and Covey, C. 1992. "Deriving Global Climate Sensitivity from Paleoclimate Reconstruction." *Nature* 360: 573–6.

Hoffert, M.I., Frei, A., and Narayanan, V.K. 1988. "Application of Solar Max Acrim Data to Analysis of Solar-Driven Climatic Variability on Earth." *Climatic Change* 13: 267–86.

Hunt, B.G. 1985. "A Model Study of Some Aspects of Soil Hydrology Relevant to Climatic Modelling." *Quarterly Journal of the Royal Meteorological Society* 111: 1071–85.

Jarvis, P.G., and McNaughton, K.G. 1986. "Stomatal Control of Transpiration: Scaling up from Leaf to Region." *Advances in Ecological Research* 15: 1–48.

Kattenberg, A., Giorgi, F., Grassl, H., Meehl, G.A., Mitchell, J.F.B., Stouffer, R.J., Tokioka, T., Weaver, A.J., and Wigley, T.M.L. 1996. "Climate Models – Projections of Future Climate." In J.T. Houghton, L.G. Meira Filho, B.A. Callander, N. Harris, A. Kattenberg, and K. Maskell, eds., *The Science of Climate Change*, Cambridge: Cambridge University Press, 285–357.

Kaufman, Y.J., Fraser, R.S., and Mahoney, R.L. 1991. "Fossil Fuel and Biomass Burning Effect on Climate – Heating or Cooling?" *Journal of Climatology* 4: 578–87.

Kelly, P.M., and Wigley, T.M.L. 1990. "The Influence of Solar Forcing on Global Mean Temperature since 1861." *Nature* 347: 460–2.

Kiehl, J.Y., and Ramanathan, V. 1982. "Radiative Heating Due to Increased CO_2: The Role of H_2O Continuum Absorption in the 12–18 μm Region." *Journal of Applied Meteorology* 39: 2923–9.

Lacis, A., Wuebbles, D., and Logan, J.A. 1990. "Radiative Forcing of Climate by Changes in the Vertical Distribution of Ozone." *Journal of Geophysical Research* 95: 9971–81.

Manabe, S. 1969. "The Atmospheric Circulation and Hydrology of the Earth's Surface." *Monthly Weather Review* 97: 739–74.

Mass, C.F., and Portman, D.A. 1989. "Major Volcanic Eruptions and Climate: A Critical Evaluation." *Journal of Climate* 2: 566–93.

Meehl, G.A., and Washington, W.M. 1988. "A Comparison of Soil-Moisture Sensitivity in Two Global Climate Models." *Journal of Atmospheric Science* 45: 1476–92.

Mitchell, J.F.B., and Warrilow, D.A. 1987. "Summer Dryness in Northern Mid-Latitudes Due to Increased CO_2." *Nature* 330: 238–40.

Mitchell, J.F.B., Wilson, C.A., and Cunnington, W.M. 1987. "On CO_2 Climate Sensitivity and Model Dependence of Results." *Quarterly Journal of the Royal Meteorological Society* 113: 293–322.

Nicholls, N., Gruza, G.V., Jouzel, J., Karl, T.R., Ogallo, L.A., and Parker, D.E. 1996. "Observed Climate Variability and Change." In J.T. Houghton, L.G. Meira Filho, B.A. Callandar, N. Harris, A. Kattenberg, and K. Maskell, eds., *Climate Change 1995: The Science of Climate Change*, Cambridge: Cambridge University Press, 133–92.

Nisbet, E.G. 1989. "Some Northern Sources of Atmospheric Methane: Production, History, and Future Implications." *Canadian Journal of Earth Sciences* 26: 1603–11.

Pitman, A.J., Henderson-Sellers, A., and Yang, Z.-L. 1990. "Sensitivity of Regional Climates to Localized Precipitation in Global Models." *Nature* 346: 734–7.

Reid, G.C. 1991. "Solar Total Irradiance Variations and the Global Sea Surface Temperature Record." *Journal of Geophysical Research* 96: 2835–44.

Rind, D. 1988. "The Doubled CO_2 Climate and the Sensitivity of the Modeled Hydrologic Cycle." *Journal of Geophysical Research* 93: 5385–412.

Ropelewski, C.F., and Halpert, M.S. 1991. "The Southern Oscillation and Northern Hemisphere Temperature Variability." In M.E. Schlesinger, ed., *Greenhouse-Gas-Induced Climatic Change: A Critical Appraisal of Simulations and Observations*, Amsterdam: Elsevier, 369–79.

Sanderson, M. 1987. *Implications of Climatic Change for Navigation and Power Generation in the Great Lakes*. Climate Change Digest 87–03. Atmospheric Environment Service, Downsview, Ont.

Schmitt, C., and Randall, D.A. 1991. "Effects of Surface Temperature and Clouds on CO_2 Forcing." *Journal of Geophysical Research* 96: 9159–68.

Shine, K.P., Derwent, R.G., Wuebbles, D.J., and Morcrette, J.-J. 1990. "Radiative Forcing of Climate." In J.T. Houghton, G.J. Jenkins, and J.J. Ephraums, eds., *Climate Change: The IPCC Scientific Assessment*, Cambridge: Cambridge University Press, 41–68.

Smit, B. 1987. *Implications of Climatic Change for Agriculture in Ontario*. Climate Change Digest No. 87–02. Atmospheric Environment Service, Downsview, Ont.

Solomon, A.M. 1986. "Transient Response of Forests to CO_2-Induced Climate Change: Simulation Modeling Experiments in Eastern North America." *Oecologia* 68: 567–79.

Verseghy, D.L. 1991. "CLASS – a Canadian Land Surface Scheme for GCMs. I. Soil Model." *International Journal of Climatology* 11: 111–33.

Verseghy, D.L., McFarlane, N.A., and Lazare, M. 1992. "CLASS – a Canadian Land Surface Scheme for GCMs. II. Vegetation Model." *International Journal of Climatology* 13: 347–70.

Warrick, R., Le Provost, C., Meier, M.F., Oerlemans, J., and Woodworth, P.L. 1996. "Changes in Sea Level." In J.T. Houghton, L.G. Meira Filho, B.A. Callander, N. Harris, A. Kattenberg, and K. Maskell, eds., *The Science of Climate Change*, Cambridge: Cambridge University Press, 359–405.

Warrilow, D.A., Sangster, A.B., and Slingo, A. 1986. *Modeling of Land-Surface Processes and Their Influence on European Climate*. Dynamics of Climatological Technology Note 38. Meteorological Office, Bracknell, Berkshire.

Washington, W.M., and Meehl, G.A. 1989. "Climate Sensitivity Due to Increased CO_2: Experiments with a Coupled Atmosphere and Ocean General Circulation Model." *Climate Dynamics* 4: 1–38.

Wetzel, P.J., and Chang, J.-T. 1987. "Concerning the Relationship between Evapotranspiration and Soil Moisture." *Journal of Climate and Applied Meteorology* 26: 18–27.

Woo, M.-K. 1992. "Impacts of Climatic Variability and Change on Canadian Wetlands." *Canadian Water Resources Journal* 17: 63–9.

Woo, M.-K., Lewcowitz, A., and Rouse, W.R. 1992. "Response of the Canadian Permafrost Environment to Climatic Change." *Physical Geography* 13: 287–317.

Xue, Y., Sellers, P.J., Kinter, J.L., and Shukla, J. 1991. "A Simplified Biosphere Model for Global Climate Studies." *Journal of Climate* 4: 345–64.

Young, G.M. 1991. "The Geologic Record of Glaciation: Relevance to the Climatic History of Earth." *Geoscience Canada* 18: 100–8.

CHAPTER FIFTEEN

Epilogue

R. TED MUNN

As previous chapters have so amply demonstrated, there is a wide spectrum of "surface climates" across Canada. These are evident not only at the relatively large scales revealed by the national network of climate-observing stations (coastal, forest, Prairie, city, and so on) but at ever-smaller scales, each embedded in a larger one (a bay, a forest clearing, a ploughed field, an urban canyon). This epilogue seeks to place the study of these climates in their historical perspective and looks ahead to the exciting challenges facing Canadian surface climatologists in the next century.

I begin with a brief historical overview of the development of the large-scale climatology of Canada and the network of observing stations that provide the data necessary for its description. This is followed by a look at the explosive growth of research studies on Canadian surface climates of the past fifty years, culminating in publication of this volume. I conclude with a brief look into the future: what are the issues that will engage the attention of Canadian climatologists in the forthcoming century?

WEATHER OBSERVING IN CANADA: A HISTORICAL PERSPECTIVE

Study of the large-scale climatology of Canada began as early as 1746, when regular weather observations were taken at Quebec City. Over the next century, enthusiasts at other eastern Canadian locations began taking observations of temperature and precipitation. In 1839 the British government founded an observatory at Toronto with full-time military weather observers. With establishment of the Meteorological Service in 1871, federal Canadian resources were provided to open volunteer temperature and precipitation stations for the purpose of delineating the Canadian climate and to provide surface data and information for government agencies, agriculture, industry/commerce, and the general public.

The Meteorological Service also organized a network of telegraphic reporting stations to obtain observations for preparation of storm warnings and general forecasts. These stations provided additional archival data to meet demands for climate data and information. Then in the late 1930s, as part of the new Air Services of the Department of Transport, the Meteorological Service began a major expansion to meet the needs of TransCanada Airlines (now Air Canada). Many new airport synoptic weather–observing stations were established, and this program was soon greatly expanded to serve the wartime Royal Canadian Air Force and the British Commonwealth Air Training Programme.

The number of volunteer observing stations taking climate observations did not increase during the Depression and the wartime 1940s, but scores of airport synoptic observing stations were added. The total number of observing stations, volunteer and synoptic, rose from 777 in 1931, to 907 in 1941, and to 1,298 in 1951, when additional resources became available for such programs after the war. By 1981 the number of stations had increased to 2,703, but, for a variety of reasons – automation, cuts in government spending, and so on – the number subsequently decreased.

Over the decades since the war, in response to increased demands for traditional climate data, the Meteorological Service/Atmospheric Environment Service has been given major resources to improve data archiving and delivery, enabling it to provide greatly improved service to a wide range of users of climate information. Further information on these historical developments is given in Thomas (1991).

SURFACE CLIMATES IN CANADA: HISTORICAL RESEARCH TRENDS

Rudolph Geiger's (1927) classic text *Climate Near the Ground* gave many examples of curious micro- and meso-climatic phenomena, but few people outside Europe were aware of the book until it was translated into English in 1950. There were a few attempts to draw Canadian analogies to Geiger's microclimatic examples. For example, foresters at Petawawa, Ontario, knew of a local frost pocket containing 100-year-old trees that were less than 1 m in height because cold-air drainage produced an annual frost-free period that lasted only a week or so.

Two Canadian pre-war studies on mesometeorology are historically noteworthy. First, in the 1930s W.E.K. Middleton designed a system to measure air temperature from a moving automobile. On several calm and cloudless nights, he drove his car from the Toronto waterfront up Yonge Street to the north side of Hogg's Hollow, a distance of about 12 km. The temperature differences recorded along this route were substantial (up to 18 Celsius degrees) and reproducible (Figure 15.1). This appears to be the first published Canadian mesometeorological study, published, curiously enough, in the *Transactions of the Royal Astronomical Society of Canada* (Middleton and Millar, 1936). For those of us who took Middleton's course on meteorological instruments in the early 1940s, this case study of climatological variability over small distances made a lasting impression.

Second, in the late 1930s and early 1940s, E.W. Hewson and G.C. Gill probed the Columbia River Valley downwind of a smelter located in Trail, BC (Dean et al., 1944; Hewson, 1945), to find out why a morning peak in concentrations of sulphur dioxide occurred at the same time all the way down the valley into the United States. The answer is that a surface inversion formed in the valley overnight. After sunrise, the ground heated, and morning convection brought the Trail smelter plume down to ground level simultaneously all along the valley. This phenomenon was termed a "fumigation" by Hewson.

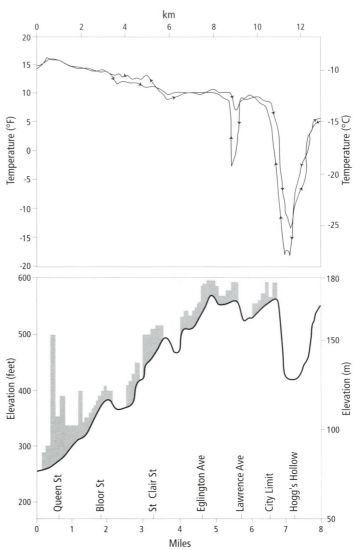

Figure 15.1
Profiles of air temperature and ground elevation observed travelling northward
from the Toronto waterfront up Yonge Street in 1936. Shaded areas represent the
heights of buildings close to Yonge Street. Air temperatures were observed at
about 0.7 m above ground on the clear night of 22 February, with a weak northerly breeze blowing. The automobile started at the inland end at 0006h, reached
the lakeshore at 0048h, and returned to the start at 0118h EST.
Source: Middleton and Millar (1936).

In Canada, microclimatology began to flourish after the Second World War, as
"surplus" government funds and trained meteorologists became available for special
assignments relating to the needs of forestry, agriculture, hydrometeorology, air pollution, building research, and so on. Many of these studies were driven by a rapidly developing community of users who needed specialized climate information far beyond

that available from standard observation networks. When I took my first meteorological courses during the war, we had lectures on climatology, and we learned about the Köppen surface-climate classification – but there was no serious discussion of the surface climates of Canada. In fact, to use a phrase of Ken Hare's, the surface layer was considered "a bit of a nuisance" for synoptic meteorologists, who preferred smooth sites when choosing weather-observing stations.

The subject of Canadian surface climates became of major interest only following the arrival at McGill in 1946 of Kenneth Hare, to whom this book is dedicated. Hare was strongly influenced by Warren Thornthwaite and his laboratory in New Jersey, Howard Penman of Rothamsted, England, and Mikhail Budyko of Leningrad. Thornthwaite's 1954 Presidential Address to the Association of American Geographers, entitled "The Task Ahead," laid out a blue-print for the physical climatology that would be undertaken in the next four decades. In the USSR, Budyko had already applied these ideas to construction of a heat balance atlas of the Soviet Union. Penman, who was a friend of Hare's in England, had just changed for ever the ways in which evaporation is estimated. The McGill group under Hare married these ideas with dynamical meteorology to start unravelling the physical and dynamical climatology of the Canadian Arctic.

One of Hare's main interests was the boreal forest, and in papers from 1950 to 1972 he tried to explain why the forest was where it was and how it related to macroclimate and to surface climates. He linked up with Reid Bryson and Heinz Lettau at the University of Wisconsin. Hare stressed the need to focus on exchanges and equilibria between the earth's surface and the atmosphere, because they are what link soil and vegetation with macro- and microclimates. He taught that in analyses of these interrelationships the physics must be sound, full account has to be taken of the physiological processes governing vegetation cover (structure, seasonality, and so on), and statistical methods, if used, must be appropriate. These ideas are epitomized in the opening chapter of this book and in Hare and Thomas's *Climate Canada* (1979).

Following the early work by Budyko and Thornthwaite, and its adaptation by Hare and Lettau, their students and others began work on the physical basis of climates of specific areas and surface types that constitute the content of this book. Soon other Canadian universities began to offer specialized courses in meteorology and climatology to students in geography, forestry, agriculture, air pollution, civil engineering, hydrology, and oceanography. By the late 1960s there was enough interest to permit the holding of two successful meetings – the first and second Canadian conferences on micrometeorology – in Toronto and at MacDonald College in Montreal, respectively, without financial support except to cover travel for the foreign guests (Hans Panofsky and Frank Pasquill in the first case, and Howard Penman in the second).

Today, almost thirty years later, surface climatology is flourishing in nearly all Canadian universities, and many engineering firms are able to offer consulting services with respect to various applications relating to surface climates. Work is greatly aided by the availability of sophisticated commercial instruments and data-acquisition systems. Further, the combination of aircraft and satellite-based remote sensing, together with geographic information systems, has made it possible to study the spatial diversity of Canada's surface landscapes and their climates as never before. Similarly, development of computers has had a huge effect on the speed with which data can be analysed and, together with advances in theoretical understanding, has made it possible to construct numerical models of the atmospheric boundary layer.

Canadians have made substantial contributions to theoretical and experimental studies of the atmospheric boundary layer, to better understanding of the effects of different types of surfaces on the lower atmosphere, and in the application of this knowledge to practical problems in such fields as hydrology, agriculture, forestry, and urban planning. They have also written basic international references in the field, such as *Descriptive Micrometeorology* (Munn, 1966) and *Boundary Layer Climates* (Oke, 1987). Many of them are people who consider themselves geographers, foresters, soil scientists, mathematicians, or engineers as much as climatologists. The authors of the present volume, although mostly geographers, reflect the mix. This melding of scientists from many different backgrounds has created a rich community of talents with a great variety of approaches. Of course, this diversity also contributes to some difficulties, such as that of finding a common "home" to meet.

FUTURE RESEARCH PROSPECTS

One of the main socioeconomic issues for the twenty-first century is climate change. So Canadian "surface-climate" specialists will not be lacking exciting work to do. As the previous chapter has demonstrated, climate-change research is a rich field, full of uncertainties, but bursting with challenges. The essential question to be answered, but in many different contexts, is: how will climate change affect the surface climates of Canada?

Recent work attempts to deal with the effects of climate change on Canadian ecosystems. In most cases, such papers begin with one or several climate-change scenarios derived from the output of general circulation models, followed by an attempt to apply the results to the regional level. These studies seek to link global and regional surface climates and thereby to lay the foundation for many practical applications.

For example, the Canadian Environmental Assessment Act, 1992, requires that cumulative effects of federally funded projects be examined within environmental-impact assessments. This extends the time-scales of interest into decades and thus forces inclusion of assessments of the effects of climate change on developments being proposed. Similar statutory requirements are being introduced in the provinces. That this is an important issue is revealed by recalling that many projects, such as suburban development, highways, flood-control systems, and power stations, have life expectancies of 50 to 100 years and in some cases require permanent commitments of land and other natural resources (Munn, 1994). By the late twenty-first century, will land surface covers and their climates move northward, or will they stay put but change radically? Will the engineering-design features of a given development be suboptimal in a changed local climate, and will adverse environmental effects ensue?

Of course, the improvement of model scenarios of global climate change will be a high-priority international task, in which Canadians will continue to play a major role. In the coming decades, however, equal emphasis must be placed on the improvement of "downscaling" techniques for applying global insights to regional climates and on estimating the influence of changed climates on ecosystems, including people. This is one of the most important tasks for the coming decades and will require the greatest skill in melding inputs from several scientific disciplines into team efforts designed to describe, explain, predict, and manage the range of Canada's ecosystems.

ACKNOWLEDGMENTS

The section on Climatology in Canada was contributed by Morley Thomas. I acknowledge helpful comments and suggestions from F.K. Hare.

REFERENCES

Dean, R.S., Swain, R.E., Hewson, E.W., and Gill, G.C. 1944. *Meteorological Investigations in the Columbia River Valley near Trail, B.C.* U.S. Bureau of Mines Bulletin No. 453. Superintendent of Documents, U.S. Government Printing Office, Washington, DC.

Geiger, R. 1927. *Das Klima der Bodennahen Luftschicht.* Braunschweig: Vieweg.

Hare, F.K., and Thomas, M.K. 1979. *Climate Canada.* Toronto: John Wiley and Sons.

Hewson, E.W. 1945. "The Meteorological Control of Atmospheric Pollution by Heavy Industry." *Quarterly Journal of the Royal Meteorological Society* 71: 266–82.

Middleton, W.E.K., and Millar, F.G. 1936. "Temperature Profiles in Toronto." *Journal of the Royal Astronomical Society of Canada* 30: 265–72.

Munn, R.E. 1966. *Descriptive Micrometeorology.* New York: Academic Press.

– ed. 1994. *Keeping Ahead: The Inclusion of Long-Term "Global" Futures in Cumulative Environmental Assessments.* CEA 9129, G31, Canadian Electrical Association, Montreal.

Oke, T.R. 1987. *Boundary Layer Climates.* 2nd edn. London: Routledge.

Thomas, M.K. 1991. *The Beginnings of Canadian Meteorology.* Toronto: ECW Press.

Glossary

ablation   process by which ice and snow are lost from the surface of a glacier, snow layer, or ice floe (for example, by evaporation, melting, runoff, wind, calving, and so on).

accumulation   process by which ice and snow are added to a glacier, snow layer, or ice floe (for example, by snowfall or sublimation); opposite of ablation.

active layer   ground layer above permafrost which thaws each summer; i.e., it is seasonally frozen and thawed.

adiabatic process   thermodynamic change of state that involves no transfer of heat with surrounding environment. For air, compression results in warming, expansion in cooling.

advection   used primarily to describe predominantly horizontal motion in the atmosphere.

albedo   ratio of the solar radiation reflected by a body to that incident upon it.

anabatic wind   upslope wind caused by local surface heating.

anthropogenic   caused by human activity.

anticyclone   large-scale atmospheric circulation system in which (in the northern hemisphere) winds rotate clockwise; interchangeable with "high" pressure system.

attenuation   any process in which the flux density of a "parallel beam" of energy decreases with increasing distance from the energy source.

baroclinic   state of the atmosphere where constant pressure and constant density surfaces intersect. In practice, it means that there are horizontal gradients of temperature and humidity, which lead to advection.

biomass   mass of biological material present at a site per unit area of the surface. In the case of plant communities, it is phytomass.

blackbody (or full radiator)   a hypothetical body that absorbs all the radiation striking it; i.e., it allows no reflection or transmission.

blending height   height at which flow changes from being in equilibrium with the local surface to being independent of horizontal position; equivalent to roughness height.

blocking   atmospheric state in which the tropospheric circulation is composed of large-amplitude waves travelling around almost stationary high-pressure systems. This pattern blocks the typical westerly flow of mid-latitudes and results in stationary and often anomalous weather patterns.

boundary layer   layer of air adjacent to a surface that is influenced by the properties of the boundary.

Bowen ratio   ratio of the turbulent transfer of sensible heat to that of latent heat.

buoyancy (or buoyant force)   upward force exerted on a parcel of fluid by virtue of the density difference between itself and the surrounding fluid.

chinook   warm, dry wind on the eastern side of the Rocky Mountains.

comma cloud   organized area of cumulus clouds associated with cyclones that look like a "comma" on a satellite image.

concentration   amount of a substance contained in unit volume.

conduction   transfer of energy in a substance by means of molecular motions without any net external motion.

convection   transport and mixing of properties in a fluid caused by mass motion; usually restricted to predominantly vertical motion in the atmosphere, as distinct from advection, which is horizontal.

convergent flow   when winds produce horizontal net inflow of air to a region.

Coriolis force   apparent force acting on objects moving in a frame of reference that is itself moving; in the atmosphere as a result of earth's rotation. It deflects airflow to the right (left) of its intended path in the northern (southern) hemisphere.

counter-gradient   situation where the direction of a flux (heat, for example) is opposite to the slope of the mean gradient of its related property (temperature, for instance); possible because the flux can be accomplished by a few large eddies whose origins are not local.

cyclone   large-scale atmospheric circulation system in which (in the northern hemisphere) winds rotate anticlockwise; interchangeable with "low" pressure system.

dewfall   condensation of water from the lower atmosphere onto objects near the ground.

dew-point   temperature to which a given parcel of air must be cooled (at constant pressure and constant water vapour content) in order for saturation to occur.

diffuse shortwave radiation   that portion of shortwave radiation reaching the earth's surface that has been scattered out of the parallel beam by molecules, particles, or cloud in the atmosphere.

direct-beam shortwave radiation   that portion of solar radiation received in a parallel beam "directly" from the sun.

diurnal   daily.

divergent flow   when winds produce horizontal net outflow of air to a region.

drag   frictional resistance offered by air to the movement of bodies moving through it (or by air flowing past stationary objects).

dryness ratio   ratio of the net radiation to the product of the latent heat and precipitation.

eddy   (a) by analogy with a molecule, a "glob" of fluid that has a life history of its own; (b) circulation in the lee of an obstacle brought about by a pressure deficit.

eddy diffusion   diffusion of atmospheric properties accomplished by turbulent motion.

emissivity   ratio of the total radiant energy emitted per unit time per unit area of a surface, at a specified wavelength and temperature, to that of a blackbody under the same conditions.

entrainment   process of mixing of fluid across a density interface bounding a region of turbulent flow. Relatively quiescent fluid is engulfed by eddies penetrating across the interface. Entrainment is an important process at the base of the inversion capping the planetary boundary layer.

equilibrium line   notional altitude line on a glacier where ablation balances accumulation. Above this line accumulation exceeds ablation; below it, the reverse holds.

evaporation (or vaporization)   process by which a liquid is transformed into a gas; in air, usually water changing to water vapour.

evapotranspiration   combined loss of water to the air by the processes of evaporation and transpiration.

fetch   distance of airflow to an object, measured in the upwind direction.

flux   rate of flow of some quantity.

flux density   flux of any quantity through unit surface area.

forced convection   fluid motion induced by mechanical forces such as deflection over an obstacle, friction, or any external force.

forcing   any process that drives flows of heat, water vapour, and momentum in a system. In the atmospheric boundary layer, the processes may be of external origin (for example, solar energy or precipitation), imposed by one layer on another (for instance, synoptic flow on the top of the boundary layer), or from the surface (for example, drag, heat flux, evaporation, or pollutant emission).

free convection   motion caused by density differences in a fluid.

front   relatively sharp interface or transition zone between two masses of air or water of different density.

fumigation   process of sudden downward mixing of pollutants from an overhead plume when convection reaches up to the height of a plume that was previously in stable air.

gradient wind   wind resulting from a balance of the horizontal pressure gradient force, Coriolis force, and the centripetal acceleration caused by motion in a curved path in a cyclone or anticyclone. The flow is horizontal and frictionless, and the isobars coincide with the streamlines.

greenhouse gas   radiatively active gases present in the atmosphere that contribute to the greenhouse effect – for example, carbon dioxide, water vapour, ozone, methane, and chlorofluorocarbons.

ground water   all subsurface water that participates in the hydrologic cycle.

growing degree-day (GDD)   Degree-day is a measure of the departure of the mean daily temperature from a base value; one degree-day is gained (lost) for each degree the daily mean is above (below) the base which is chosen to be meaningful to plant growth, such as germination. GDDs (or heat units) are accumulated over a season, so the total is a measure of the past effect.

heat capacity   amount of heat absorbed (or released) by unit volume of a system for a corresponding temperature rise (or fall) of one degree.

hydraulic conductivity   constant of proportionality in Darcy's law that relates the water flow through soil to the water head. It reflects the ease with which water flows through a porous medium such as soil.

interception   process whereby precipitation is trapped on vegetation and other surfaces before reaching the ground.

inversion   departure from the usual decrease or increase of atmospheric properties with height. Most commonly it refers to a temperature inversion, where temperatures increase rather than lapse with height.

irradiation   total radiant flux received by unit area of a surface.

katabatic wind   any wind blowing down a slope, most often due to the drainage of cold air because of gravity.

laminar boundary layer   layer next to a fixed boundary where flow is laminar.

laminar flow   smooth fluid flow in parallel sheets; non-turbulent flow.

lapse rate   decrease of an atmospheric property with height; usually refers to temperature unless otherwise specified.

latent heat   heat released or absorbed by unit mass of a system when it changes among solid, liquid, and gas states.

leaf area index   measure of the surface area of exposed leaves (one side only) over the height of a plant per unit area of the underlying ground.

local scale   class of atmospheric phenomena with horizontal dimensions from 100 m to 50 km, with lifetimes of less than one day; usually associated with topographic features, such as land/sea and mountain/valley winds and urban heat islands.

macroscale   class of atmospheric phenomena with horizontal dimensions of hundreds to thousands of kilometres and a lifetime of weeks to years; includes continental to global size features such as jet streams, weather systems, monsoons, and global climate distributions.

mesoscale   class of atmospheric phenomena with horizontal dimensions of 20 to 200 km, such as tornadoes, squall lines, and thunderstorms.

microscale   class of atmospheric phenomena with horizontal dimensions from millimetres to one km and lifetimes of minutes to hours. Examples include climates of leaves, animals, gardens, buildings, fields, and rivers.

mixed layer   layer of air (usually sub-inversion) within which pollutants are mixed by turbulence.

mixing height (or mixing depth)   the thickness of the layer measured from the surface upward, through which pollutants are presumed to mix as a result of convection caused by daytime heating of the surface.

momentum   property of a particle given by the product of its mass and its velocity.

nocturnal   nighttime.

orographic   influences associated with lifting caused by airflow over topographic barriers; hence orographic cloud and orographic precipitation.

pan evaporation   amount of water lost to the air from an open pan of water.

parameterization   an approximation to nature; usually involves replacing a fully physical equation with a relation that is an intuitively or physically reasonable approximation or surrogate.

permafrost   soil and rock where temperatures persist below 0°C for at least two consecutive winters and the intervening summer.

planetary boundary layer   atmospheric boundary (or Ekman) layer, which extends from the ground surface to the level where its frictional influence is absent.

polynya   pool of open water within pack ice.

potential evaporation   amount of water that would be evaporated from a surface with an unrestricted supply of water, as a result of the evaporative capacity of the atmosphere and the available surface heat.

potential temperature   temperature that a parcel of dry air would have if brought adiabatically from its present altitude to a standard pressure of 100 kPa.

profile   graph of an atmospheric quantity versus a horizontal or vertical distance, or time-scale.

psychrometer   instrument used to measure atmospheric water vapour content using wet- and dry-bulb thermometers.

radiation   process by which electromagnetic radiation is propagated through free space by virtue of joint undulatory variations in the electric and magnetic fields in space.

resistance   measure of the difficulty for entities such as heat, water vapour, and momentum to be transported within a system (air, water, soil, plant and animal tissue, building materials, etc.); based on the analogy between atmospheric fluxes and flow of electrons in an electrical circuit (Ohm's law). Inverse of resistance is called conductance.

ridge   elongated area of high atmospheric pressure.

runoff   surface water that ultimately reaches stream channels.

runoff ratio   ratio of runoff to mean precipitation.

saltation   hopping movement of snow particles or sand grains as they are carried bouncing across the surface by the wind.

scalar   any physical quantity that can be described by its magnitude alone, as opposed to a vector, which requires both a magnitude and a direction for its proper description. Temperature and water vapour are examples of scalars; winds are vectors.

screen-level   refers to the height of the instruments in a weather screen (approximately 1.5 m above ground).

sensible heat   heat energy able to be sensed (for example, with a thermometer); used in contrast to "latent heat."

sky view factor   ratio of the amount of the sky "seen" from a given point on a surface to that potentially available (i.e., proportion of the sky hemisphere subtended by a horizontal surface).

snow water equivalent (SWE)   depth of water that would result if a snow sample were melted; value depends on density of snow.

snowline   altitude on the landscape separating areas where snow is lost in summer from those where it remains throughout the year.

solar altitude   vertical direction of the sun above the horizon expressed in degrees.

solar azimuth   horizontal direction of the sun relative to a reference direction (usually true north), expressed in degrees.

solar zenith angle   vertical direction of the sun relative to the zenith, expressed in degrees; reciprocal of solar altitude.

specific heat   amount of heat absorbed (or released) by unit mass of a system for a corresponding temperature rise (or fall) of one degree.

stability (atmospheric)   tendency of air parcels after being moved vertically through the atmosphere. After forcible movement up or down, due to buoyancy a parcel tends to: (a) accelerate upward or downward (it is unstable), (b) resist displacement and return to its original position (it is stable), or (c) remain at the new level (it is neutral).

stomata   minute pores on the surface of leaves allowing them to exchange mass with the atmosphere (carbon dioxide for photosynthesis, and water by transpiration).

streamline   line whose tangent at any point in a fluid is parallel to the instantaneous velocity of the fluid. A map of streamlines gives an instantaneous "snap-shot" of the flow.

sublimation   transition of a substance directly from the solid to the vapour phase, or vice versa.

subsidence   slow sinking of air, usually associated with high pressure areas.

suspension   movement of snow particles or sand grains at speeds close to that of the transporting wind. Such particles are derived from the saltation layer next to the surface.

synoptic   referring to the use of meteorological data obtained simultaneously over a wide area for the purpose of presenting a comprehensive and nearly instantaneous picture of the state of the atmosphere.

thermal admittance   surface thermal property that governs the ease with which it will take up or release heat. It is the square root of the product of the thermal conductivity and heat capacity.

thermal conductivity   physical property of a substance describing its ability to conduct heat by molecular motion.

thermal diffusivity   ratio of the thermal conductivity to the heat capacity of a substance; determines the rate of heating caused by a given temperature distribution in a substance.

thermocline   layer of water in a lake or ocean in which there is a sharper gradient of temperature than above or below. It has an important control on vertical exchange. In some oceans it is a permanent feature, whereas in lakes it is usually seasonal.

trace gas   atmospheric gases that occur at very small concentrations. Notwithstanding this, several trace gases have important radiative effects.

trajectory   curve in space tracing the points successively occupied by a particle in motion.

transpiration   process by which water in plants is transferred as water vapour to the atmosphere.

treeline   upper altitudinal (or latitudinal) limit for tree growth.

troposphere   lowest 10–20 km of the atmosphere. This layer is characterized by decreasing temperature with height, appreciable water vapour, and vertical motion, and it contains most weather activity.

trough   elongated area of low atmospheric pressure.

turbidity   any condition of air or water that reduces its transparency to solar radiation.

turbulence   state of fluid flow in which the instantaneous velocities exhibit irregular and apparently random fluctuations so that in practice only statistical properties can be recognized. Capable of transporting atmospheric properties (such as heat, water vapour, and momentum) at rates far in excess of molecular processes.

upper air sounding   vertical profile of an atmospheric variable (such as temperature) gathered by a radiosonde attached to a balloon which reaches into the upper atmosphere.

urban heat island   characteristic warmth of a city, often approximated by urban–rural temperature differences.

zenith   point in the sphere surrounding an observer that lies directly overhead.

SOURCES

Ahrens, C.D. 1994. *Meteorology Today.* 5th edn. Minneapolis, Minn.: West Publishing Co.

Goudie, A., Atkinson, B.W., Gregory, K.J., Simmons, I.G., Stoddart, D.R., and Sugden, D. 1994. *The Encyclopedic Dictionary of Physical Geography.* 2nd edn. Oxford: Basil Blackwell Ltd.

Huschke, R.E., ed. 1959. *Glossary of Meteorology.* Boston: American Meteorological Society.

Oke, T.R. 1987. *Boundary Layer Climates.* 2nd edn. London: Routledge.

Contributors

BRIAN D. AMIRO, Research Scientist, Health and Environmental Sciences Division, Environmental Science Branch, Atomic Energy of Canada Limited, Whiteshell Laboratories, Pinawa, Manitoba, R0E 1L0

W.G. BAILEY, Professor, Department of Geography, Simon Fraser University, Burnaby, British Columbia, V5A 1S6

RICHARD L. BELLO, Associate Professor, Department of Geography, York University, North York, Ontario, M3T 1P3

TERRY J. GILLESPIE, Professor, Department of Land Resource Science, University of Guelph, Guelph, Ontario, N1G 2W1

BARRY E. GOODISON, Chief, Climate Processes and Earth Observations Division, Atmospheric Environment Service, Downsview, Ontario, M3H 5T4

F. KENNETH HARE, University Professor Emeritus, University of Toronto, 301 Lakeshore Road West, Oakville, Ontario, L6K 1G2

L.D. DANNY HARVEY, Associate Professor, Department of Geography, University of Toronto, Toronto, Ontario, M5S 3G3

OWEN HERTZMAN, Assistant Professor (Adjunct), Department of Oceanography, Dalhousie University, Halifax, Nova Scotia, B3H 4J1

PETER M. LAFLEUR, Associate Professor, Department of Geography, Trent University, Peterborough, Ontario, K9J 7B8

J. HARRY MCCAUGHEY, Professor, Department of Geography, Queen's University, Kingston, Ontario, K7L 3N6

LINDA MORTSCH, Research Scientist, National Water Research Institute, Environment Canada, Burlington, Ontario, L7R 4A6

R. TED MUNN, Institute for Environmental Studies, University of Toronto, Toronto, Ontario, M5S 3E8

D. SCOTT MUNRO, Professor, Department of Geography, University of Toronto, Mississauga, Ontario, L5L 1C6

ATSUMU OHMURA, Professor, Geographisches Institute ETH, Swiss Federal Institute of Technology, Winterthurestrasse 190, CH-8057 Zurich, Switzerland

TIMOTHY R. OKE, Professor, Department of Geography, University of British Columbia, Vancouver, British Columbia, V6T 1Z2

JOHN W. POMEROY, Research Scientist, National Hydrology Research Institute, Environment Canada, Saskatoon, Saskatchewan, S7N 3H5

ALEXANDER W. ROBERTSON, Faculty of Engineering and Applied Science, Memorial University of Newfoundland, St John's, Newfoundland, A1B 3X5

NIGEL T. ROULET, Associate Professor, Department of Geography, and Director, Centre for Climate and Global Change Research, McGill University, Montreal, Quebec, H3A 2K6

WAYNE R. ROUSE, Professor, Department of Geography, McMaster University, Hamilton, Ontario, L8S 4K1

IAN R. SAUNDERS, Research Scientist, Department of Geography, Simon Fraser University, Burnaby, British Columbia, V5A 1S6

WILLIAM M. SCHERTZER, Research Scientist, National Water Research Institute, Environment Canada, Burlington, Ontario, L7R 4A6

HANS-PETER SCHMID, Assistant Professor, Department of Geography, Indiana University, Bloomington, Indiana 47405–6101, United States of America

DAVID L. SPITTLEHOUSE, Research Scientist, British Columbia Ministry of Forests, Victoria, British Columbia, V8W 3E7

DOUW G. STEYN, Professor, Department of Geography, University of British Columbia, Vancouver, British Columbia, V6T 1Z2

JOHN L. WALMSLEY, Research Branch, Atmospheric Environment Service, Toronto, Ontario, M3H 5T4

JOHN D. WILSON, Professor, Department of Earth and Atmospheric Science, University of Alberta, Edmonton, Alberta, T6G 2H4

MING-KO WOO, Professor, Department of Geography, McMaster University, Hamilton, Ontario, L8S 4K1

Index

Primary index entries are organized by topic; secondary entries occur according to surface environment.

aerodynamic resistance, 33, 38
agricultural activities in Canada, 280–1
agricultural surfaces, 277–301; Bowen ratio, 285; climate change, 299–301, 343–5; cold protection, 293–5; energy balance, 277–9, 288–9; forage and field crops, 285–7; horticultural crops, 289–92; Ohm's law, 277–9; Prairie crops, 288–9; radiation balance, 277; surface modifications, 292–3
agroclimatic classification, 280–4
airmass, 8–9
albedo, 27, 41–2
alpine environments, 222–45; albedo, 231–2; Bowen ratio, 234; characteristics, 222–5; energy balance, 233–41; evapotranspiration, 242; momentum, 225–7; net radiation, 232–4; physiographic control of climate, 222; precipitation, 240–2; radiation balance, 228–34; runoff and drainage, 243–4; surface storage, 242–3; water balance, 240–4; wind flow, 225–7
alpine glaciers, 229–44; albedo, 232; energy balance, 235–8; evaporation, 242; radiation balance, 230, 233; surface storage, 242–3

anabatic wind, 61, 225–7
anticyclones, 7–10, 25–6
arctic glaciers, 189–95; climate change, 194–5; climates, 189–92; energy balance, 192; equilibrium line climate, 192–3
Arctic Islands, 172–95; albedo, 183–4, 187; distribution, 173; energy balance, 183–4; precipitation, 177–9; radiation balance, 174–5; snow distribution, 183; snowmelt, 182, 185–7; temperature, 175–8, 188–9; temperature profiles, 179–80; vegetation zonality, 181
arctic tundra, 202–8; Bowen ratio, 203–4, 207; carbon fluxes, 207–8; energy balance, 202–7; evaporation, 203–4; radiation balance, 202–7; water balance, 204–7
atmospheric forcing, 46
atmospheric scales, 44–5
atmospheric stability, 33

biogeochemical cycles, 346
boreal forest, 198
BOREAS project, 272–3
boundary layers, 21–4
Bowen ratio, 15, 34–6, 203–4, 207, 211–12

Canadian Cordillera, physiography, 226

canopy layer, 23
carbon, global storage, 167–8
carbon dioxide, 328–48; flux, 35, 39; profiles, 36–7, 41
chinooks, 70
CLASS, 339–40
climate: Cretaceous epoch, 330; future research, 356
climate change, 328–48, 356; agriculture, 299–301, 343–5; arctic glaciers, 194–5; coastal zone, 345; effects on Canadian surfaces, 343–6; forests, 343–4, 371–2; lakes, 345; natural causes, 331–2; permafrost, 345; scenarios for Canada, 334–7; urban, 323–4, 346; wetlands, 166–8
climate change models, 334, 336–41
climatic moisture index, 280–3
climatology, Atlantic coast, 108–9; coastal regions, 114–18; spatial variability, 44–65
combination model, 38, 203, 205, 235, 279, 317
comma cloud, 10
conduction, 26
cyclones, 7–10, 25–6

deep ocean circulation, 104
drainage flows, 58
dryness ratio, 15; distribution for Canada, 19
dynamic climatology, 3–10, 25

ecoclimate distribution for Canada, 344
eddy correlation, 32, 35
emissivity, 26
energy and mass exchange, 11
energy balance, 14, 29, 33–4, 40
entrainment, 50
entrainment zone, 22–3, 28
equilibrium layer, 47
evaporation, 14, 25, 37–8, 40; distribution for Canada, 17–18

feedbacks, 46
flux-gradient equation, 32, 35
fog, 118–20
forests, 247–73; acid precipitation, 270–1; albedo, 249; Bowen ratio, 251; canopy turbulence, 256–9; carbon fluxes, 253–6; climate change, 271–2, 343–4; distribution in Canada, 248; ecosystem stress, 271; energy balance, 251–3; evapotranspiration, 263–7; latent heat, 252; macroclimate control, 247; net radiation, 250–1; photosynthesis, 253–5; precipitation, 260–1; radiation balance, 249–51; runoff, 261–3; sensible heat, 252; snow accumulation, 262; temperature profiles, 257–8; water balance, 260–7; water storage, 261–3; wind flow, 267–8; wind profiles, 257–8
fronts, 8–10
fumigation, 51, 353

general circulation models, 333–4
greenhouse gases, 328
growing degree days, 284

Hare, F.K., v, 3–20, 355
history, climatology in Canada, 352–3; surface climate research, 353–6
humidity profiles, 36–7, 41

infiltration, 37–8
inhomogeneity, 44–8
internal boundary layer, 47–56; mechanical, 47; thermal, 47, 50
inversion, 22–3, 25, 28, 33, 50

irrigation, 296–7

jet stream, 3–10

katabatic wind, 62, 226–7, 291

lakes, 124–45; advection, 136; albedo, 129–31; Bowen ratio, 138–9; climate change, 345; distribution of drainage basins in Canada, 125; distribution of large lakes in Canada, 125; energy balance, 135–40; evaporation, 140–4; evaporation distribution for Canada, 144–5; form and type, 126; heat storage, 136–7; latent heat, 138–9; momentum, 128; radiation balance, 129–35; seasonal thermal cycles, 127; sensible heat, 138–9; surface roughness, 128; water balance, 135–45
laminar boundary layer, 23
land-sea breezes, 58–60, 64, 119
latent heat, 14, 33–7
leading edge, 47, 49
Lettau's law, 15–16
logarithmic wind profile, 30, 49
longwave radiation, 13–4, 28–9
low arctic and subarctic region, 198–219; climate change, 218–19; location, 199; regional climate, 200–2

McNaughton-Jarvis coupling parameter, 39, 252
mechanical convection, 22
methane, 328
microclimate profiles, 36–7, 41
misery index, 303–4
mixing layer, 22–3, 25
momentum, 24, 30–2, 35
mountain wind, 60, 62, 226–7

net radiation, 13–15, 28–9; distribution for Canada, 13–15

oceans, 5, 101–21; advection, 112; air temperature, 114–18; albedo, 110; Bowen ratio, 113; currents, 104–6; energy balance, 112–13; net radiation, 112; radiation balance, 108–12; storm tracks, 107;

upwelling, 106; water balance, 113–14
Ohm's law, 32, 35
open subarctic forest, 208–13; Bowen ratio, 211–12; energy balance, 209–13; evaporation, 209–13; radiation balance, 208–10

patchy surfaces, 52–6
peat, thermal properties, 159
permafrost, 177, 198–9, 214; distribution in Canada, 199
photosynthesis, 39, 280, 285–6
physical climatology, 10–43
physiographic control, 5–7
planetary boundary layer, 22–3, 36, 41–2, 53
plumes, 54
pollutants, 63–4
pollution, 51, 297–8
polynyas, 52, 176, 182
potential evaporation, 38
precipitation, 14, 37–8; cyclonic, 9–10; distribution for Canada, 16–17
Priestley-Taylor model, 39, 218–19, 234

radiation balance, 28–9
radiative damping parameter, 329–30
respiration, 39
roughness layer, 22–3, 54
roughness length, 30–1
runoff, 14, 37–8
runoff ratio, 15; distribution for Canada, 17–19

sea ice, 346–7; northern hemisphere distribution, 101–2
sea surface temperature, 101, 108–9
sensible heat, 14, 33–7
shelter effects, 60–1
shelterbelts, 60–2
slope breeze, 60
snow, 41, 68–97; accumulation in forests, 83–4; accumulation in open environments, 84–9; aerodynamic characteristics, 70; albedo, 70, 93; creep, 79–80; density, 72, 75; energy balance, 90–2, 95–6; interception, 76–8; measurement, 74;

measurement errors, 75–6; net radiation, 93–4; physical properties, 70–1; radiation balance, 93–4; saltation, 80–1, 88; sublimation, 78–9, 82, 87–9; suspension, 80–1, 88; water retention, 87

snowcover measurement, 89–90; distribution for North America, 71–2; temperature relationships, 72

snowfall, distribution for Canada, 73–4; distribution for southern Ontario, 74

snowmelt, 94–7

snowpack, ground heat flux, 91–2; latent heat, 91–2; sensible heat, 91–2

soil heat, 14, 33–7

soil temperature, 34

solar radiation, 11–14, 25–9; distribution for Canada, 11–13

spatial scales, 24, 46

Stefan-Boltzmann law, 28

stratosphere, 5–6

subarctic lakes and ponds, 213–18; energy balance, 215–18; evaporation, 215–18; radiation balance, 215; temperature regimes, 216–17

surface conductance, 317

surface forcing, 46

surface layer, 22–3, 41–2

surface resistance, 38

surface roughness length, 24, 26, 30, 47

synoptic control of climate, 24–6

temperature, 25–6; global trends, 331–2; profiles, 23, 30, 36–7, 41; transect of Toronto, 353–4; transects, 291; trends for Canada, 332–3

temporal scales, 24

thermal conductivity, 27, 34

thermal convection, 22, 25

thermal diffusivity, 27

thermoclines, 127

topographic forcing, 62

treeline, 199

troposphere, 3–6

tundra, energy balance, 237–9, 241

turbulence, 22

urban environments, 303–24; airflow, 305–7; atmospheric structure, 305; Bowen ratio, 312–14; climate change, 323–4, 346; energy balance, 310–16; evaporation, 314–19; heat storage, 311–12; humidity, 317–19; latent heat, 312–14; radiation balance, 308–10; sensible heat, 312–14; water balance, 314–19

urban heat island, 319–23; Canadian cities, 320–1

urban surfaces, 55

valley winds, 60, 62, 226–7

vapour pressure deficit, 25–6

water, physical properties, 126

water balance, 14, 29, 37–8, 40

westerlies, 3–7

wetlands, 149–68; albedo, 156–7; Bowen ratio, 158, 161–2; carbon balance, 166–8; characteristics, 149–54; classification, 153; climate change, 166–8; distribution, 149–54; distribution for Canada, 151; energy balance, 158–64; evaporation, 163–6; latent heat, 160–2; net radiation, 157–8; radiation balance, 154–8; sensible heat, 160–2, 164; surface resistance, 164–6

wind profiles, 31, 33, 36–7, 42, 49, 56, 287

windbreaks, 60–2, 295–6

windflow, uneven terrain, 56–8

winter season, 68–97

winter, precipitation regime for Canada, 68–9; temperature regime for Canada, 68–9

zero-plane displacement, 24